Pathogenesis and Pharmacological Applications of Bacterial Toxins

Pathogenesis and Pharmacological Applications of Bacterial Toxins

Pathogenesis and Pharmacological Applications of Bacterial Toxins

Editor: Abigale Bellisio

AMERICAN
MEDICAL PUBLISHERS
www.americanmedicalpublishers.com

AMERICAN
MEDICAL PUBLISHERS
www.americanmedicalpublishers.com

Cataloging-in-Publication Data

Pathogenesis and pharmacological applications of bacterial toxins / edited by Abigale Bellisio.
p. cm.
Includes bibliographical references and index.
ISBN 978-1-63927-914-2
1. Bacterial toxins. 2. Pathogenic bacteria. 3. Pharmacology. 4. Physiology, Pathological. I. Bellisio, Abigale.
QP632.B3 P38 2023
615.952 99--dc23

American Medical Publishers,
41 Flatbush Avenue,
1st Floor, New York,
NY 11217, USA

ISBN 978-1-63927-914-2 (Hardback)

Contents

Contents

Preface

In my initial years as a student, I used to run to the library at every possible instance to grab a book and learn something new. Books were my primary source of knowledge and I would not have come such a long way without all that I learnt from them. Thus, when I was approached to edit this book; I became understandably nostalgic. It was an absolute honor to be considered worthy of guiding the current generation as well as those to come. I put all my knowledge and hard work into making this book most beneficial for its readers.

Bacterial toxins operate as independent molecular devices, targeting particular cells in an organism. They punch holes in the cell membrane and modify their intracellular components. Many bacterial toxins are strong human poisons, such as diphtheria, botulinum, Shiga and tetanus toxins. The lethality of a majority of the bacterial exotoxins can be compared with the lethality of endotoxin, snake venom and strychnine. They are multifunctional and self-programmed for reaching their target cells and organs. Due to this reason, they are significant pharmacological tools that can be used to treat various diseases. Diagnosis of various diseases depends on the identification and detection of toxins. Vaccines are made from chemically detoxified bacterial toxins and certain properties of these toxins make them appropriate for therapeutic applications. Different pharmacological applications, evaluations, and advanced studies on bacterial toxins have been included in this book. It presents researches and studies performed by experts across the globe. This book is appropriate for students seeking detailed information on bacterial toxins and their pathogenesis.

I wish to thank my publisher for supporting me at every step. I would also like to thank all the authors who have contributed their researches in this book. I hope this book will be a valuable contribution to the progress of the field.

Editor

1

Targeting Type II Toxin–Antitoxin Systems as Antibacterial Strategies

Marcin Równicki [1,2,*], Robert Lasek [2], Joanna Trylska [1] and Dariusz Bartosik [2,*]

[1] Centre of New Technologies, University of Warsaw, ul. S. Banacha 2c, 02-097 Warsaw, Poland; joanna@cent.uw.edu.pl

[2] Department of Bacterial Genetics, Institute of Microbiology, Faculty of Biology, University of Warsaw, Mecznikowa 1, 02-096 Warsaw, Poland; lasek@biol.uw.edu.pl

* Correspondence: marcin.rownicki1@gmail.com (M.R.); bartosik@biol.uw.edu.pl (D.B.)

Abstract: The identification of novel targets for antimicrobial agents is crucial for combating infectious diseases caused by evolving bacterial pathogens. Components of bacterial toxin–antitoxin (TA) systems have been recognized as promising therapeutic targets. These widespread genetic modules are usually composed of two genes that encode a toxic protein targeting an essential cellular process and an antitoxin that counteracts the activity of the toxin. Uncontrolled toxin expression may elicit a bactericidal effect, so they may be considered "intracellular molecular bombs" that can lead to elimination of their host cells. Based on the molecular nature of antitoxins and their mode of interaction with toxins, TA systems have been classified into six groups. The most prevalent are type II TA systems. Due to their ubiquity among clinical isolates of pathogenic bacteria and the essential processes targeted, they are promising candidates for the development of novel antimicrobial strategies. In this review, we describe the distribution of type II TA systems in clinically relevant human pathogens, examine how these systems could be developed as the targets for novel antibacterials, and discuss possible undesirable effects of such therapeutic intervention, such as the induction of persister cells, biofilm formation and toxicity to eukaryotic cells.

Keywords: toxin–antitoxin systems; toxin activation; antibacterial agents; bacterial persistence

Key Contribution: Bacterial toxin–antitoxin (TA) systems are promising targets for the development of novel antimicrobial strategies.

---

1. Introduction

The emergence of antibiotic-resistant bacteria is a critical issue in modern medicine [1]. Considerable efforts are being made to identify novel antimicrobials to combat infectious diseases caused by evolving multi-resistant pathogens. Bacterial toxin–antitoxin (TA) systems represent promising targets for such compounds. These small genetic modules encode two components: a stable toxin (always a protein) that recognizes a specific cellular target, and a labile antitoxin (protein or RNA molecule), produced in excess, that counteracts the activity of the toxin [2]. TAs have been classified into several groups based on the molecular nature of the antitoxin and their mode of interaction with their cognate toxin. The most prevalent and most extensively studied are type II TA systems, encoding proteic antitoxins which neutralize the toxin by forming TA complexes under optimal growth conditions [3,4]. These two-gene *loci* are organized in operons, whose expression is tightly regulated by both the antitoxin and toxin–antitoxin complexes [5,6]. Degradation of the antitoxin by cellular proteases liberates the toxin, which elicits a bacteriostatic or bactericidal effect [7–10].

TA *loci* were initially identified within bacterial plasmids, where they function as post-segregational cell killing systems (PSK), providing stable maintenance for their carrier replicon in a bacterial population due to the elimination of plasmid-less cells [11]. A surprising observation was that TA systems are also highly abundant and widespread in the chromosomes of free-living bacteria (both Gram-negative and Gram-positive), as well as in archaea [12]. More detailed studies have revealed that these TAs may be involved in important biological processes, such as (i) the stringent response, helping cells to survive stressful conditions by limiting various metabolic activities [13], (ii) programmed cell death [14] and (iii) biofilm formation [15]. In the light of these observations, it seems likely that chromosomal TA *loci* have been transferred to extrachromosomal replicons and adopted as plasmid stabilization systems.

The toxins of TA systems target essential cellular processes of bacteria; therefore, they may be considered intracellular "molecular time bombs", which are activated under certain environmental conditions [16]. Compounds that can artificially activate TA toxins may form a new class of antimicrobials that could represent an alternative to antibiotics [17]. Type II TA systems are ubiquitous in bacteria [18], and their mechanisms are fairly well characterized [18], so they would appear to be excellent candidates for testing the merits of this idea.

The potential use of TA systems in combatting bacterial infections has been examined in several valuable reviews [19–22]. Here, we describe the most promising antibacterial strategies employing type II TA systems and discuss possible undesirable effects of their application, such as the induction of persister cells, biofilm formation and toxicity to eukaryotic cells. In addition, the distribution of type II TA systems among clinically relevant human pathogens is described, which may help to identify appropriate TA candidates for fighting particular infections.

2. Type II TA Systems in Pathogenic Bacteria

Since 2011, the studies of TA systems have been assisted by the Toxin–Antitoxin Database (TADB) [23], updated to version 2.0 in 2017 [24]. Currently, TADB 2.0 lists 6194 type II TA *loci* [24] categorized, where possible, in two partially interdependent ways. The first classification scheme uses a system based on the structural and functional characteristics of the toxin proteins [25,26]. There are 11 families of two-component type II TA systems: *ccd*, *hicBA*, *hipBA*, *mazEF*, *parD(PemKI)*, *parDE*, *phd-doc*, *relBE* (with 5 subfamilies—*relBE*, *higBA*, *yfeM-yoeB*, *ygiTU(mqsAR)* and *prlF-yhaV*), *vapBC*, *mosAT* and *yeeU*. The less numerous three-component type II TA systems are *ω-ε-ζ*, *pasABC* and *paaR-paaA-parE*. The second classification scheme is based on a set of 44 conserved toxin–antitoxin protein domain pairings and better reflects the versatility and modularity of TA systems [27]. Figure 1 summarizes the variety of type II TA systems identified in bacterial species commonly associated with pathogenesis in humans. Among type II TA system (sub)families, four are most frequently encountered: *vapBC*, *relBE*, *mazEF* and *higBA*, and these account for ~80% of the listed *loci*. (An expanded list of type II TA systems of bacterial pathogens, organized into toxin–antitoxin domain pair groupings, is given in Supplementary Figure S1.)

Due to their widespread occurrence within the accessory genomes of human pathogens, their probable role in pathogenicity and potential as therapeutic targets, type II TA systems have been a subject of growing interest in recent years. Several systematic reviews have focused on the distribution and roles of the TA systems in clinically relevant pathogenic bacteria, including *Escherichia coli*, *Mycobacterium tuberculosis*, *Neisseria gonorrheae*, *Streptococcus* spp., *Burkholderia* spp. and species of the ESKAPE group (*Enterococcus faecium*, *Staphylococcus aureus*, *Klebsiella pneumoniae*, *Acinetobacter baumannii*, *Pseudomonas aeruginosa* and *Enterobacter* spp.) [19,28–32]. An increasing number of studies are being undertaken to describe the repertoire of TA systems in pathogenic bacteria at the level of a given strain or of a whole taxon. Not only do such approaches shed light on the importance of overall TA system networks with respect to the physiology, virulence and evolution of pathogens, but thanks to the large-scale genome comparisons involved, they also lead to the discovery of novel TA *loci*

Species	Total	vapBC	relBE	mazEF	higBA	parDE	hicBA	phd-doc	ccd	parD	yefM-yoeB	ygiTU(mqsAR)	ω-ε-ζ	paaRA-parE	prlF-yhaV	mosAT	yeeUV	Class
Mycobacterium bovis	47	38	3	4	1	1												Actinobacteria
Mycobacterium tuberculosis	89	73	4	8	1	3												
Corynebacterium diphtheriae	1							1										
Nocardia farcinica	4	2		1	1													
Propionibacterium acnes	3	1					2											
Bacteroides fragilis	4		1		2		1											Bacteroidia
Staphylococcus aureus	18		11	7														Bacilli
Staphylococcus epidermidis	3			2				1										
Staphylococcus simulans	1		1															
Streptococcus agalactiae	5		2			2	1											
Streptococcus mutans	2		1	1														
Streptococcus pneumoniae	11		6		1		3	1										
Streptococcus pyogenes	12					9	1						2					
Streptococcus suis	2		1							1								
Bacillus anthracis	2			2														
Bacillus cereus	5			3								2						
Bacillus thuringiensis	1			1														
Listeria monocytogenes	1			1														
Enterococcus faecalis	3		1	1				1										
Enterococcus faecium	2										2							
Clostridium perfringens	1			1														Clostridia
Brucella melitensis	2		1					1										Alpha-proteobacteria
Brucella suis	2		1					1										
Bartonella henselae	12		3	1	5	1	2											
Rickettsia conorii	10	6	1	2	1													
Bordetella bronchiseptica	2	1			1													Beta-proteobacteria
Bordetella parapertussis	4	1			1		2											
Burkholderia pseudomallei	3		2				1											
Neisseria meningitidis	3	1		2														
Escherichia coli	26	1	2	4	5	2	2		3	1	1			2	2		1	Gamma-proteobacteria
Salmonella entercia	20	7	6		4			3										
Shigella flexneri	5		3	1						1								
Yersinia pestis	9		2		4		2	1										
Yersinia pseudotuberculosis	4		2		2													
Vibrio cholerae	11		5		2	3										1		
Vibrio parahaemolyticus	4		2			2												
Vibrio vulnificus	5		4			1												
Coxiella burnetii	5	1	1		2	1												
Legionella pneumophila	6		2		4													
Haemophilus ducreyi	1						1											
Haemophilus influenzae	6	2	1		3													
Pseudomonas aeruginosa	8		7	1														
Helicobacter hepaticus	1	1																Epsilon-proteobacteria
Helicobacter pylori	6		5				1											
Campylobacter jejuni	2		1					1										
Leptospira interrogans	9	6		3														Spirochaetes
Treponema denticola	27	12	6	2	2		5											
Total:	410	153	88	48	42	25	24	11	3	3	3	2	2	2	2	1	1	

Figure 1. Type II toxin–antitoxin (TA) systems identified in human pathogenic bacteria according to the toxin family classification system [25,26] as collected in the Toxin–Antitoxin Database (TADB) 2.0 [24]. The number given for an individual species is the sum of the values for all strains of a given taxon. This list was manually curated to represent strains and/or species frequently associated with infections in humans (including opportunistic pathogens).

M. tuberculosis is notable for its abundance of TA systems, especially of the *vapBC*, *mazEF* and *relBE* families (Figure 1), and their activity has been linked to the regulation of adaptive responses to stress caused by interaction with the host and drug treatment [33]. Predictably, *M. tuberculosis* has been the object of studies on the common expression patterns of TA *loci* [34–36], as well as cross-activation between homologous systems [37,38], and putative interaction between non-cognate toxin–antitoxin pairs, which remains controversial [39–41]. Furthermore, comparative studies of the distribution of TA systems in mycobacteria have demonstrated the variability of TA *loci* between different *M. tuberculosis* lineages, and led to the discovery of putative novel TA systems [42,43]. Similarly, the body of knowledge about the TA systems of *S. aureus* [44] was recently expanded by several studies providing insights into the association between the TA systems landscape and strain phenotypes [45–48]. A recent in-depth analysis of the diversity and distribution of type II and type IV TA systems in the genomes of

K. pneumoniae species complex identified several novel toxins and demonstrated the co-occurrence of TA *loci* and clinically relevant genes [49].

Comprehensive strategies for the identification of novel TA *loci* are required in order to gain a greater understanding of the role of TA networks in the biology of bacterial cells. In response to this need, Akarsu and colleagues created the "discovery-oriented" database TASmania in 2019 [50]. Using this newly developed pipeline, they annotated a set of over 41,000 assemblies from the EnsemblBacteria database, resulting in the identification of $>2 \times 10^6$ candidate TA *loci* [50]. The greater flexibility of TASmania compared to the TAFinder search tool in TADB 2.0 allowed for the identification of a higher number of putative TA *loci*, thus providing a starting point for experimental analyses [50]. Moreover, the "guilt-by-association" strategy used throughout the annotation process, i.e., targeting *loci* directly neighboring orphan toxin or antitoxin genes, facilitated the discovery of new TA protein families. In the case of *Listeria monocytogenes*, TASmania-assisted large-scale genomic comparisons led to the identification of 14 putative TA genes [51].

3. Strategies for the Artificial Activation of Toxin–Antitoxin Systems

The growth-inhibitory and lethal consequences of the activity of TA systems led to the proposal that artificial activation of the toxins could provide an effective antibacterial strategy [17,52]. Type II TA systems seem to be most convenient for developing such strategies because many toxins of these systems have been thoroughly characterized and their cellular targets are known [53–55]. Moreover, as shown in Figure 1, the presence and conservation of type II TAs have been confirmed in major human-associated bacterial pathogens. Importantly, TA systems have no human homologs, and no pre-existing resistance against TA toxins has been observed. To date, several TA-based antibacterial strategies have been proposed [19,20], but in most cases they are not supported by significant experimental data.

In this review, we focus our attention on the most promising approaches: (i) the direct activation of TA systems by the use of specific molecules that interfere with TAs and imbalance the delicate stoichiometry of active toxin and antitoxin in bacterial cells, and (ii) the indirect activation of TAs by triggering other cellular components whose functions are interconnected with the TA system. Another novel and interesting approach that we consider involves the use of engineered species-specific toxins that can selectively kill selected strains of pathogenic bacteria.

3.1. *Direct Activation of TA Systems*

3.1.1. Disruption or Preventing the Formation of TA Complexes

The most straightforward strategy for direct activation of the antibacterial effect of type II TA systems requires disruption of the protein toxin and antitoxin complexes, leading to toxin liberation (Figure 2a). This strategy has been validated by several research groups who focused their efforts on identifying high affinity peptide inhibitors that can efficiently displace toxins from their cognate antitoxins.

Lioy and co-workers selected the ε-ζ TA system from *Streptococcus pyogenes* as their research model because the free ζ toxin was shown to trigger a loss of cell proliferation similar to that caused by known antimicrobials [56]. Using high-throughput methods, they screened several oligopeptide libraries for the ability to impair the assembly of ε–ζ complexes. This led to the identification of a library containing a mixture of 17-amino acid-long oligopeptides that interfered with the TA interaction. However, further subfractionation of this library resulted in a diminished effect. The authors suggested that the disruption of ε–ζ complexes that was originally observed might have been a consequence of the concerted action of several weak-binding oligopeptides. Nevertheless, this study provided a proof-of-concept for the antimicrobial potential of this strategy [56].

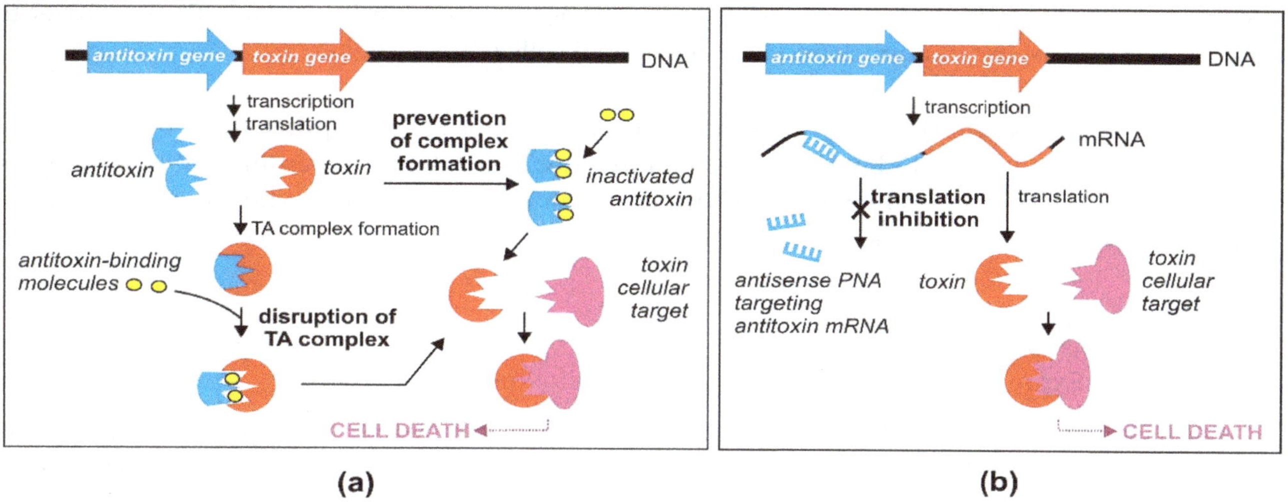

Figure 2. Proposed antibacterial strategies based on the direct activation of toxins of TA systems; (**a**) disruption of the protein toxin and antitoxin complex and/or prevention of protein complex formation; (**b**) inhibition of antitoxin translation (see text for details).

A similar approach was applied in the case of two related TA systems—*pemIK* and *moxXT* of *Bacillus anthracis*. The former module encodes the toxin PemK, a ribonuclease whose overexpression exerts a toxic effect in *B. anthracis* cells characterized by the drastic inhibition of protein synthesis [57]. Based on in silico protein structural modeling, several peptides were designed to mimic the *C*-terminal domain of the antitoxin PemI, which is involved in toxin binding. In vitro experiments indicated the effectiveness of the designed molecules and demonstrated the feasibility of disrupting the TA interaction using octapeptides [58]. Similar results were obtained in studies on the related TA system *moxXT*. Based on the crystal structure of the MoxX–MoxT complex, Verma and colleagues [59] designed a series of peptides that effectively disturbed the interaction of MoxT with antitoxin MoxX and stimulated MoxT ribonuclease activity [60].

Analogously, following a detailed analysis of *vapBC* TA systems of *Mycobacterium tuberculosis*, Lee and colleagues [61] designed VapB- and VapC-based peptides that effectively disrupted the TA complexes, causing activation of the VapC toxin ribonuclease [61].

3.1.2. Inhibition of Antitoxin Translation

Toxin release can also be caused by a reduction in the amount of antitoxin molecules in a bacterial cell. One way to achieve such an effect is by blocking antitoxin translation (Figure 2b) [20]. The toxin and antitoxin genes, co-transcribed as a single mRNA, possess separate Shine–Dalgarno sequences [62]. Therefore, inhibition of antitoxin translation should not influence the translation of the toxin protein. This strategy was tested by Równicki and co-workers who designed a peptide nucleic acid (PNA)-based treatment to inhibit translation of the antitoxins of the *mazEF* and *hipBA* TA systems of *E. coli* [63].

The sequence-specific antisense PNAs targeted either *mazE* or *hipB* antitoxin mRNAs and as a result lowered the cellular levels of these transcripts, which caused an effective inhibition of *E. coli* growth [63]. Importantly, the PNA treatment did not change the relative levels of the *mazF* or *hipA* toxin mRNAs. The crucial role of the MazF and HipA toxins in the observed growth inhibition was confirmed by showing that *E. coli* mutants lacking the genes encoding these proteins were "resistant" to treatment with the antitoxin-specific PNAs.

Another PNA-based strategy, described in the same report, used PNA oligomers directed at the cellular target of the toxin, thus bypassing the involvement of the TA system. It was demonstrated that PNAs can mimic the action of the HipA toxin by silencing the *gltX* gene encoding its cellular target, glutamyl-tRNA synthase [63]. For these experiments, the PNA oligomers were conjugated with a cell penetrating peptide—$(KFF)_3K$, which indicated the importance of employing an efficient carrier to introduce these antisense oligonucleotides into bacterial cells [63].

The above results confirmed experimentally that TA systems are susceptible to sequence-specific antisense agents and provided a basis for their further exploitation in antimicrobial strategies. Importantly, PNA oligomers exhibit nuclease and protease resistance, high binding affinity to natural nucleic acids, and negligible toxicity to eukaryotic cells [64,65].

3.2. Indirect Activation of TA Systems

3.2.1. Enhanced Expression of Proteases Degrading Antitoxins

As previously mentioned, the antitoxins of type II TA systems are more susceptible to degradation by host cytoplasmic proteases than their cognate toxins. In most cases, bacterial Lon protease is involved in antitoxin degradation; however, the two-component protease ClpP, acting in cooperation with the chaperones ClpA or ClpX, is used by some TA systems [66–68].

The depletion of antitoxin molecules results in the liberation of the toxin proteins from TA complexes and relieves transcriptional repression of TA operons, causing increased production of TA transcripts. These events alter the stoichiometry of TA molecules in a cell and ultimately have a lethal effect [69]. Therefore, increasing the level of cellular proteases or designing specific molecules that activate these proteases could be a promising strategy for the indirect activation of toxins (Figure 3a).

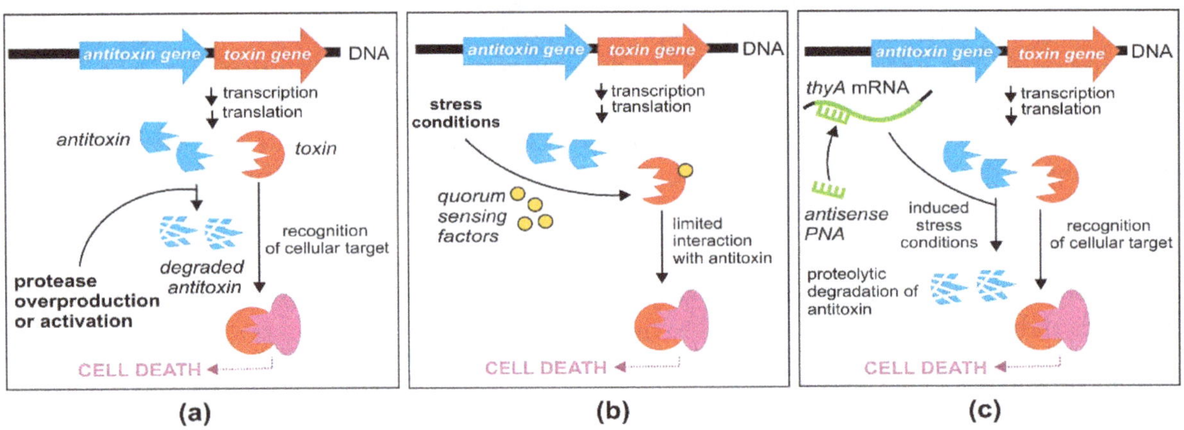

Figure 3. Proposed antibacterial strategies based on the indirect activation of toxins of TA systems: (**a**) activation of the Lon or ClpP proteases that degrade antitoxins; (**b**) triggering TA systems by quorum sensing factors; (**c**) triggering TAs by artificial induction of the stringent response (see text for details).

Increased protease expression can be achieved by introducing a plasmid carrying a cloned protease gene. Overproduction of the Lon protease is known to be lethal for *E. coli* cells [70]. However, by employing an inducible Lon overproduction system, Christensen and colleagues overcame this problem to demonstrate that the lethality is partially dependent on the *yefM-yoeB* TA system [66]. They showed that overproduction of Lon triggered YoeB-dependent mRNA cleavage, leading to translation inhibition. This, in turn, activated the YoeB toxin by preventing the synthesis of its unstable YefM antidote, which was eventually lethal to the host cells [66].

Proteolysis is a tightly controlled process which can be significantly influenced by specific molecules targeting proteases. For example, acyldepsipeptides (ADEPs) are compounds with antibiotic properties that specifically activate the bacterial protease ClpP. Uncontrolled proteolysis induced in this way inhibits bacterial cell division and results in cell death, possibly with the participation of activated toxins [71].

3.2.2. Triggering of TA Systems by Quorum-Sensing Factors

Another interesting antimicrobial strategy was developed by Kumar and Engelberg-Kulka [72,73]. Their approach targeted *mazEF* TA systems, which are among the most abundant bacterial TA *loci* (Figure 1). Toxin MazF is a sequence-specific endoribonuclease that initiates a programmed cell death pathway in response to environmental stress conditions [74], while MazE is a labile antitoxin that is preferentially degraded by the serine protease ClpAP [14].

The proposed strategy involves the use of a newly discovered group of pentapeptides secreted by bacteria, called extracellular death factors (EDFs), which act in quorum sensing and enhance MazF activity under stressful conditions (Figure 3b) [72]. Interestingly, it was shown that EDFs bind directly to MazF in a sequence-specific manner, and this binding is likely to limit interaction of the toxin with its cognate antitoxin MazE [73,75]. EDFs can also stimulate activation of *mazEF* in heterologous hosts, which might broaden the potential application of this strategy [73,75]. Although an early study on MazF classified its toxicity as lethal to cells [14], this statement has since been revisited, suggesting that MazF is involved in growth arrest rather than cell death [76,77].

3.2.3. Induction of the Stringent Response

Another antibacterial strategy involving the indirect activation of toxins is based on induction of the stringent response, a conserved mechanism that allows bacteria to adapt their metabolism in response to stressful environmental conditions, e.g., nutrient deprivation. Many chromosomal type II TA systems, including the aforementioned *mazEF loci*, are transcriptionally upregulated under stressful conditions [78], and their activity leads to remodeling of cellular metabolism and/or programmed cell death, affecting part of the bacterial community [74].

The stringent response (mediated by the alarmone guanosine 3,5 bispyrophosphate, ppGpp) is activated by different natural starvation and stress signals [79], and this can also be achieved by the application of artificial factors. Równicki and colleagues used sequence-specific PNAs targeting the *thyA* gene of *E. coli*, conjugated with a $(KFF)_3K$ peptide as a carrier, to trigger MazF toxin production by inducing thymine starvation [63] (Figure 3c). The *thyA* gene encodes thymidylate synthase, an enzyme involved in folic acid metabolism, which normally interferes with *mazEF*-mediated growth inhibition [74]. As shown for *E. coli*, thymine starvation leads to accumulation of ppGpp in bacterial cells, which reduces global transcription [80]. As a consequence, the inhibition of transcription of *mazEF* leads to activation of the MazF toxin [81]. The significantly reduced level of *thyA* mRNA after treatment with a complementary anti-*thyA* PNA and the resulting growth inhibition confirmed the effectiveness of this silencing strategy [63].

3.3. *Engineered TA Systems in a Targeted Killing Strategy*

An innovative strategy for the targeted killing of selected pathogenic bacteria, without harming beneficial members of the microbiota inhabiting eukaryotic host organisms, was recently proposed by López-Igual et al. [82] (Figure 4). This strategy is based on the use of the *ccdA-ccdB* type II TA system, encoding the toxin CcdB that poisons DNA gyrase—an enzyme responsible for the negative supercoiling of bacterial DNA [83]. CcdB interferes with the activity of DNA gyrase, inducing it to form a covalent GyrA–DNA complex that cannot be resolved, thus promoting DNA breakage and cell death. This mechanism is closely related to the action of quinolone antibiotics, which also target DNA gyrase [84].

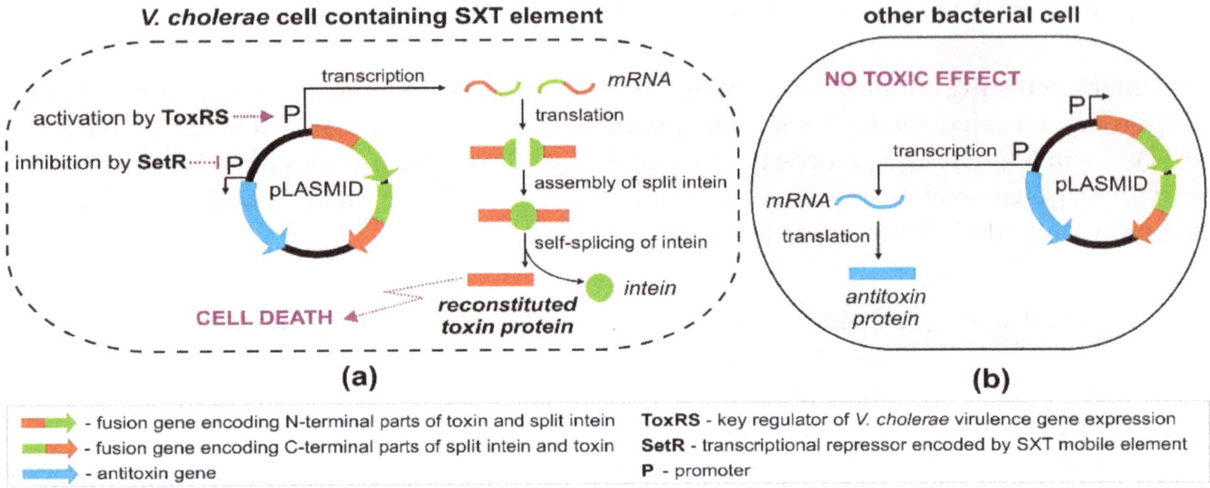

Figure 4. Engineered CcdB toxin as a *V. cholerae*-specific antimicrobial agent. pLASMID—carrier plasmid molecule containing engineered genes of the *ccdAB* TA system. (**a**) In *V. cholerae* cells, by producing ToxRS and SetR transcription factors, transcription of the antitoxin gene is repressed, and expression of engineered toxin genes is activated, leading to cell death; (**b**) in other bacterial cells (lacking both transcription factors), the antitoxin gene is preferentially expressed.

To better control CcdB production in vivo, the gene for this toxin was divided into two parts, and each was fused with DNA encoding split inteins. The generation of a functional toxin occurs in three stages: (i) expression of the two polypeptides, (ii) their association and ligation into a single fusion protein, and (iii) self-splicing of the intein (Figure 4). An important step in this strategy was the construction of a mobilizable plasmid containing genes encoding the antitoxin and the engineered toxin–intein, whose expression could be independently regulated in bacteria harboring specific transcription factors. Although the plasmid could be readily transferred by conjugation from *E. coli* to other bacteria, the toxic effect was observed exclusively in strains of *Vibrio cholerae*. This host specificity was achieved by cloning the engineered *ccdB* toxin–intein genes downstream of a promoter regulated by the transcriptional activator ToxRS, a cholera toxin-associated activator characteristic for *V. cholerae* [85] (Figure 4). Bacterial strains lacking ToxRS were unaffected, including the *E. coli* donor strain and non-pathogenic *Vibrio* spp.

The second component of the TA module, the antitoxin gene *ccdA*, was placed under the transcriptional control of the repressor SetR. The presence of *setR* is considered a hallmark of SXT—an integrative and conjugative mobile element (ICE) of *V. cholerae* that often includes various antibiotic resistance genes [86]. Therefore, expression of the antitoxin gene is repressed in pathogenic multidrug-resistant *V. cholerae* cells containing the *setR* gene, allowing the toxin to poison gyrase and cause a bactericidal effect (Figure 4). However, in cells that lack *setR*, the antitoxin CcdA is produced, neutralizing the effects of the toxin. Thus, the described approach allows species-specific killing of antibiotic-resistant *V. cholerae* strains without affecting the growth of other bacteria present in the mixed populations. The effectiveness of this strategy was confirmed in vivo, in the microbiota of zebrafish and crustacean larvae, where *Vibrio* spp. naturally occur [82].

4. Toxicity to Eukaryotic Cells

Although TA *loci* do not occur in eukaryotic genomes, most toxins of TA systems are endoribonucleases that cleave mRNAs irrespective of their origin, be that prokaryotic or eukaryotic. Therefore, before using these toxins in potential antibacterial strategies, it is important to consider possible cytotoxicity and side effects on human cells. Unfortunately, only a few studies have addressed this issue to date. Notably, toxin cytotoxicity was exploited in one study that employed an engineered version of the MazF toxin to reduce solid tumors in mice [87]. Similarly, the VapC toxin from a TA system of *M. tuberculosis* demonstrated pro-apoptotic activity in human cancer cells, regardless of the expression system used. In another study, Chono and colleagues found that the MazF toxin dosage is a critical factor in determining its activity and cytotoxicity in eukaryotic cells [88,89]. The above examples show that mammalian cells are sensitive to the ribonuclease activity of toxins. However, the ability of bacterial TA-system toxins to penetrate eukaryotic cells is currently unknown, and strategies that specifically target pathogens or utilize toxins that lack targets in human cells might prevent any deleterious effects.

5. Role of Type II TA Modules in Biofilm Formation and Bacterial Persistence

Two potentially undesirable effects of artificial activation of TAs are bacterial persistence and biofilm formation. Persister cells are a subpopulation of slow-growing or growth-arrested bacterial cells that have a decreased susceptibility to killing by bactericidal antibiotics within an otherwise susceptible clonal population [90]. It has been shown that increased tolerance of biofilms to antibiotics is due to the higher amounts of persister cells within the biofilm community [91,92]. While antibiotics kill the majority of biofilm cells, persisters remain viable and repopulate biofilms when the level of antibiotics drops [93]. Thus, persister cells appear to play a central role in the recalcitrance of chronic and biofilm-related infections [94].

One of the first items of evidence linking type II TA modules to biofilm formation comes from a well-characterized chromosomal TA system *mqsRA* in *E. coli* [95,96]. The *mqsRA* is a unique type II TA system where the toxin gene *msqR* precedes the antitoxin *msqA* [97]. It has been shown that a Tn*5* insertion mutant of the toxin *mqsR* formed less adherent biomass [96]. On the contrary, later reanalysis of the *msqRA* revealed that this TA module does not affect biofilm formation in nutrient-rich conditions [98]. The authors identified two new promoters located in the toxin coding sequence that allow the constitutive expression of *mqsA*, thereby allowing a constant and steady level of the MqsA antitoxin compared to the MqsR toxin. This work, quite understandably, has opened up the debate about the role of TA systems in persistence and biofilm formation [78,99]. This result disproves the role of *mqsRA* (and nine others) TA modules in spontaneous biofilm formation, but does not address the question of a role of TA systems in growth control and biofilm formation under stress conditions. The first persistence-related gene to be identified was *hipA*, encoding the toxin of the *E. coli hipBA* TA system—a serine/threonine kinase that inhibits cell growth by inactivating the glutamyl-tRNA synthetase GltX [100]. This discovery linked TA modules and bacterial persistence for the first time. Since then, correlations between antibiotic persistence and TA systems have been extensively studied. Evidence both for and against the participation of TA systems in persister cell formation has been accumulated, making their contribution to this phenomenon unclear [99,101–104]. For example, the deletion of 10 ribonuclease-encoding TA systems from the *E. coli* genome was found to decrease the number of persisters. However, it was subsequently shown that this result was influenced by the presence of φ80 bacteriophage contamination [105]. Moreover, reconstruction of this mutant strain demonstrated that deletion of the 10 TA systems did not affect susceptibility to ofloxacin or ampicillin [106]. These contradictory findings have given rise to considerable debate [78,107–109], and the relevance of TA systems to bacterial persistence remains unclear.

In contrast, the results of several other studies support the contribution of TA systems to bacterial persistence. For example, overexpression of the toxins RelE or MazF was shown to increase the survival of *E. coli* under antibiotic exposure [9,110]. In another study, the *dinJ/yafQ* TA module was found to be involved in tolerance to cephalosporin and aminoglycoside antibiotics [111]. Further evidence linking type II TA modules and bacterial persistence was obtained in a study on uropathogenic isolates of *E. coli* [112]. TA systems have also been linked to bacterial persistence in *Salmonella*, where the *shpB1* allele, carrying a mutation in the antitoxin of the *shpAB* TA module, was associated with the salmonella high persister phenotype [113]. In addition to the *shpAB* system, 13 other type II TA modules were shown to contribute to persistence in *Salmonella* triggered by stresses encountered during macrophage infection [114]. Furthermore, the overexpression of three acetyltransferase toxins (TacT, TacT2, TacT3—acetylating tRNA) in *Salmonella enterica* was found to increase the level of persisters in the population, and their deletion resulted in a decrease in the proportion of persister cells [115,116]. Since the general mechanism of persistence is still unclear, it is crucial to determine whether the activation of TAs influences persister or biofilm formation before they are considered for use as antimicrobial targets. Besides the type II TA modules, other TA systems have been shown to be involved in persistence in *E. coli*. For further information see important recent publications by Fisher et al. [117], Ronneau and Helaine [78], Wilmaerts et al. [118], Dorr et al. [119], and papers from the Wood group [120,121].

While persistence is among the possible side effects of using TAs as antibacterial targets, these systems can also serve in antipersister strategies [122]. Conlon and colleagues [123] reasoned that a compound capable of striking a target in dormant cells will kill persisters. They used an acyldepsipeptide antibiotic (ADEP4) to globally activate the protease ClpP and showed that this becomes fairly non-specific and kills persisters by degrading over 400 proteins, which forces cells to self-digest. Subsequently, this strategy was used to eradicate a biofilm in an animal model, which confirmed its potential as the basis of therapies to treat chronic infections. In a different approach, Li and co-workers [124] identified a novel inhibitor of the *E. coli* HipA toxin which interfered with persister formation in an antibiotic-independent manner. A comprehensive review describing current antipersister strategies was recently published by Defraine et al. [122].

6. Conclusions

Although TA-based antimicrobial strategies have great potential, it is still too early to assess the therapeutic value of toxin activation in clinical settings. Drug discovery and development are time consuming and costly processes, so selecting the right approach will be very important if efforts to produce new antimicrobials based on TA activators are to progress. It is also crucial to select the correct TA targets, and such decisions should be based on data concerning their clinical relevance (i.e., prevalence in antibiotic-resistant clinical isolates), effect exerted on bacterial cells (bactericidal, bacteriostatic), influence on persister cell and biofilm formation, cytotoxicity to human cells and lastly the necessary mode of drug delivery. It would appear advantageous to combine the use of TA activators with conventional antibiotics, since such a strategy could be more effective against a broader spectrum of multidrug-resistant strains [63]. The use of engineered toxins is another promising avenue of research [82,125]. According to the type and properties of such recombinant proteins, they might be applied in targeted antimicrobial strategies or directed against human cancer cells in novel anti-cancer therapies.

References

1. Ventola, C.L. The antibiotic resistance crisis: Causes and threats. *Pharm. Ther. J.* **2015**, *40*, 277–283.
2. Gerdes, K.; Nielsen, A.; Thorsted, P.; Wagner, E.G.H. Mechanism of killer gene activation. Antisense RNA-dependent RNase III cleavage ensures rapid turn-over of the stable Hok, SrnB and PndA effector messenger RNAs. *J. Mol. Biol.* **1992**, *226*, 637–649. [CrossRef]
3. Fraikin, N.; Goormaghtigh, F.; Van Melderen, L. Type II toxin-antitoxin systems: Evolution and revolutions. *J. Bacteriol.* **2020**, *202*, e00763-19. [CrossRef]
4. Page, R.; Peti, W. Toxin-antitoxin systems in bacterial growth arrest and persistence. *Nat. Chem. Biol.* **2016**, *12*, 208–214. [CrossRef]
5. Coussens, N.P.; Daines, D.A. Wake me when it's over-Bacterial toxin-antitoxin proteins and induced dormancy. *Exp. Biol. Med.* **2016**, *241*, 1332–1342. [CrossRef]

6. Yamaguchi, Y.; Inouye, M. Regulation of growth and death in *Escherichia coli* by toxin–antitoxin systems. *Nat. Rev. Microbiol.* **2011**, *9*, 779–790. [CrossRef]
7. Piscotta, F.J.; Jeffrey, P.D.; James Link, A. ParST is a widespread toxin–antitoxin module that targets nucleotide metabolism. *Proc. Natl. Acad. Sci. USA* **2019**, *116*, 826–834. [CrossRef]
8. Unterholzner, S.J.; Poppenberger, B.; Rozhon, W. Toxin–antitoxin systems. Biology, identification, and application. *Mob. Genet. Elem.* **2013**, *3*, 1–13. [CrossRef]
9. Keren, I.; Shah, D.; Spoering, A.; Kaldalu, N.; Lewis, K. Specialized persister cells and the mechanism of multidrug tolerance in *Escherichia coli*. *J. Bacteriol.* **2004**, *186*, 8172–8180. [CrossRef]
10. Robson, J.; McKenzie, J.L.; Cursons, R.; Cook, G.M.; Arcus, V.L. The vapBC operon from Mycobacterium smegmatis is an autoregulated toxin-antitoxin module that controls growth via inhibition of translation. *J. Mol. Biol.* **2009**, *390*, 353–367. [CrossRef]
11. Dziewit, L.; Jazurek, M.; Drewniak, L.; Baj, J.; Bartosik, D. The SXT conjugative element and linear prophage N15 encode toxin-antitoxin-stabilizing systems homologous to the tad-ata module of the Paracoccus aminophilus plasmid pAMI2. *J. Bacteriol.* **2007**, *189*, 1983–1997. [CrossRef]
12. Pandey, D.P.; Gerdes, K. Toxin-antitoxin loci are highly abundant in free-living but lost from host-associated prokaryotes. *Nucleic Acids Res.* **2005**, *33*, 966–976. [CrossRef]
13. Christensen, S.K.; Mikkelsen, M.; Pedersen, K.; Gerdes, K. RelE, a global inhibitor of translation, is activated during nutritional stress. *Proc. Natl. Acad. Sci. USA* **2001**, *98*, 14328–14333. [CrossRef]
14. Aizenman, E.; Engelberg-Kulka, H.; Glaser, G. An *Escherichia coli* chromosomal "addiction module" regulated by 3′,5′-bispyrophosphate: A model for programmed bacterial cell death. *Proc. Natl. Acad. Sci. USA* **1996**, *93*, 6059–6063. [CrossRef]
15. Wen, Y.; Behiels, E.; Devreese, B. Toxin-antitoxin systems: Their role in persistence, biofilm formation, and pathogenicity. *Pathog. Dis.* **2014**, *70*, 240–249. [CrossRef]
16. Kędzierska, B.; Lian, L.Y.; Hayes, F. Toxin-antitoxin regulation: Bimodal interaction of YefM-YoeB with paired DNA palindromes exerts transcriptional autorepression. *Nucleic Acids Res.* **2007**, *35*, 325–339. [CrossRef]
17. Williams, J.; Hergenrother, P. Artificial activation of toxin-antitoxin systems as an antibacterial strategy. *Trends Microbiol.* **2014**, *20*, 291–298. [CrossRef]
18. Harms, A.; Brodersen, D.E.; Mitarai, N.; Gerdes, K. Toxins, targets, and triggers: An overview of toxin-antitoxin biology. *Mol. Cell* **2018**, *70*, 768–784. [CrossRef]
19. Lee, K.Y.; Lee, B.J. Structure, biology, and therapeutic application of toxin-antitoxin systems in pathogenic bacteria. *Toxins (Basel)* **2016**, *8*, 305. [CrossRef]
20. Chan, W.T.; Balsa, D.; Espinosa, M. One cannot rule them all: Are bacterial toxins-antitoxins druggable? *FEMS Microbiol. Rev.* **2015**, *39*, 522–540. [CrossRef]
21. Bravo, A.; Ruiz-Cruz, S.; Alkorta, I.; Espinosa, M. When humans met superbugs: Strategies to tackle bacterial resistances to antibiotics. *Biomol. Concepts* **2018**, *9*, 216–226. [CrossRef] [PubMed]
22. Carlos, L.; Barbosa, B.; Sânder, A.; Cangussu, R.; Garrido, S.S.; Marchetto, R. Toxin-antitoxin systems and its biotechnological applications. *Afr. J. Biotechnol.* **2014**, *13*, 11–17.
23. Shao, Y.; Harrison, E.M.; Bi, D.; Tai, C.; He, X.; Ou, H.Y.; Rajakumar, K.; Deng, Z. TADB: A web-based resource for type 2 toxin-antitoxin loci in bacteria and archaea. *Nucleic Acids Res.* **2011**, *39*, D606–D611. [CrossRef]
24. Xie, Y.; Wei, Y.; Shen, Y.; Li, X.; Zhou, H.; Tai, C.; Deng, Z.; Ou, H.Y. TADB 2.0: An updated database of bacterial type II toxin-antitoxin loci. *Nucleic Acids Res.* **2018**, *46*, D749–D753. [CrossRef] [PubMed]
25. Gerdes, K.; Christensen, S.K.; Løbner-Olesen, A. Prokaryotic toxin-antitoxin stress response loci. *Nat. Rev. Microbiol.* **2005**, *3*, 371–382. [CrossRef] [PubMed]
26. Van Melderen, L.; De Bast, M.S. Bacterial toxin-antitoxin systems: More than selfish entities? *PLoS Genet.* **2009**, *5*, e1000437. [CrossRef]
27. Makarova, K.S.; Wolf, Y.I.; Koonin, E.V. Comprehensive comparative-genomic analysis of Type 2 toxin-antitoxin systems and related mobile stress response systems in prokaryotes. *Biol. Direct* **2009**, *4*, 19. [CrossRef]
28. Fernandez-Garcia, L.; Blasco, L.; Lopez, M.; Bou, G.; Garcia-Contreras, R.; Wood, T.; Tomas, M. Toxin-antitoxin systems in clinical pathogens. *Toxins (Basel)* **2016**, *8*, 227. [CrossRef]
29. Lobato-Márquez, D.; Díaz-Orejas, R.; García-del Portillo, F. Toxin-antitoxins and bacterial virulence. *FEMS Microbiol. Rev.* **2016**, *40*, 592–609. [CrossRef]
30. Kędzierska, B.; Hayes, F. Emerging roles of toxin-antitoxin modules in bacterial pathogenesis. *Molecules* **2016**, *21*, 790. [CrossRef]

31. Yang, Q.E.; Walsh, T.R. Toxin-antitoxin systems and their role in disseminating and maintaining antimicrobial resistance. *FEMS Microbiol. Rev.* **2017**, *41*, 343–353. [CrossRef] [PubMed]
32. Kang, S.M.; Kim, D.H.; Jin, C.; Lee, B.J. A systematic overview of type II and III toxin-antitoxin systems with a focus on druggability. *Toxins (Basel)* **2018**, *10*, 515. [CrossRef] [PubMed]
33. Slayden, R.A.; Dawson, C.C.; Cummings, J.E. Toxin-antitoxin systems and regulatory mechanisms in *Mycobacterium tuberculosis*. *Pathog. Dis.* **2018**, *76*, fty039. [CrossRef] [PubMed]
34. Korch, S.B.; Malhotra, V.; Contreras, H.; Clark-Curtiss, J.E. The *Mycobacterium tuberculosis* relBE toxin:antitoxin genes are stress-responsive modules that regulate growth through translation inhibition. *J. Microbiol.* **2015**, *53*, 783–795. [CrossRef] [PubMed]
35. Gupta, A.; Venkataraman, B.; Vasudevan, M.; Gopinath Bankar, K. Co-expression network analysis of toxin-antitoxin loci in *Mycobacterium tuberculosis* reveals key modulators of cellular stress. *Sci. Rep.* **2017**, *7*, 5868. [CrossRef] [PubMed]
36. Thakur, Z.; Saini, V.; Arya, P.; Kumar, A.; Mehta, P.K. Computational insights into promoter architecture of toxin-antitoxin systems of *Mycobacterium tuberculosis*. *Gene* **2018**, *641*, 161–171. [CrossRef]
37. Yang, M.; Gao, C.; Wang, Y.; Zhang, H.; He, Z.G. Characterization of the interaction and cross-regulation of three *Mycobacterium tuberculosis* RelBE modules. *PLoS ONE* **2010**, *5*, e10672. [CrossRef]
38. Agarwal, S.; Tiwari, P.; Deep, A.; Kidwai, S.; Gupta, S.; Thakur, K.G.; Singh, R. System-wide analysis unravels the differential regulation and in vivo essentiality of virulence-associated proteins B and C toxin-antitoxin systems of *Mycobacterium tuberculosis*. *J. Infect. Dis.* **2018**, *217*, 1809–1820. [CrossRef]
39. Zhu, L.; Sharp, J.D.; Kobayashi, H.; Woychik, N.A.; Inouye, M. Noncognate *Mycobacterium tuberculosis* toxin-antitoxins can physically and functionally interact. *J. Biol. Chem.* **2010**, *285*, 39732–39738. [CrossRef]
40. Ramirez, M.V.; Dawson, C.C.; Crew, R.; England, K.; Slayden, R.A. MazF6 toxin of *Mycobacterium tuberculosis* demonstrates antitoxin specificity and is coupled to regulation of cell growth by a Soj-like protein. *BMC Microbiol.* **2013**, *13*, 240. [CrossRef]
41. Tandon, H.; Melarkode Vattekatte, A.; Srinivasan, N.; Sandhya, S. Molecular and structural basis of cross-reactivity in M. tuberculosis toxin–antitoxin systems. *Toxins (Basel)* **2020**, *12*, 481. [CrossRef] [PubMed]
42. Solano-Gutierrez, J.S.; Pino, C.; Robledo, J. Toxin-antitoxin systems shows variability among *Mycobacterium tuberculosis* lineages. *FEMS Microbiol. Lett.* **2019**, *366*, fny276. [CrossRef] [PubMed]
43. Tandon, H.; Sharma, A.; Wadhwa, S.; Varadarajan, R.; Singh, R.; Srinivasan, N.; Sandhya, S. Bioinformatic and mutational studies of related toxin–antitoxin pairs in *Mycobacterium tuberculosis* predict and identify key functional residues. *J. Biol. Chem.* **2019**, *294*, 9048–9063. [CrossRef]
44. Schuster, C.F.; Bertram, R. Toxin-antitoxin systems of Staphylococcus aureus. *Toxins (Basel)* **2016**, *8*, 140. [CrossRef] [PubMed]
45. Bukowski, M.; Hyz, K.; Janczak, M.; Hydzik, M.; Dubin, G.; Wladyka, B. Identification of novel mazEF/pemIK family toxin-antitoxin loci and their distribution in the Staphylococcus genus. *Sci. Rep.* **2017**, *7*, 13462. [CrossRef] [PubMed]
46. Habib, G.; Zhu, Q.; Sun, B. Bioinformatics and functional assessment of toxin-antitoxin systems in Staphylococcus aureus. *Toxins (Basel)* **2018**, *10*, 473. [CrossRef]
47. Kato, F.; Yoshizumi, S.; Yamaguchi, Y.; Inouye, M. Genome-wide screening for identification of novel toxin-antitoxin systems in Staphylococcus aureus. *Appl. Environ. Microbiol.* **2019**, *85*, e00915-19. [CrossRef]
48. Sierra, R.; Viollier, P.; Renzoni, A. Linking toxin-antitoxin systems with phenotypes: A Staphylococcus aureus viewpoint. *Biochim. Biophys. Acta Gene Regul. Mech.* **2019**, *1862*, 742–751. [CrossRef]
49. Horesh, G.; Fino, C.; Harms, A.; Dorman, M.J.; Parts, L.; Gerdes, K.; Heinz, E.; Thomson, N.R. Type II and type IV toxin-antitoxin systems show different evolutionary patterns in the global *Klebsiella pneumoniae* population. *Nucleic Acids Res.* **2020**, *48*, 4357–4370. [CrossRef]
50. Akarsu, H.; Bordes, P.; Mansour, M.; Bigot, D.J.; Genevaux, P.; Falquet, L. TASmania: A bacterial toxin-antitoxin systems database. *PLoS Comput. Biol.* **2019**, *15*, e1006946. [CrossRef]
51. Agüero, J.A.; Akarsu, H.; Aguilar-Bultet, L.; Oevermann, A.; Falquet, L. Large-scale comparison of toxin and antitoxins in *Listeria monocytogenes*. *Toxins (Basel)* **2020**, *12*, 29. [CrossRef] [PubMed]
52. Shapiro, S. Speculative strategies for new antibacterials: All roads should not lead to Rome. *J. Antibiot. (Tokyo)* **2013**, *66*, 371–386. [CrossRef] [PubMed]
53. Germain, E.; Castro-Roa, D.; Zenkin, N.; Gerdes, K. Molecular mechanism of bacterial persistence by HipA. *Mol. Cell* **2013**, *52*, 248–254. [CrossRef] [PubMed]

54. Zhang, S.-P.; Wang, Q.; Quan, S.-W.; Yu, X.-Q.; Wang, Y.; Guo, D.-D.; Peng, L.; Feng, H.-Y.; Yong-Xing, H. Type II toxin–antitoxin system in bacteria: Activation, function, and mode of action. *Biophys. Rep.* **2020**, *6*, 68–79. [CrossRef]
55. Ramisetty, B.C.M.; Santhosh, R.S. Endoribonuclease type II toxin-antitoxin systems: Functional or selfish? *Microbiology* **2017**, *163*, 931–939. [CrossRef]
56. Lioy, V.S.; Rey, O.; Balsa, D.; Pellicer, T.; Alonso, J.C. A toxin-antitoxin module as a target for antimicrobial development. *Plasmid* **2010**, *63*, 31–39. [CrossRef]
57. Agarwal, S.; Agarwal, S.; Bhatnagar, R. Identification and characterization of a novel toxin-antitoxin module from Bacillus anthracis. *FEBS Lett.* **2007**, *581*, 1727–1734. [CrossRef]
58. Agarwal, S.; Mishra, N.K.; Bhatnagar, S.; Bhatnagar, R. PemK toxin of Bacillus anthracis is a ribonuclease: An insight into its active site, structure, and function. *J. Biol. Chem.* **2010**, *285*, 7254–7270. [CrossRef]
59. Verma, S.; Kumar, S.; Gupta, V.P.; Gourinath, S.; Bhatnagar, S.; Bhatnagar, R. Structural basis of Bacillus anthracis MoxXT disruption and the modulation of MoxT ribonuclease activity by rationally designed peptides. *J. Biomol. Struct. Dyn.* **2015**, *33*, 606–624. [CrossRef]
60. Chopra, N.; Agarwal, S.; Verma, S.; Bhatnagar, S.; Bhatnagar, R. Modeling of the structure and interactions of the B. anthracis antitoxin, MoxX: Deletion mutant studies highlight its modular structure and repressor function. *J. Comput. Aided. Mol. Des.* **2011**, *25*, 275–291. [CrossRef]
61. Lee, I.G.; Lee, S.J.; Chae, S.; Lee, K.Y.; Kim, J.H.; Lee, B.J. Structural and functional studies of the *Mycobacterium tuberculosis* VapBC30 toxin-antitoxin system: Implications for the design of novel antimicrobial peptides. *Nucleic Acids Res.* **2015**, *43*, 7624–7637. [CrossRef] [PubMed]
62. Chan, W.T.; Moreno-Córdoba, I.; Yeo, C.C.; Espinosa, M. Toxin-antitoxin genes of the Gram-positive pathogen Streptococcus pneumoniae: So few and yet so many. *Microbiol. Mol. Biol. Rev.* **2012**, *76*, 773–791. [CrossRef] [PubMed]
63. Równicki, M.; Pieńko, T.; Czarnecki, J.; Kolanowska, M.; Bartosik, D.; Trylska, J. Artificial activation of Escherichia coli mazEF and hipBA toxin–antitoxin systems by antisense peptide nucleic acids as an antibacterial strategy. *Front. Microbiol.* **2018**, *9*, 2870. [CrossRef] [PubMed]
64. Good, L.; Awasthi, S.K.; Dryselius, R.; Larsson, O.; Nielsen, P.E. Bactericidal antisense effects of peptide-PNA conjugates. *Nat. Biotechnol.* **2001**, *19*, 360–364. [CrossRef] [PubMed]
65. Wojciechowska, M.; Równicki, M.; Mieczkowski, A.; Miszkiewicz, J.; Trylska, J. Antibacterial peptide nucleic acids—Facts and perspectives. *Molecules* **2020**, *25*, 559. [CrossRef] [PubMed]
66. Christensen, S.K.; Maenhaut-Michel, G.; Mine, N.; Gottesman, S.; Gerdes, K.; Van Melderen, L. Overproduction of the Lon protease triggers inhibition of translation in Escherichia coli: Involvement of the yefM-yoeB toxin-antitoxin system. *Mol. Microbiol.* **2004**, *51*, 1705–1717. [CrossRef]
67. Brzozowska, I.; Zielenkiewicz, U. The ClpXP protease is responsible for the degradation of the Epsilon antidote to the Zeta toxin of the streptococcal pSM19035 plasmid. *J. Biol. Chem.* **2014**, *289*, 7514–7523. [CrossRef]
68. Van Melderen, L.; Thi, M.H.D.; Lecchi, P.; Gottesman, S.; Couturier, M.; Maurizi, M.R. ATP-dependent degradation of CcdA by Lon protease. Effects of secondary structure and heterologous subunit interactions. *J. Biol. Chem.* **1996**, *271*, 27730–27738. [CrossRef]
69. Muthuramalingam, M.; White, J.C.; Bourne, C.R. Toxin-antitoxin modules are pliable switches activated by multiple protease pathways. *Toxins (Basel)* **2016**, *8*, 214. [CrossRef]
70. Goff, S.A.; Goldberg, A.L. An increased content of protease La, the lon gene product, increases protein degradation and blocks growth in Escherichia coli. *J. Biol. Chem.* **1987**, *262*, 4508–4515.
71. Brötz-Oesterhelt, H.; Beyer, D.; Kroll, H.P.; Endermann, R.; Ladel, C.; Schroeder, W.; Hinzen, B.; Raddatz, S.; Paulsen, H.; Henninger, K.; et al. Dysregulation of bacterial proteolytic machinery by a new class of antibiotics. *Nat. Med.* **2005**, *11*, 1082–1087. [CrossRef] [PubMed]
72. Kumar, S.; Engelberg-Kulka, H. Quorum sensing peptides mediating interspecies bacterial cell death as a novel class of antimicrobial agents. *Curr. Opin. Microbiol.* **2014**, *21*, 22–27. [CrossRef]
73. Kumar, S.; Kolodkin-Gal, I.; Engelberg-Kulka, H. Novel quorum-sensing peptides mediating interspecies bacterial cell death. *MBio* **2013**, *4*, e00314-13. [CrossRef] [PubMed]
74. Engelberg-Kulka, H.; Hazan, R.; Amitai, S. mazEF: A chromosomal toxin-antitoxin module that triggers programmed cell death in bacteria. *J. Cell Sci.* **2005**, *118*, 4327–4332. [CrossRef] [PubMed]

75. Nigam, A.; Kumar, S.; Engelberg-Kulka, H. Quorum sensing extracellular death peptides enhance the endoribonucleolytic activities of *Mycobacterium tuberculosis* MazF toxins. *MBio* **2018**, *9*, e00685-18. [CrossRef] [PubMed]
76. Ramisetty, B.C.M.; Raj, S.; Ghosh, D. Escherichia coli MazEF toxin-antitoxin system does not mediate programmed cell death. *J. Basic Microbiol.* **2016**, *56*, 1398–1402. [CrossRef] [PubMed]
77. Fu, Z.; Tamber, S.; Memmi, G.; Donegan, N.P.; Cheung, A.L. Overexpression of MazF in *Staphylococcus aureus* induces bacteriostasis by selectively targeting mRNAs for cleavage. *J. Bacteriol.* **2009**, *191*, 2051–2059. [CrossRef]
78. Ronneau, S.; Helaine, S. Clarifying the link between toxin–antitoxin modules and bacterial persistence. *J. Mol. Biol.* **2019**, *431*, 3462–3471. [CrossRef]
79. Jimmy, S.; Saha, C.K.; Kurata, T.; Stavropoulos, C.; Oliveira, S.R.A.; Koh, A.; Cepauskas, A.; Takada, H.; Rejman, D.; Tenson, T.; et al. A widespread toxin–antitoxin system exploiting growth control via alarmone signaling. *Proc. Natl. Acad. Sci. USA* **2020**, *117*, 10500–10510. [CrossRef]
80. Török, I.; Kari, C. Accumulation of ppGpp in a relA mutant of Escherichia coli during amino acid starvation. *J. Biol. Chem.* **1980**, *255*, 3838–3840.
81. Ramisetty, B.C.M.; Natarajan, B.; Santhosh, R.S. MazEF-mediated programmed cell death in bacteria: "What is this?". *Crit. Rev. Microbiol.* **2015**, *41*, 89–100. [CrossRef] [PubMed]
82. López-Igual, R.; Bernal-Bayard, J.; Rodríguez-Patón, A.; Ghigo, J.M.; Mazel, D. Engineered toxin–intein antimicrobials can selectively target and kill antibiotic-resistant bacteria in mixed populations. *Nat. Biotechnol.* **2019**, *37*, 755–760. [CrossRef] [PubMed]
83. Collin, F.; Karkare, S.; Maxwell, A. Exploiting bacterial DNA gyrase as a drug target: Current state and perspectives. *Appl. Microbiol. Biotechnol.* **2011**, *92*, 479–497. [CrossRef] [PubMed]
84. Bernard, P.; Kézdy, K.E.; Van Melderen, L.; Steyaert, J.; Wyns, L.; Pato, M.L.; Higgins, P.N.; Couturier, M. The F plasmid CcdB protein induces efficient ATP-dependent DNA cleavage by gyrase. *J. Mol. Biol.* **1993**, *234*, 534–541. [CrossRef]
85. Childers, B.M.; Klose, K.E. Regulation of virulence in Vibrio cholerae: The ToxR regulon. *Future Microbiol.* **2007**, *2*, 335–344. [CrossRef]
86. Beaber, J.W.; Hochhut, B.; Waldor, M.K. SOS response promotes horizontal dissemination of antibiotic resistance genes. *Nature* **2004**, *427*, 72–74. [CrossRef]
87. Shimazu, T.; Mirochnitchenko, O.; Phadtare, S.; Inouye, M. Regression of solid tumors by induction of MazF, a bacterial mRNA endoribonuclease. *J. Mol. Microbiol. Biotechnol.* **2014**, *24*, 228–233. [CrossRef]
88. Chono, H.; Saito, N.; Tsuda, H.; Shibata, H.; Ageyama, N.; Terao, K.; Yasutomi, Y.; Mineno, J.; Kato, I. In vivo safety and persistence of endoribonuclease gene-transduced CD4+ t cells in cynomolgus macaques for HIV-1 gene therapy model. *PLoS ONE* **2011**, *6*, e23585. [CrossRef]
89. Chono, H.; Matsumoto, K.; Tsuda, H.; Saito, N.; Lee, K.; Kim, S.; Shibata, H.; Ageyama, N.; Terao, K.; Yasutomi, Y.; et al. Acquisition of HIV-1 resistance in T lymphocytes using an ACA-specific *E. coli* mRNA interferase. *Hum. Gene Ther.* **2011**, *22*, 35–43. [CrossRef]
90. Lewis, K. Persister cells. *Annu. Rev. Microbiol.* **2010**, *64*, 357–372. [CrossRef]
91. Lewis, K. Persister cells and the riddle of biofilm survival. *Biochemistry* **2005**, *70*, 267–274. [CrossRef] [PubMed]
92. Lewis, K. Multidrug tolerance of biofilms and persister cells. *Curr. Top. Microbiol. Immunol.* **2008**, *322*, 107–131. [PubMed]
93. Wood, T.K.; Knabel, S.J.; Kwan, B.W. Bacterial persister cell formation and dormancy. *Appl. Environ. Microbiol.* **2013**, *79*, 7116–7121. [CrossRef] [PubMed]
94. Lebeaux, D.; Ghigo, J.-M.; Beloin, C. Biofilm-related infections: Bridging the gap between clinical management and fundamental aspects of recalcitrance toward antibiotics. *Microbiol. Mol. Biol. Rev.* **2014**, *78*, 510–543. [CrossRef]
95. Soo, V.W.C.; Wood, T.K. Antitoxin MqsA represses curli formation through the master biofilm regulator CsgD. *Sci. Rep.* **2013**, *3*, 3186. [CrossRef]
96. González Barrios, A.F.; Zuo, R.; Hashimoto, Y.; Yang, L.; Bentley, W.E.; Wood, T.K. Autoinducer 2 controls biofilm formation in *Escherichia coli* through a novel motility quorum-sensing regulator (MqsR, B3022). *J. Bacteriol.* **2006**, *188*, 305–316. [CrossRef]

97. Kasari, V.; Kurg, K.; Margus, T.; Tenson, T.; Kaldalu, N. The Escherichia coli mqsR and ygiT genes encode a new toxin-antitoxin pair. *J. Bacteriol.* **2010**, *192*, 2908–2919. [CrossRef]
98. Fraikin, N.; Rousseau, C.J.; Goeders, N.; Van Melderen, L. Reassessing the role of the type II MqsRA toxin-antitoxin system in stress response and biofilm formation: MqsA is transcriptionally uncoupled from mqsR. *MBio* **2019**, *10*, e02678-19. [CrossRef]
99. Ramisetty, B.C.M.; Ghosh, D.; Chowdhury, M.R.; Santhosh, R.S. Corrigendum: What is the link between stringent response, endoribonuclease encoding type II toxin-antitoxin systems and persistence? *Front. Microbiol.* **2017**, *8*, 458. [CrossRef]
100. Moyed, H.S.; Bertrand, K.P. hipA, a newly recognized gene of Escherichia coli K-12 that affects frequency of persistence after inhibition of murein synthesis. *J. Bacteriol.* **1983**, *155*, 768–775. [CrossRef]
101. Van Melderen, L.; Wood, T.K. Commentary: What is the link between stringent response, endoribonuclease encoding type II toxin-antitoxin systems and persistence? *Front. Microbiol.* **2017**, *14*, 191. [CrossRef] [PubMed]
102. Shan, Y.; Gandt, A.B.; Rowe, S.E.; Deisinger, J.P.; Conlon, B.P.; Lewis, K. ATP-Dependent persister formation in Escherichia coli. *MBio* **2017**, *8*, e02267-16. [CrossRef] [PubMed]
103. Ramisetty, B.C.M.; Ghosh, D.; Chowdhury, M.R.; Santhosh, R.S. What is the link between stringent response, endoribonuclease encoding type II toxin-antitoxin systems and persistence? *Front. Microbiol.* **2016**, *7*, 1882. [CrossRef] [PubMed]
104. Harms, A.; Maisonneuve, E.; Gerdes, K. Mechanisms of bacterial persistence during stress and antibiotic exposure. *Science* **2016**, *16*, aaf4268. [CrossRef] [PubMed]
105. Maisonneuve, E.; Castro-Camargo, M.; Gerdes, K. Retraction notice to: (p)ppGpp controls bacterial persistence by stochastic induction of toxin-antitoxin activity. *Cell* **2018**, *172*, 1135. [CrossRef]
106. Goormaghtigh, F.; Fraikin, N.; Putrinš, M.; Hallaert, T.; Hauryliuk, V.; Garcia-Pino, A.; Sjödin, A.; Kasvandik, S.; Udekwu, K.; Tenson, T.; et al. Reassessing the role of type II toxin-antitoxin systems in formation of escherichia coli type II persister cells. *MBio* **2018**, *9*, e00640-18. [CrossRef]
107. Goeders, N.; Van Melderen, L. Toxin-antitoxin systems as multilevel interaction systems. *Toxins (Basel)* **2013**, *6*, 304–324. [CrossRef]
108. Van Melderen, L. Toxin-antitoxin systems: Why so many, what for? *Curr. Opin. Microbiol.* **2010**, *13*, 781–785. [CrossRef]
109. Pacios, O.; Blasco, L.; Bleriot, I.; Fernandez-Garcia, L.; Ambroa, A.; López, M.; Bou, G.; Cantón, R.; Garcia-Contreras, R.; Wood, T.K.; et al. (p)ppGpp and its role in bacterial persistence: New challenges. *Antimicrob. Agents Chemother.* **2020**. [CrossRef]
110. Tripathi, A.; Dewan, P.C.; Siddique, S.A.; Varadarajan, R. MazF-induced growth inhibition and persister generation in Escherichia coli. *J. Biol. Chem.* **2014**, *289*, 4191–4205. [CrossRef]
111. Harrison, J.J.; Wade, W.D.; Akierman, S.; Vacchi-Suzzi, C.; Stremick, C.A.; Turner, R.J.; Ceri, H. The chromosomal toxin gene yafQ is a determinant of multidrug tolerance for *Escherichia coli* growing in a biofilm. *Antimicrob. Agents Chemother.* **2009**, *53*, 2253–2258. [CrossRef] [PubMed]
112. Norton, J.P.; Mulvey, M.A. Toxin-Antitoxin Systems Are Important for Niche-Specific Colonization and Stress Resistance of Uropathogenic *Escherichia coli*. *PLoS Pathog.* **2012**, *8*, e1002954. [CrossRef] [PubMed]
113. Slattery, A.; Victorsen, A.H.; Brown, A.; Hillman, K.; Phillips, G.J. Isolation of highly persistent mutants of *Salmonella enterica* serovar typhimurium reveals a new toxin-antitoxin module. *J. Bacteriol.* **2013**, *195*, 647–657. [CrossRef] [PubMed]
114. Helaine, S.; Cheverton, A.M.; Watson, K.G.; Faure, L.M.; Matthews, S.A.; Holden, D.W. Internalization of salmonella by macrophages induces formation of nonreplicating persisters. *Science* **2014**, *343*, 204–208. [CrossRef] [PubMed]
115. Cheverton, A.M.; Gollan, B.; Przydacz, M.; Wong, C.T.; Mylona, A.; Hare, S.A.; Helaine, S. A Salmonella toxin promotes persister formation through acetylation of tRNA. *Mol. Cell* **2016**, *63*, 86–96. [CrossRef] [PubMed]
116. Rycroft, J.A.; Gollan, B.; Grabe, G.J.; Hall, A.; Cheverton, A.M.; Larrouy-Maumus, G.; Hare, S.A.; Helaine, S. Activity of acetyltransferase toxins involved in Salmonella persister formation during macrophage infection. *Nat. Commun.* **2018**, *9*, 1993. [CrossRef]
117. Fisher, R.A.; Gollan, B.; Helaine, S. Persistent bacterial infections and persister cells. *Nat. Rev. Microbiol.* **2017**, *15*, 453–464. [CrossRef]

118. Wilmaerts, D.; Dewachter, L.; De Loose, P.-J.; Bollen, C.; Verstraeten, N.; Michiels, J. HokB Monomerization and Membrane Repolarization Control Persister Awakening. *Mol. Cell* **2019**, *75*, 1031–1042. [CrossRef]
119. Dörr, T.; Vulic, M.; Lewis, K. Ciprofloxacin causes persister formation by inducing the TisB toxin in *Escherichia coli*. *PLoS Biol.* **2010**, *8*, e1000317. [CrossRef]
120. Cheng, H.Y.; Soo, V.W.C.; Islam, S.; McAnulty, M.J.; Benedik, M.J.; Wood, T.K. Toxin GhoT of the GhoT/GhoS toxin/antitoxin system damages the cell membrane to reduce adenosine triphosphate and to reduce growth under stress. *Environ. Microbiol.* **2014**, *16*, 1741–1754. [CrossRef]
121. Wang, X.; Lord, D.M.; Hong, S.H.; Peti, W.; Benedik, M.J.; Page, R.; Wood, T.K. Type II toxin/antitoxin MqsR/MqsA controls type V toxin/antitoxin GhoT/GhoS. *Environ. Microbiol.* **2013**, *15*, 1734–1744. [CrossRef] [PubMed]
122. Defraine, V.; Fauvart, M.; Michiels, J. Fighting bacterial persistence: Current and emerging anti-persister strategies and therapeutics. *Drug Resist. Updates* **2018**, *38*, 12–26. [CrossRef] [PubMed]
123. Conlon, B.P.; Rowe, S.E.; Gandt, A.B.; Nuxoll, A.S.; Donegan, N.P.; Zalis, E.A.; Clair, G.; Adkins, J.N.; Cheung, A.L.; Lewis, K. Persister formation in *Staphylococcus aureus* is associated with ATP depletion. *Nat. Microbiol.* **2016**, *1*, 16051. [CrossRef] [PubMed]
124. Li, T.; Yin, N.; Liu, H.; Pei, J.; Lai, L. Novel inhibitors of toxin HipA reduce multidrug tolerant persisters. *ACS Med. Chem. Lett.* **2016**, *7*, 449–453. [CrossRef] [PubMed]
125. Solecki, O.; Mosbah, A.; Baudy Floc'h, M.; Felden, B. Converting a *Staphylococcus aureus* toxin into effective cyclic pseudopeptide antibiotics. *Chem. Biol.* **2015**, *22*, 329–335. [CrossRef]

2

Chloroquine Analog Interaction with C2- and Iota-Toxin in Vitro and in Living Cells

Angelika Kronhardt [1,†], Christoph Beitzinger [1,†], Holger Barth [2] and Roland Benz [3,*]

1. Rudolf Virchow Center, Research Center for Experimental Biomedicine, University of Würzburg, Versbacher Straße 9, 97078 Würzburg, Germany; ange.kronhardt@web.de (A.K.); christoph.beitzinger@web.de (C.B.)
2. Institute of Pharmacology and Toxicology, University of Ulm Medical Center, Albert-Einstein-Allee 11, 89081 Ulm, Germany; holger.barth@uni-ulm.de
3. Department of Life Sciences and Chemistry, Jacobs-University Bremen, Campus-Ring 1, 28759 Bremen, Germany

* Correspondence: r.benz@jacobs-university.de.

† These authors contributed equally to this work.

Abstract: C2-toxin from *Clostridium botulinum* and Iota-toxin from *Clostridium perfringens* belong both to the binary A-B-type of toxins consisting of two separately secreted components, an enzymatic subunit A and a binding component B that facilitates the entry of the corresponding enzymatic subunit into the target cells. The enzymatic subunits are in both cases actin ADP-ribosyltransferases that modify R177 of globular actin finally leading to cell death. Following their binding to host cells' receptors and internalization, the two binding components form heptameric channels in endosomal membranes which mediate the translocation of the enzymatic components Iota a and C2I from endosomes into the cytosol of the target cells. The binding components form ion-permeable channels in artificial and biological membranes. Chloroquine and related 4-aminoquinolines were able to block channel formation in vitro and intoxication of living cells. In this study, we extended our previous work to the use of different chloroquine analogs and demonstrate that positively charged aminoquinolinium salts are able to block channels formed in lipid bilayer membranes by the binding components of C2- and Iota-toxin. Similarly, these molecules protect cultured mammalian cells from intoxication with C2- and Iota-toxin. The aminoquinolinium salts did presumably not interfere with actin ADP-ribosylation or receptor binding but blocked the pores formed by C2IIa and Iota b in living cells and in vitro. The blocking efficiency of pores formed by Iota b and C2IIa by the chloroquine analogs showed interesting differences indicating structural variations between the types of protein-conducting nanochannels formed by Iota b and C2IIa.

Keywords: C2-toxin; iota-toxin; binding components; chloroquine; black lipid bilayer; aminoquinolinium salts

---

1. Introduction

Binary A-B type protein toxins are potent virulence factors of certain gram-positive bacteria (for reviews see refs [1–3]). The most prominent example of this type of toxins is the anthrax toxin produced by *Bacillus anthracis*, which is also known as a possible biological weapon [4–6]. Other prominent examples are C2-toxin of *Clostridium botulinum* and Iota-toxin of *Clostridium perfringens*. Both toxins consist of two distinct components that are secreted separately into the extracellular media: an enzymatically active component A—which acts as an actin-specific ADP-ribosyltransferase—and a separate component

B, which is the binding/translocation subunit needed for binding of the toxins to target cells and responsible for translocation of the enzymatic subunits into the cytosol of target cells [4,7–12].

After the proteolytic activation the B components C2II of *Clostridium botulinum* and also Iota b of *Clostridium perfringens* form ring-shaped heptamers similar to the B component of the anthrax toxin PA [11,13–16]. These heptamers (C2IIa, Iota b) are the biologically active species of the B components and mediate two different functions during cellular uptake of the toxins: First, they bind to their receptors on the surface of target cells and form complexes with their A components. These complexes are subsequently taken up into cells via receptor-mediated endocytosis and thereby reach early endosomal vesicles. The acidic conditions in such endosomes trigger a conformational change of the compound B heptamers, which insert into endosomal membranes to form trans-membrane pores. These pores serve as translocation channels for the subsequent transport of the unfolded A components of these toxins from the endosomal lumen into the host cell cytosol. Treatment of cells with bafilomycin (Baf) A1, a compound that prevents acidification of the endosomes, inhibits pore-formation by the B components, and therefore the translocation of the A components across endosomal membranes into the cytosol and thus protects cells from intoxication with these toxins [1,17–20]. Such a translocation mechanism is common to other binary toxins, including anthrax toxin from *Bacillus anthracis* [1,21].

The enzymatic components develop their activity in the cytosol of the target cells where they ADP-ribosylate monomeric G-actin at position arginine 177 with NAD as co-substrate leading to actin depolymerization, cell rounding, and eventually cell death [1,22–26]. Similarly, other members of the family of binary toxins act also as ADP-ribosylating toxins. These are CDT (*Clostridium difficile* binary toxin) of *Clostridium difficile* [27–29], *Clostridium spiroforme* toxin [30], and the vegetative insecticidal proteins (VIPs) of *Bacillus cere*us and *Bacillus thuringiensis* [31,32].

The inhibition of channel function by binding components and intoxication of target cells by compounds that bind to the binding components is of considerable interest because of the possible use of A-B type of toxins as biological weapons. Possible candidates are tailored azolopyridinium salts and tailored cyclic dextrines [33–36]. In previous studies, we have demonstrated that low concentrations of chloroquine were able to inhibit intoxication of target cells by C2-toxin in cell-based assays and pore-formation by C2IIa in lipid bilayer membranes [37,38]. Similarly, blockage of iota b channels by chloroquine was also observed in reconstitution experiments with lipid bilayers but at much higher concentrations than those needed in experiments with C2IIa [39,40]. The binding site for chloroquine and related compounds in the channel formed by C2IIa was identified in the vestibule on the cis-side of the mushroom-sized heptamers that corresponds to the cell surface exposed side [41]. It is presumably the same binding site that also interacts also with the positively charged N-terminus of the enzymatic subunits C2I and Iota b and directs them to the channel lumen and further on into the cytosol of the target cells [1,3,40]. This means that binding is the prerequisite for transport. Site-directed mutagenesis of E399, D426, and F428 (corresponding to the Φ–clamp in PA [42,43]) in C2IIa has clearly demonstrated that these three amino acids are elements of the binding site within the vestibule of the channel formed by C2II [41]. These amino acids are also present in the primary sequence of Iota b in similar positions (D386, D413, and F415) and there exists no doubt that they are also involved in the binding site of the heptameric Iota b channel [40]. Besides these amino acids that are directly involved in binding of Iota a and chloroquine the sequence of Iota b also contains several threonines (T292 and T320) that are probably involved in the structure and stability of the pore-forming heptamers of Iota b. Their replacement by other amino acids leads to misfolded Iota b channels that have completely different properties than the ones formed by wildtype Iota b [40].

In this study, we investigated the binding of different chloroquine analogs to the channels formed by the binding components C2IIa and Iota b. The interaction between the protein-conducting nanopores and the different ligands was performed by titration experiments with artificial membranes containing C2IIa and Iota b channels. This type of investigation using the dose-dependent decrease of membrane conductance allowed a rapid and meaningful investigation of the affinity of the different chloroquine analogs to the binding site inside the vestibule of the heptameric channels. Similarly, we investigated the effect of the chloroquine analog with the highest affinity for binding to C2IIa and Iota b on the pH-dependent trans-membrane transport of the A components of C2- and Iota-toxin through the trans-membrane pores formed by the B components of these toxins in living cells. The results suggested indeed that this compound blocked the trans-membrane transport of these binary toxins with much higher efficiency than chloroquine.

2. Results

2.1. Binding of Different Aminoquinolinium Salts to the Channels Formed by the Binding Components C2IIa and Iota b

The channels formed by the binding components C2IIa and Iota b are fully oriented in artificial and presumably also in biological membranes when they are added to only one side of an artificial and biological membrane [37,39]. Most of the water-soluble part of the mushroom-sized heptamer is localized on the *cis*-side of the membrane (the side of addition of the binding components). Only a few amino acids at the end of the beta-barrel cylinder of 14 beta-strands are directed to the trans-side of the membrane. The structure is similar to that of α-toxin of *Staphylococcus aureus* and the recently elucidated 3D-structure of the membrane-spanning form of the PA_{63}-channel, which forms also a heptamer with some sort of vestibule on the *cis*-side [16,44]. In previous studies we demonstrated that reconstituted C2IIa channels as well as Iota b channels can be blocked in lipid bilayer membranes by the addition of 4-aminoquinolines [38,39,45] and identified the binding site for chloroquine to C2IIa channels on the *cis* side of the C2IIa heptamer within the vestibule of the channels [41]. The binding affinity strongly depends on negatively charged amino acids and also on the Φ-clamp within the vestibule of the C2IIa and Iota b channels. The stability constant *K* for ligand binding to the C2IIa and the Iota b channels was calculated by multi-channel titration experiments [37–40]. Similar experiments were performed here with the chloroquine analogs C 23, C 164, C 268, and C 280. Figure 1 shows an experiment of this type. Activated Iota b was added in a concentration of about 20 ng/mL to the *cis*-side (the side of the applied potential) of a black lipid bilayer membrane while stirring. The reconstitution of Iota b channels led to a substantial increase of membrane conductivity by several orders of magnitude caused by insertion of Iota b channels in the membrane monitored by a strip chart recorder. After about 30 min to several hours, when the membrane conductance was virtually stationary, the titration experiments started. Small amounts of concentrated solution of C 164 were added after about two hours after the start of the experiment to the aqueous phase on the *cis*-side of the membrane while stirring to allow equilibration. Subsequently, the Iota b channels were blocked and the dose-dependent decrease of conductance was measured as a function of time (see Figure 1).

The analysis of the data of Figure 1 indicated that the Iota b channels were not fully blocked by the addition of C 164 at a concentration of 4.03 mM. This was caused by the problem to reach sufficiently high concentrations of the chloroquine analog, which are limited by its solubility in aqueous salt solution. The fit of the titration data shown in Figure 2A suggest in principle that the Iota b channel was only blocked by about 50%. However, when the concentration of C 164 was extrapolated to higher ones (see Figure 2B), then it was clear that compound C 164 was also able to almost fully block the Iota b channel. The stability constant of binding of C 164 to the Iota b channels was about (348 ± 48) 1/M and the channel block was at maximum 92% ± 7%. This was a very low stability constant for binding of a channel blocker to one of the binding component channels. However, the data of binding of all chloroquine analogs and chloroquine itself demonstrates that the iota b channel is not a good target for binding of chloroquine and the different chloroquine analogs (aminoquinolinium salts) (see Table 1). Only chloroquine itself and C 280 had a reasonably high affinity to the Iota b channel. Chloroquine analog C 280 that has a permanent positive charge and a bulky side chain was used for the study of the inhibition of cell cytotoxicity by Iota-toxin (see Figure 3).

The affinity of chloroquine and the chloroquine analogs to the channels formed by C2IIa was definitely more substantial as the summary of stability constants and half saturation constants of Table 1 clearly demonstrates. Chloroquine and all aminoquinolinium salts used in this study have a higher affinity to the C2IIa channel than to that formed by Iota b. With the aminoquinolinium salts, the affinity to the C2IIa channels increased in the series C 23, C 164, C 268, and C 280 by factor of more than 4000. C 268 had already a binding constant to C2IIa channels that was about twofold higher than chloroquine. The highest stability constant for binding to the C2IIa channel was C 280, which had a half saturation constant of 0.16 μM. This low K_S value for binding of C 280 to C2IIa is about a factor of 60 smaller than that of chloroquine binding which is remarkable and suggested indeed that C 280 could serve as an inhibitor of intoxication by C2-toxin.

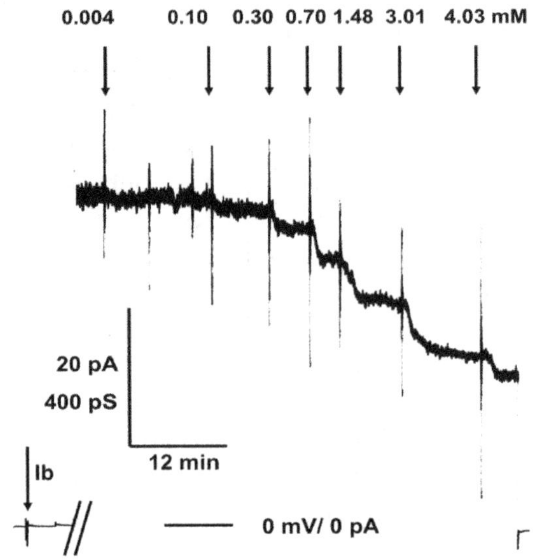

Figure 1. Titration experiment of Iota b-induced membrane conductance with C 164. The membrane was formed from diphytanoyl phosphatidylcholine/n-decane. The aqueous phase contained 20 ng/mL Iota b (Ib) protein (added to the *cis*-side of the membrane at the time of the left side arrow), 150 mM KCl, 10 mM MES, pH 6.0. The temperature was constantly 20 °C and the applied voltage was 50 mV. The two bars at the base line indicate the time interval of about 1h and 50 min between the addition of Ib and the start of the titration experiment. The membrane contained about 70 Iota b-channels (single channel conductance G = 15 pS) when C 164 was added at the indicated concentrations to the aqueous phase. The bottom line represents zero level of conductance (at begin of the experiment) or also zero level of current when voltage was switched off (at the end of titrations). Note that the high noise of the current recording during the titration experiment was caused by stirring in the membrane cell to allow rapid equilibration of C 164 in the aqueous phase.

Table 1. Stability constants *K* for the block of binding components channels formed by C2II and Iota b by chloroquine and related aminoquinolinium salts in lipid bilayer membranes [a]. * The results of similar titration experiments performed with C2IIa channels and chloroquine are given for comparison.

Chloroquine Analog	$K/10^3$ M^{-1}	K_S/μM	$K/10^3$ M^{-1}	K_S/μM
-	C2II		Iota b	
Chloroquine	110 *	9.1 *	7.1 ± 1.7	140
C 23	1.5 ± 0.4	710	0.82 ± 0.21	1200
C 164	18.5 ± 2.5	54	0.39 ± 0.13	2400
C 268	198 ± 15	5.1	2.5 ± 0.4	400
C 280	6200 ± 40	0.16	12.5 ± 2.1	80

[a] The data represent means ± SD of at least three individual titration experiments. The membranes were formed from diphytanoyl phosphatidylcholine/n-decane. The aqueous phase contained 150 mM KCl, 10 mM MES-KOH, pH 6, and about 10 ng/mL activated C2II or about 20 ng/mL Iota b; T = 20 °C. * The stability constant *K* for binding of chloroquine to C2IIa channels is given for comparison and was taken from Bachmeyer et al. (2003) [45].

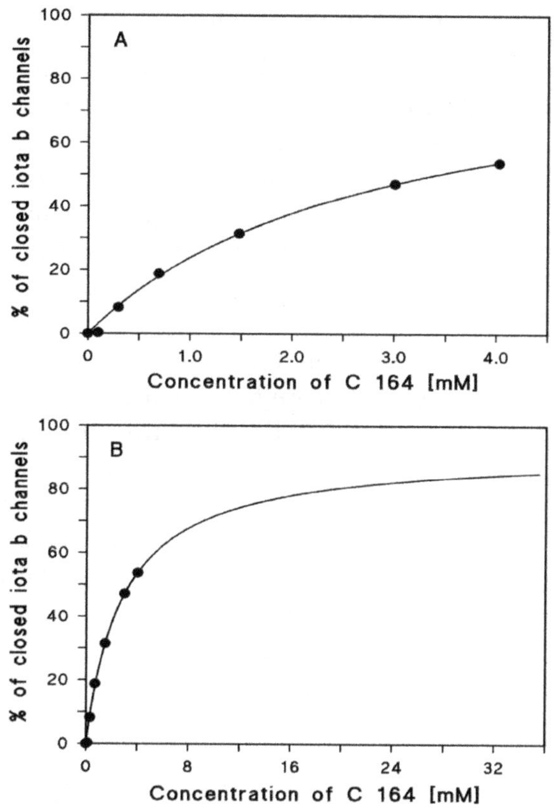

Figure 2. (**A**) Langmuir isotherm of the inhibition of Iota b-induced membrane conductance (about 70 Iota b-channels) by the aminoquinolinium salt C 164. The fit line corresponds to the data points taken from the titration experiment in Figure 1. The fit of the data was performed using Equation (2). The stability constant, K, for binding of C 164 to the Iota b-channels was (348 ± 48) 1/M (The channel block was at maximum 92% ± 7%; K_S = 2.9 mM (r^2 = 0.997645)); (**B**) Because of the low degree of inhibition in C 164 concentration range, we extrapolated its concentration to about 33 mM and used the same fit parameters as in A. The fit curve indicates that high concentration of C 164 almost fully blocked the Iota b channels.

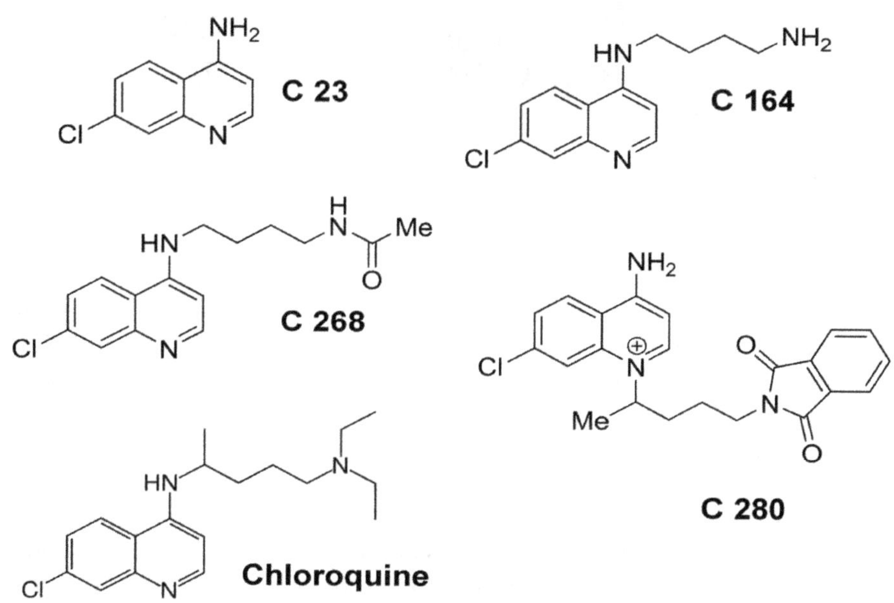

Figure 3. Structures of chloroquine and the chloroquine analogs (aminoquinolinium salts) used in this study). The chloroquine analogs were designated as suggested by Lödige (2013) [46].

2.2. C 280 Inhibited pH-Dependent Membrane Translocation of C2 and Iota Toxin

Prompted by the observation that C 280 interfered with the C2IIa and Iota b pores in vitro, we addressed the question whether C 280 also inhibits translocation of the enzyme components C2I and Iota a through their respective toxin pores across the membranes of intact cells. To this end, we used an established assay, which mimics the acidic conditions of endosomes on the surface of intact cells and allows direct translocation of the C2I and Iota a enzyme components into the cytosol through CIIa and Iota b pores, respectively, which were inserted in the plasma membrane under acidic conditions [34]. All steps of this assay were performed in the presence of Baf A1 to block the normal uptake of C2 and Iota toxins into the cytosol via acidified endosomes.

Vero cells were exposed to either C2 or Iota toxin under acidic conditions in the presence and absence of C 280 and subsequently the cells were further incubated under neutral conditions also with or without C 280. As shown in Figure 4, C2- and Iota-treated cells rounded up consequently to the acidic pulse. However, significantly less cells were rounded in the presence of C 280, indicating that less C2I or Iota a reached the cytosol. This result strongly suggests that C 280 inhibits membrane translocation of C2I and Iota a through the lumen of the C2IIa and Iota b pores, respectively. This result suggested that chloroquine-analogs similar to the structure of C 280 might provide interesting tools for the further study of aminoquinolinium salts as blockers for intoxication by C2-toxin. The block of intoxication of Iota-toxin by the same molecules does not look so promising because of the possible different structure of the Iota b channel (see Discussion).

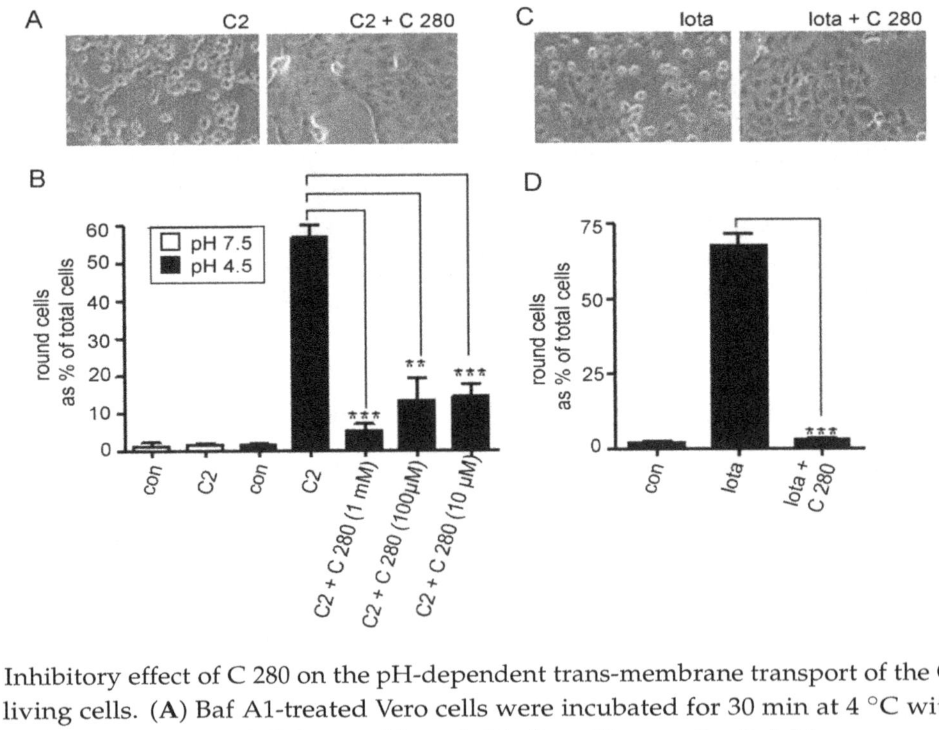

Figure 4. Inhibitory effect of C 280 on the pH-dependent trans-membrane transport of the C2 and iota toxins in living cells. (**A**) Baf A1-treated Vero cells were incubated for 30 min at 4 °C with C2 toxin (100 ng/mL C2I + 200 ng/mL C2IIa) to enable toxin binding. Noteworthy, Baf A1 was present to inhibit in a later step the "normal" transport of the A components of the internalized toxins into the cytosol via acidified endosomes, which is a prerequisite to investigate the toxin transport across the cytoplasmic membrane in this approach. For control (con), cells were incubated without toxin. Subsequently, cells were exposed for 5 min at 37 °C to acidic medium (pH 4.5 or to neutral medium pH 7.5 for control) and subsequently incubated at 37 °C in neutral medium containing Baf A1. During the acidic pulse, the B components insert as pores into the cytoplasmic membrane and the A components translocate through these pores into the cytosol of the cells and induce ADP-ribosylation of actin and cell-rounding. In this approach, the toxin-induced cell-rounding serves as an established specific and sensitive endpoint to monitor the uptake of the A components into the cytosol in the presence and absence of the inhibitor.

To test the effect of C 280 on toxin translocation, the indicated concentrations of C 280 were present in the medium during acidic pulse and the subsequent incubation periods. Pictures were taken after 1 and 2 h to document cell rounding, i.e., intoxication of cells (shown in A for 2 h and 1 mM C 280); (**B**) The percentage of intoxicated cells was determined after 1 h, and values are given as mean ± S.D. ($n = 3$). Significance was tested between cells, which have been treated with C2 toxin either in the absence or presence of C 280 by using the student's t-test (*** $p < 0.0005$, ** $p < 0.005$); (**C**) To test the influence of C 280 on membrane translocation of iota toxin, Baf A1-treated Vero cells were exposed for 15 min at ph 4.0 to Iota toxin (100 ng/mL Ia + 200 ng/mL Ib) in the presence or absence of 1 mM C 280. Cells were incubated for further 2 h at 37 °C in neutral medium containing Baf A1 in the presence or absence of C 280. Pictures were taken and the percentage of round cells was determined (D) Values are given as mean ± S.D. ($n = 3$). Significance was tested between cells, which have been treated with iota toxin either in the absence or presence of C 280 by using the student's t-test (*** $p < 0.0005$, ** $p < 0.005$).

3. Discussion

3.1. The Structure of the Aminoquinolinium Salts Allows an Interesting Insight in Structural Elements Required for Efficient Binding Protein Channel Blocking

The experiments with different aminoquinolinium salts to block the C2IIa channel allow an interesting insight in the structural requirement for efficient channel blocking, which can be considered as important tool for further in vivo and in vitro studies. The simplest chloroquine analog in the study here is C 23 (4-amino-7-chloroquinoline) which represents only the heterocyclic (bicyclic) part of the chloroquine molecule without side chain. C 23 has the smallest binding affinity to the C2IIa channels with a half saturation constant of about 700 μM for binding to C2IIa. The addition of n-butylamine to the amino group at the bicyclic molecule C 23 decreased the half saturation constant K_S for binding to C2IIa by a factor of more than 10 to 54 μM. The further addition of an acetyl group to the amino group of C 164 led to an additional decrease of the half saturation constant for the resulting C 268 molecule to about 5 μM. The half saturation constant is already below that of chloroquine, which means that C 268 is already a more efficient blocker of C2IIa channels than chloroquine. Another big step to improve binding of the aminoquinolinium salts to the C2IIa channel is the attachment of the bulky side chain to the nitrogen in the pyridine ring of C 23 (see Figure 3). The resulting chloroquine analog (C 280) has the highest affinity for binding to C2IIa but also for binding to Iota b. The half saturation constant for C 280 binding to C2IIa drops down to 0.16 μM, which is about 60-times lower than the half saturation constant for chloroquine binding.

The situation is in the case of binding of the aminoquinolinium salts to the Iota b channel not such straightforward as in the case of C2IIa. The reason for this is that the half saturation constant for binding shows some increase from C 23 to C 164 and then starts to become smaller in the series C 164, C 268, and C 280 (see Table 1). However, all half saturation constants for binding of the aminoquinolinium salts to Iota b were considerably higher than those for their binding to C2IIa. Even C 280, which had the highest binding affinity to Iota b with a half saturation constant of 80 μM showed a half saturation constant that was about 500 times higher than the corresponding constant for binding to C2IIa, which is remarkable. Similarly, the half saturation constant for binding of C 280 to the Iota b channel (80 μM) is only little smaller that for binding of chloroquine to the channel (140 μM), which looks a little strange when the situation is again compared to that of C2IIa. Nevertheless, the binding of C 280 to the Iota b channel is still strong enough that inhibition of the translocation of Iota a through Iota b is blocked in cell-based assays. This means that also certain aminoquinolinium salts could serve as blockers for intoxication of cells by Iota-toxin.

3.2. What Could Be the Reason That Blockers of Channel Function Have a Much Smaller Affinity to Iota b Than to C2IIa?

The results presented here and in previous studies demonstrated that chloroquine had much lower binding affinity to Iota b channels as compared to binding to C2IIa heptamers [37–40]. Similarly, the affinity of the aminoquinolinium salts to both binding protein channels differed considerably (see Table 1). The reason for this discrepancy is not quite clear because many structural elements that probably contribute to the binding site in the vestibule of the two channels are present in both primary sequences. There are the two rings of—at maximum—seven negatively charged residues in the binding site of the heptamer (D386 in Iota b corresponding to E399 in C2IIa and D413 in Iota b corresponding to D426 in C2IIa [41]). The only difference is that Iota b contains aspartate in position 386 whereas C2IIa has in the corresponding position 399 a glutamate. However, this very conservative exchange should not interfere much with the properties of the binding site in the vestibule of the two heptamers. Similarly, both cannels contain an Φ-clamp (F415 in Iota b and F428 in C2IIa), which is also an important structural element in both channels [40,41]. Thus, it is only slightly understandable that they differ so substantially in binding of the aminoquinolinium salts. The only remarkable difference between Iota b and C2IIa is the number of negatively charged groups within the channel-forming domain itself [41]. Whereas the membrane-spanning beta-sheet structure of C2IIa contains one glutamate (E307), there are no charges in the membrane spanning part of Iota b. This could represent the difference. However, experiments with a C2IIa mutant where E307 was replaced by lysine demonstrated that the E307K C2IIa mutant channel had approximately the same affinity for chloroquine as wildtype C2IIa [47]. This means presumably that charges within the pore-forming domain of the binding protein heptamers are most likely not essential for aminoquinolinium salt binding. This has to do with the strong image force along the channel that is created by the many negatively charged groups within the vestibule of the channels [47]. Taken together, it seems moreover that the structure of the binding site within the vestibule of the Iota b heptamers shows some structural differences to those of C2IIa and PA that result in a lower affinity for binding of chloroquine and its analogs [41,48].

3.3. The Aminoquinolinium Salts Inhibit the Trans-Membrane Transport of the A Components of the Binary C2 and Iota Toxins Through the Pores Formed under Acidic Conditions by the B Components in Membranes of Living Cells and Protect Cells from Intoxication with These Toxins

As expected from the in vitro results with lipid bilayers, the compounds inhibited the pH-triggered trans-membrane transport of the enzyme components of C2-toxin and Iota-toxin through the pores formed by the B-components of these toxins in cell membranes under acidic conditions, thereby protecting the cells from intoxication. This was tested in a well-established experimental approach where the situation of acidic endosomal vesicles was experimentally mimicked on the surface of living cultured epithelial cells. Vero cells were pre-treated with Baf A1 to inhibit endosomal acidification and thereby prevent the "normal" uptake of the binary C2 and iota toxins via acidic endosomes [17–19]. Then, the cells were incubated for 30 min at 4 °C with the respective toxin to enable toxin binding to the cell surface receptors but not receptor-mediated endocytosis, which does not occur at 4 °C. Subsequently, the cells were exposed to a short acidic pulse at 37 °C to trigger the conformational change of the cell-bound B components, which then form pores in the cytoplasmic membrane and mediate the transport of their bound A components through these pores across the cytoplasmic membrane into the cytosol of the cells. This transport of the A components into the cytosol results in ADP-ribosylation of actin and cell-rounding, which serves as a highly specific and sensitive endpoint to monitor the uptake of the A components into the cytosol in the presence and absence of the inhibitor. No cell-rounding was observed without toxin, indicating that the acidic conditions alone had no effect on cell morphology, or when cells were exposed to neutral medium because in this case, Baf A1 prevented the toxin uptake into the cytosol (Figure 4). As shown in Figure 4, cells only rounded up when the toxin was bound prior to the acidic pulse, indicating translocation of the A components into the cytosol. Significantly fewer cells rounded up under such conditions in the presence of C 280, clearly indicating the inhibitory effect of this compound on the membrane transport of the binary toxins C2 and iota in living cells. The aminoquinolinium salts had no effect on cell morphology under the experimental conditions used in this study. They have also the unique advantage that their toxicology is known from malaria treatment. In conclusion, these compounds should represent attractive lead compounds for development of novel pharmacological inhibitors against binary clostridial actin ADP-ribosylating toxins and may further be used against related binary toxins from pathogenic bacteria.

4. Experimental Procedures

4.1. Materials

The recombinant components of C2 toxin, C2I and C2II, were expressed as GST fusion proteins in *Escherichia (E.) coli* BL21 cells and purified as described [8]. To obtain biologically active C2IIa, C2II was treated with trypsin as reported earlier [8]. Iota a and Iota b were kind gifts of Dr. Michel R. Popoff (Institut Pasteur, Paris, France) [11]. Iota a and Iota b were activated by α-chymotrypsin as described previously [11,40,49]. The heterocyclic chloroquine analogs (aminoquinolinium salts) C 23, C 164, C 268, and C 280 (see Figure 3) were kind gifts of Dr. Gerhard Bringmann and Dr. Melanie Lödige, Institute for Organic Chemistry, University of Würzburg, 97074 Würzburg, Germany. The chloroquine analogs were termed and synthesized as was described in detail recently [46]. The chloroquine analogs were dissolved in ultrapure water supplemented with 10% (*v/v*) ETOH. Cell culture media (DMEM, MEM) and fetal calf serum were obtained from Invitrogen (Karlsruhe, Germany) and cell culture materials from Techno Plastic Products. (Trasadingen, Switzerland). Complete® protease inhibitor and streptavidin-peroxidase were from Roche (Mannheim, Germany), Baf A1 from Calbiochem (Bad Soden, Germany), and biotinylated NAD^+ from R&D Systems GmbH (Wiesbaden-Nordenstadt, Germany).

All salts (analytical grade) were obtained from Sigma-Aldrich Chemie GmbH (München, Germany) and were dissolved in ultrapure H_2O (Milli-Q® systems, Merck Millipore, Darmstadt, Germany). Diphytanoyl phosphatidylcholine (DiPhPC) was obtained from Avanti Polar Lipids Alabaster AL and n-decane (analytical grade from Merck, Darmstadt, Germany).

4.2. Methods

4.2.1. Cell Culture and Cytotoxicity Tests

African green monkey kidney (Vero) cells were cultivated at 37 °C and 5% CO_2 in MEM containing 10% FCS, 1.5 g/L sodium bicarbonate, 1 mM sodium-pyruvate, 2 mM L-glutamine, 0.1 mM non-essential amino acids. Vero cells were reseeded twice a week for, at most, 15–20 times. The macrophage-like murine J774A.1 cells were cultivated at 37 °C and 5% CO_2 in DMEM containing 10% FCS and 4 mM L-glutamate. For cytotoxicity experiments with C2-toxin or Iota-toxin, Vero cells were incubated at 37 °C in 1 mL serum-free medium containing both components of C2 toxin (200 ng/mL C2IIa + 100 ng/mL C2I) or iota toxin (200 ng/mL Iota b + 100 ng/mL Iota a). After different incubation periods, the toxin-induced cell-rounding was documented with a Zeiss Axiovert 40CFl microscope (Zeiss, Oberkochen, Germany) containing a Jenoptik progress C10 CCD camera (Carl Zeiss GmbH, Jena, Germany) and the percentage of round cells was determined from the pictures [19]. Inhibitory effects of C 280 were analyzed by incubating the cells with toxin in the presence of C 280.

4.2.2. Toxin-translocation Assay with Intact Vero Cells

The pH-dependent translocation of C2 toxin across the cytoplasmic membranes of intact Vero cells was performed as described earlier [8]. In brief, Vero cells were pretreated for 30 min at 37 °C with Baf A1 (100 nM) to prevent normal internalization of the toxin via acidified endosomes. Subsequently, cells were incubated at 4 °C in serum-free medium with C2IIa (200 ng/mL) and C2I (100 ng/mL) to enable toxin binding. Cells were washed and exposed for 5 min to warm acidic medium (37 °C, pH 4.5, Baf A1) to trigger insertion of cell-bound C2IIa into the cytoplasmic membrane and subsequent translocation of C2I through the pores across the membrane. Subsequently, the cells were further incubated at 37 °C in complete medium under neutral conditions in the presence of Baf A1 and C2I-toxin induced cell-rounding was documented by photography. The pH-driven translocation of cell-bound Iota toxin across the cytoplasmic membrane of Vero cells was performed as described earlier by Blöcker et al. (2001) [50]. Baf A1-treated cells were exposed for 15 min at 37 °C to acidic medium (pH 4.0) containing Iota toxin (100 ng/mL Iota a + 200 ng/mL Iota b) and subsequently incubated at 37 °C in neutral medium containing Baf A1. The number of round cells was determined to document the cytopathic action of Iota toxin. To test an inhibitory effect of C 280 on membrane translocation of C2- and Iota-toxins, C 280 was applied to the medium during the acidic pulse and the subsequent incubation periods, pictures from the cells were taken and the number of round cells was determined from the pictures.

4.2.3. Lipid Bilayer Experiments

The experiments with planar lipid bilayers were performed as has been described previously in detail [51]. In brief membranes were formed by the painting method using DiPhPC dissolved to 1% (*w/v*) in n-decane. The membrane hole had an area of about 0.4 mm^2 in the thin wall separating two 5 mL compartments in a Teflon cell. The different binding components (C2IIa and Iota b) were added from concentrated solutions to the aqueous phase either immediately before membrane formation or after the membranes had turned black in concentrations of about 1 to 10 ng/mL. The temperature was maintained at 20 °C during all experiments. The membrane conductance induced by channels formed by the binding components C2IIa and Iota b was measured after application of fixed membrane potentials with a pair of silver/silver chloride electrodes with salt bridges inserted into the aqueous compartments on both sides of the DiPhPC membranes. The electrodes were connected in series to a voltage source and a homemade current-to-voltage converter made with a Burr Brown operational amplifier. The amplified signal was monitored on a digital storage oscilloscope (OWON) and recorded on a strip chart recorder.

4.2.4. Titration Experiments with the Different Aminoquinolinium Salts

The binding of the compounds C 23, C 164, C 268, and C 280 (see Figure 3) to the channels formed by the binding components was investigated with titration experiments similar to those used previously to study the binding of carbohydrates to the LamB-channel of Escherichia coli [52] and the binding of tailored azolopyridinium salts to channels formed by protective antigen (PA) [33] and C2IIa [34]. About 30 min after start of the reconstitution of the binding components into lipid bilayer membranes, their reconstitution rate in the membranes became very small. Then concentrated solutions of different aminoquinolinium salts were added to both sides of the membranes while stirring to allow equilibration. The results of the titration experiments were analyzed in a similar way as has been performed previously for the binding of azolopyridinium salts to channels formed by protective antigen (PA) [33] and C2IIa [34]. The conductance, *G(c)* of the membrane at a given concentration, *c*, of the different aminoquinolinium salts C 23, C 164, C 268, and C 280 relative to the initial conductance, *Gmax* (in the absence of the ligands), was analyzed using the following equation, which corresponds to Langmuir adsorption isotherms [33,34]:

$$\%\text{ of fraction of blocked channels} = \frac{(G_{\max} - G(c))}{G_{\max}} = \frac{100 \cdot K \cdot c}{(K \cdot c + 1)} \quad (1)$$

K is the stability constant for the binding of aminoquinolinium salts to channels formed by Iota b and C2IIa [33]. The half saturation constant, K_S of this process is given by the inverse stability constant $1/K$. K can be derived from the titration experiments by a fit of the experimental data to Equation (1). We did not observe full channel blockage in all titration experiments. In cases of only partial blockage of the C2IIa and Iota b channels, Equation (1) had to be modified to account for the reduced maximum blockage given by A in percent (the maximum degree A of blockage was in all cases between 85% to 100%):

$$\%\text{ of fraction of blocked channels} = \frac{(G_{\max} - G(c))}{G_{\max}} = \frac{K \cdot c}{(K \cdot c + 1)} \quad (2)$$

Acknowledgments: The authors would like to thank Gerhard Bringmann and Melanie Lödige, Institute for Organic Chemistry, University of Würzburg for the kind gift of the aminoquinolinium salts C 23, C 164, C 268, and C 280 and for their contribution at the early stage of this work. We thank Michel R. Popoff, Institut Pasteur Paris, France for providing the Iota-toxin. We thank Ulrike Binder, University of Ulm, and Jennifer Rausch, University of Würzburg for expert technical assistance. This work was financially supported by the Deutsche Forschungsgemeinschaft (DFG, grant BA 2087/2-2).

Author Contributions: Angelika Kronhardt, Christoph Beitzinger, Holger Barth, and Roland Benz conceived and designed the experiments; Angelika Kronhardt, Christoph Beitzinger, and Holger Barth performed the experiments; Holger Barth and Roland Benz analyzed the data; Holger Barth and Roland Benz wrote the paper.

References

1. Barth, H.; Aktories, K.; Popoff, M.R.; Stiles, B.G. Binary bacterial toxins: Biochemistry, biology, and applications of common Clostridium and Bacillus proteins. *Microbiol. Mol. Biol. Rev.* **2004**, *68*, 373–402. [CrossRef] [PubMed]
2. Aktories, K.; Barth, H. New insights into the mode of action of the actin ADP-ribosylating virulence factors *Salmonella enterica* SpvB and *Clostridium botulinum* C2 toxin. *Eur. J. Cell Biol.* **2011**, *90*, 944–950.
3. Knapp, O.; Benz, R.; Popoff, M.R. Pore-forming activity of clostridial binary toxins. *Biochim. Biophys. Acta* **2016**, *1858*, 512–525. [CrossRef] [PubMed]
4. Friedlander, A.M. Macrophages are sensitive to anthrax lethal toxin through an acid-dependent process. *J. Biol. Chem.* **1986**, *261*, 7123–7126. [PubMed]
5. Mock, M.; Fouet, A. Anthrax. *Annu. Rev. Microbiol.* **2001**, *55*, 647–671. [CrossRef] [PubMed]
6. Collier, R.J.; Young, J.A. Anthrax toxin. *Annu. Rev. Cell. Dev. Biol.* **2003**, *19*, 45–70. [CrossRef] [PubMed]
7. Ohishi, I.; Iwasaki, M.; Sakaguchi, G. Purification and characterization of two components of botulinum C2 toxin. *Infect. Immun.* **1980**, *30*, 668–673. [PubMed]
8. Barth, H.; Blöcker, D.; Behlke, J.; Bergsma-Schutter, W.; Brisson, A.; Benz, R.; Aktories, K. Cellular uptake of *Clostridium botulinum* C2 toxin requires oligomerization and acidification. *J. Biol. Chem.* **2000**, *275*, 18704–18711. [CrossRef] [PubMed]
9. Stiles, B.G.; Wilkins, T.D. Purification and characterization of *Clostridium perfringens* iota toxin: Dependence on two nonlinked proteins for biological activity. *Infect. Immun.* **1986**, *54*, 683–688. [PubMed]
10. Popoff, M.R.; Boquet, P. Clostridium spiroforme toxin is a binary toxin which ADP-ribosylates cellular actin. *Biochem. Biophys. Res. Commun.* **1988**, *152*, 1361–1368. [CrossRef]
11. Gibert, M.; Petit, L.; Raffestin, S.; Okabe, A.; Popoff, M.R. *Clostridium perfringens* iota-toxin requires activation of both binding and enzymatic components for cytopathic activity. *Infect Immun.* **2000**, *68*, 3848–3853. [CrossRef] [PubMed]
12. Popoff, M.R.; Stiles, B.G. Bacterial toxins and virulence factors targeting the actin cytoskeleton and intercellular junctions. In *The Comprehensive Sourcebook of Bacterial Toxins*; Alouf, J.E., Popoff, M.R., Eds.; Academic Press: London, UK, 2006.
13. Schleberger, C.; Hochmann, H.; Barth, H.; Aktories, K.; Schulz, G.E. Structure and action of the binary C2 toxin from *Clostridium botulinum*. *J. Mol. Biol.* **2006**, *364*, 705–715. [CrossRef] [PubMed]
14. Petosa, C.; Collier, R.J.; Klimpel, K.R.; Leppla, S.H.; Liddington, R.C. Crystal structure of the anthrax toxin protective antigen. *Nature* **1997**, *385*, 833–838. [CrossRef] [PubMed]
15. Nguyen, T.L. Three-dimensional model of the pore form of anthrax protective antigen. Structure and biological implications. *J. Biomol. Struct. Dyn.* **2004**, *22*, 253–265. [CrossRef] [PubMed]
16. Jiang, J.; Pentelute, B.L.; Collier, R.J.; Zhou, Z.H. Atomic structure of anthrax protective antigen pore elucidates toxin translocation. *Nature* **2015**, *521*, 545–549. [CrossRef] [PubMed]
17. Haug, G.; Leemhuis, J.; Tiemann, D.; Meyer, D.K.; Aktories, K.; Barth, H. The host cell chaperone Hsp90 is essential for translocation of the binary *Clostridium botulinum* C2 toxin into the cytosol. *J. Biol. Chem.* **2003**, *278*, 32266–32274. [CrossRef] [PubMed]
18. Kaiser, E.; Pust, S.; Kroll, C.; Barth, H. Cyclophilin A facilitates translocation of the *Clostridium botulinum* C2 toxin across membranes of acidified endosomes into the cytosol of mammalian cells. *Cell. Microbiol.* **2009**, *11*, 780–795. [CrossRef] [PubMed]

19. Kaiser, E.; Kroll, C.; Ernst, K.; Schwan, C.; Popoff, M.; Fischer, G.; Buchner, J.; Aktories, K.; Barth, H. Membrane translocation of binary actin-ADP-ribosylating toxins from *Clostridium difficile* and *Clostridium perfringens* is facilitated by cyclophilin A and Hsp90. *Infect. Immun.* **2011**, *79*, 3913–3921. [CrossRef] [PubMed]
20. Zhang, S.; Finkelstein, A.; Collier, R.J. Evidence that translocation of anthrax toxin's lethal factor is initiated by entry of its N terminus into the protective antigen channel. *Proc. Natl. Acad. Sci. USA* **2004**, *101*, 16756–16761. [CrossRef] [PubMed]
21. Young, J.A.; Collier, R.J. Anthrax toxin: Receptor binding, internalization, pore formation, and translocation. *Annu. Rev. Biochem.* **2007**, *76*, 243–265. [CrossRef] [PubMed]
22. Aktories, K.; Bärmann, M.; Ohishi, I.; Tsuyama, S.; Jakobs, K.H.; Habermann, E. Botulinum C2 toxin ADP-ribosylates actin. *Nature* **1986**, *322*, 390–392. [CrossRef] [PubMed]
23. Vandekerckhove, J.; Schering, B.; Bärmann, M.; Aktories, K. Botulinum C2 toxin ADP-ribosylates cytoplasmic beta/gamma-actin in arginine 177. *J. Biol. Chem.* **1988**, *263*, 696–700. [PubMed]
24. Schering, B.; Barmann, M.; Chhatwal, G.S.; Geipel, U.; Aktories, K. ADP-ribosylation of skeletal muscle and non-muscle actin by *Clostridium perfringens* iota toxin. *Eur. J. Biochem.* **1988**, *171*, 225–229. [CrossRef] [PubMed]
25. Heine, K.; Pust, S.; Enzenmüller, S.; Barth, H. ADP-ribosylation of actin by *Clostridium botulinum* C2 toxin in mammalian cells results in delayed caspase-dependent apoptotic cell death. *Infect. Immun.* **2008**, *76*, 4600–4608. [CrossRef] [PubMed]
26. Popoff, M.R.; Bouvet, P. Clostridial toxins. *Future Microbiol.* **2009**, *4*, 1021–1064. [CrossRef] [PubMed]
27. Gülke, I.; Pfeifer, G.; Liese, J.; Fritz, M.; Hofmann, F.; Aktories, K.; Barth, H. Characterization of the enzymatic component of the ADP-ribosyltransferase toxin CDTa from *Clostridium difficile*. *Infect. Immun.* **2001**, *69*, 6004–6011. [CrossRef] [PubMed]
28. Perelle, S.; Gibert, M.; Bourlioux, P.; Corthier, G.; Popoff, M.R. Production of a complete binary toxin (actin-specific ADP-ribosyltransferase) by *Clostridium difficile* CD196. *Infect. Immun.* **1997**, *65*, 1402–1407. [PubMed]
29. Popoff, M.R.; Rubin, E.J.; Gill, D.M.; Boquet, P. Actin-specific ADP-ribosyltransferase produced by a *Clostridium difficile* strain. *Infect. Immun.* **1988**, *56*, 2299–2306. [PubMed]
30. Popoff, M.R.; Boquet, P. *Clostridium spiroforme* toxin is a binary toxin which ADP-ribosylates cellular actin. *Biochem. Biophys. Res. Commun.* **1988**, *152*, 1361–1368. [CrossRef]
31. Han, S.; Craig, J.A.; Putnam, C.D.; Carozzi, N.B.; Tainer, J.A. Evolution and mechanism from structures of an ADP-ribosylating toxin and NAD complex. *Nat. Struct. Biol.* **1999**, *6*, 932–936. [PubMed]
32. Leuber, M.; Orlik, F.; Schiffler, B.; Sickmann, A.; Benz, R. Vegetative insecticidal protein (Vip1Ac) of *Bacillus thuringiensis* HD201: Evidence for oligomer and channel formation. *Biochemistry* **2006**, *45*, 283–288. [CrossRef] [PubMed]
33. Beitzinger, C.; Bronnhuber, A.; Duscha, K.; Riedl, Z.; Huber-Lang, M.; Benz, R.; Hajós, G.; Barth, H. Designed azolopyridinium salts block protective antigen pores in vitro and protect cells from anthrax toxin. *PLoS ONE* **2013**, *8*. [CrossRef] [PubMed]
34. Bronnhuber, A.; Maier, E.; Riedl, Z.; Hajós, G.; Benz, R.; Barth, H. Inhibitions of the translocation pore of *Clostridium botulinum* C2 toxin by tailored azolopyridinium salts protects human cells from intoxication. *Toxicology* **2014**, *316*, 25–33. [CrossRef] [PubMed]
35. Nestorovich, E.M.; Karginov, V.A.; Popoff, M.R.; Bezrukov, S.M.; Barth, H. Tailored ß-cyclodextrin blocks the translocation pores of binary exotoxins from *C. botulinum* and *C. perfringens* and protects cells from intoxication. *PLoS ONE* **2011**, *6*. [CrossRef] [PubMed]
36. Roeder, M.; Nestorovich, E.M.; Karginov, V.A.; Schwan, C.; Aktories, K.; Barth, H. Tailored cyclodextrin pore blocker protects mammalian cells from *Clostridium difficile* binary toxin CDT. *Toxins (Basel)* **2014**, *6*, 2097–2114. [CrossRef] [PubMed]
37. Schmid, A.; Benz, R.; Just, I.; Aktories, K. Interaction of *Clostridium botulinum* C2 toxin with lipid bilayer membranes. Formation of cation-selective channels and inhibition of channel function by chloroquine. *J. Biol. Chem.* **1994**, *269*, 16706–16711. [PubMed]
38. Bachmeyer, C.; Benz, R.; Barth, H.; Aktories, K.; Gilbert, M.; Popoff, M.R. Interaction of *Clostridium botulinum* C2 toxin with lipid bilayer membranes and Vero cells: Inhibition of channel function by chloroquine and related compounds in vitro and intoxification in vivo. *FASEB J.* **2001**, *15*, 1658–1660. [CrossRef] [PubMed]

39. Knapp, O.; Benz, R.; Gibert, M.; Marvaud, J.C.; Popoff, M.R. Interaction of *Clostridium perfringens* iota-toxin with lipid bilayer membranes. Demonstration of channel formation by the activated binding component Ib and channel block by the enzyme component Ia. *J. Biol. Chem.* **2002**, *277*, 6143–6152. [CrossRef] [PubMed]
40. Knapp, O.; Maier, E.; Waltenberger, E.; Mazuet, C.; Benz, R.; Popoff, M.R. Residues involved in the pore-forming activity of the *Clostridium perfringens* iota toxin. *Cell Microbiol.* **2015**, *17*, 288–302. [CrossRef] [PubMed]
41. Neumeyer, T.; Schiffler, B.; Maier, E.; Lang, A.E.; Aktories, K.; Benz, R. *Clostridium botulinum* C2 toxin. Identification of the binding site for chloroquine and related compounds and influence of the binding site on properties of the C2II channel. *J. Biol. Chem.* **2008**, *283*, 3904–3914. [CrossRef] [PubMed]
42. Krantz, B.A.; Melnyk, R.A.; Zhang, S.; Juris, S.J.; Lacy, D.B.; Wu, Z.; Finkelstein, A.; Collier, R.J. A phenylalanine clamp catalyzes protein translocation through the anthrax toxin pore. *Science* **2005**, *309*, 777–781. [CrossRef] [PubMed]
43. Melnyk, R.A.; Collier, R.J. A loop network within the anthrax toxin pore positions the phenylalanine clamp in an active conformation. *Proc. Natl. Acad. Sci. USA* **2006**, *103*, 9802–9807. [CrossRef] [PubMed]
44. Song, L.; Hobaugh, M.R.; Shustak, C.; Cheley, S.; Bayley, H.; Gouaux, J.E. Structure of staphylococcal alpha-hemolysin, a heptameric transmembrane pore. *Science* **1996**, *274*, 1859–1866. [CrossRef] [PubMed]
45. Bachmeyer, C.; Orlik, F.; Barth, H.; Aktories, K.; Benz, R. Mechanism of C2-toxin inhibition by fluphenazine and related compounds: Investigation of their binding kinetics to the C2II-channel using the current noise analysis. *J. Mol. Biol.* **2003**, *333*, 527–540. [CrossRef] [PubMed]
46. Lödige, M. Synthese und Evaluierung Neuartiger Wirkstoffklassen Gegen Infektionskrankheiten (Synthesis and Evaluation of Novel Drug Classes Against Infectious Diseases). Ph.D. Thesis, University of Würzburg, Würzburg, Germany, 29 July 2013.
47. Blöcker, D.; Bachmeyer, C.; Benz, R.; Aktories, K.; Barth, H. Channel formation by the binding component of *Clostridium botulinum* C2 toxin: Glutamate 307 of C2II affects channel properties in vitro and pH-dependent C2I translocation in vivo. *Biochemistry* **2003**, *42*, 5368–5377. [CrossRef] [PubMed]
48. Orlik, F.; Schiffler, B.; Benz, R. Anthrax toxin protective antigen: Inhibition of channel function by chloroquine and related compounds and study of binding kinetics using the current noise analysis. *Biophys. J.* **2005**, *88*, 1715–1724. [CrossRef] [PubMed]
49. Ohishi, I. Activation of botulinum C2 toxin by trypsin. *Infect. Immun.* **1987**, *55*, 1461–465. [PubMed]
50. Blöcker, D.; Behlke, J.; Aktories, K.; Barth, H. Cellular uptake of the binary *Clostridium perfringens* iota toxin. *Infect. Immun.* **2001**, *69*, 2980–2987. [CrossRef] [PubMed]
51. Benz, R.; Janko, K.; Boos, W.; Lauger, P. Formation of large, ion-permeable membrane channels by the matrix protein (porin) of *Escherichia coli*. *Biochim. Biophys. Acta* **1978**, *511*, 305–319. [CrossRef]
52. Benz, R.; Schmid, A.; Vos-Scheperkeuter, G.H. Mechanism of sugar transport through the sugar-specific LamB channel of *Escherichia coli* outer membrane. *J. Membr. Biol.* **1987**, *100*, 21–29. [CrossRef] [PubMed]

3

Antitoxin ε Reverses Toxin ζ-Facilitated Ampicillin Dormants

María Moreno-del Álamo, Chiara Marchisone and Juan C. Alonso *

Department of Microbial Biotechnology, Centro Nacional de Biotecnología, CNB-CSIC, 3 Darwin Str., 28049 Madrid, Spain; mmoreno@cnb.csic.es (M.M.-d.Á.); cmarchisone@cnb.csic.es (C.M.)

* Correspondence: jcalonso@cnb.csic.es

Abstract: Toxin-antitoxin (TA) modules are ubiquitous in bacteria, but their biological importance in stress adaptation remains a matter of debate. The inactive ζ-ε_2-ζ TA complex is composed of one labile ε_2 antitoxin dimer flanked by two stable ζ toxin monomers. Free toxin ζ reduces the ATP and GTP levels, increases the (p)ppGpp and c-di-AMP pool, inactivates a fraction of uridine diphosphate-*N*-acetylglucosamine, and induces reversible dormancy. A small subpopulation, however, survives toxin action. Here, employing a genetic orthogonal control of ζ and ε levels, the fate of bacteriophage SPP1 infection was analyzed. Toxin ζ induces an active slow-growth state that halts SPP1 amplification, but it re-starts after antitoxin expression rather than promoting abortive infection. Toxin ζ-induced and toxin-facilitated ampicillin (Amp) dormants have been revisited. Transient toxin ζ expression causes a metabolic heterogeneity that induces toxin and Amp dormancy over a long window of time rather than cell persistence. Antitoxin ε expression, by reversing ζ activities, facilitates the exit of Amp-induced dormancy both in *rec*$^+$ and *recA* cells. Our findings argue that an unexploited target to fight against antibiotic persistence is to disrupt toxin-antitoxin interactions.

Keywords: toxin-antitoxin system; cell wall inhibition; persistence; nucleotide hydrolysis; uridine diphosphate-*N*-acetylglucosamine

Key Contribution: The understanding of bacterial persistence might enable us to devise treatments against such threat. Toxin ζ, which is a UNAG-dependent ATPase, interferes with purine nucleotide homeostasis, and this physiological heterogeneity induces reversible dormancy and a subpopulation of survivors. Transient ζ expression indirectly reduces Amp persistence without massive autolysis. Antitoxin ε expression reverses toxin action and induces the exit of cells from an Amp dormant state. A deeper understanding of the metabolic state to inhibit toxin action, which sensitizes bacterial cells to different antibiotics, and disruption of antitoxin-mediated reversion of toxic action, are crucial to overcome antibiotic persistence.

---

1. Introduction

Bacteria have evolved complex regulatory controls and a diverse repertoire of cell transition states in response to various environmental stresses. In order to survive, cells slow down their growth rate and redirect their metabolic resources, until conditions improve and growth can increase [1]. It was shown early in *Staphylococcus aureus* that a large bacterial fraction is unable to survive to lethal doses of penicillin but a subpopulation survives, showing a biphasic inactivation curve [2,3]. It was proposed that in the presence of phenotypic heterogeneity (noise at the level of transcription of key genes during the life cycle), stochastic and antibiotic-induced stress leads to a non-genetic phenotypic "antibiotic insensitive" state, which was termed as persistence [3–6].Persisters have been featured throughout due to their important role in bacterial infections. In the Proteobacteria Phylum

(*Escherichia coli* being the best-characterized representative) starvation-induced (p)ppGpp contributes to antibiotic persistence, whereas in bacteria of the Firmicutes phylum, the persistence effectors are more complex [7–10]. It has been proposed that in *S. aureus* the *bona fide* persistence effector is ATP depletion [11], but treatment with ATP synthesis inhibitors also induces (p)ppGpp accumulation and decreases GTP levels [12]. In *Bacillus subtilis*, which is the best-characterized representative of the Firmicutes phylum, the synthase-hydrolase RelA, and the two small alarmone synthases SasA (also termed RelP or YwaC) and SasB (RelQ or YjbM) downregulate ATP and GTP levels, increase (p)ppGpp, and indirectly the cyclic 3,5-diadenosine monophosphate (c-di-AMP) pools [13,14]. Antibiotic persistence might be controlled by deregulating (p)ppGpp and c-di-AMP pools [10,15]. In response to antibiotics that target cell wall biosynthesis, as the bactericidal β-lactam ampicillin (Amp), (p)ppGpp and c-di-AMP synthesis are induced in *B. subtilis* cells [16]. Furthermore, the *sasA* transcripts are increased by certain classes of cell-wall-active antibiotics (Amp among them) that trigger a membrane stress, and increase (p)ppGpp and c-di-AMP levels [13,17].

Bacterial toxin-antitoxin (TA) modules, which are ubiquitous across a broad range of extrachromosomal elements and are present in bacterial and archaeal chromosomes, are operons encoding a toxin that interferes with vital cell processes and an antitoxin that counteracts toxin action [18–21]. The type I to VI TA modules, which are classified according to the nature of the antitoxin and to their mode of toxin inhibition, are implicated in the fine tuning of multiple cellular functions and are associated with cell survival under different stress conditions [20,21]. The largest group is formed by type II TAs, which were first described as loci that enforce the maintenance of plasmids via post-segregational killing [22], and their ability to stabilize their mobile elements is considered as a selfish property [21,23]. In a type II TA system, which comprises a pair of genes coding for two proteins, the antitoxin forms a tight complex with the toxin to avoid its action [18–21,23]. Under certain stress conditions, the unstable antitoxin is rapidly degraded by host ATP-dependent proteases, freeing the proteolytically stable toxin [18,20]. The free toxin, which targets essential cellular processes such as RNA, DNA and protein synthesis, cell division, general stress response, etc., triggers a biphasic cell inactivation curve [24].

A variant of the HipAB type II TA system, rather than deletion of *hipBA* operon, has been shown to contribute to the formation of antibiotic persistence in *E. coli* cells. The *hipA7* mutation, which maps in the HipA toxin, produces a high fraction of persisters [7,25]. HipA activates the stringent response, and by increasing (p)ppGpp levels, leads to persister formation [26–28]. However, a strong link between the induction of TA systems and antibiotic persistence is missing, and further work is needed to establish whether physiological levels of a toxin contribute to antibiotic persistence [23,29–31]. Deletion of *E. coli*- and *B. subtilis*-encoded chromosomal type II TA systems have no effect on antibiotic persistence [12,31].

The type II TAs of the ζ-ε superfamily, which is among one of the most broadly distributed TA modules in nature, are found in major human and plant pathogens. TAs of this superfamily are composed of two toxin monomers (with ζ and PezT as representatives), that interact with their cognate dimeric ε_2/PezA$_2$ antitoxin forming an inactive $\zeta\varepsilon_2\zeta$ or PezT-PezA$_2$-PezT complex [32,33]. ζ/PezT interacts with its activator/substrate, the peptidoglycan precursor uridine diphosphate-*N*-acetylglucosamine (UNAG) [34]. UNAG is an essential precursor of bacterial cell wall biosynthesis [34]. The structures of the inactive $\zeta\varepsilon_2\zeta$ or PezT-PezA$_2$-PezT complex, alone or UNAG bound, have been reported [32–35]. The interaction of the antitoxin with the toxin sterically blocks ATP-Mg^{2+} binding by the ζ/PezT toxin, but not with UNAG [32–35]. Therefore, the interactions with ATP and ε_2/PezA$_2$ are mutually exclusive [32,33]. The structure of the inactive $\zeta\varepsilon_2\zeta$ or PezT-PezA$_2$-PezT complex, both alone or UNAG bound, has been reported [32–35]. Toxin ζ hydrolyses ATP in the presence of UNAG and phosphorylates a small fraction of UNAG, rendering it irreversibly inactive [15,34,36,37].

Transient toxin ζ expression, at or near physiological concentrations, leads to several cellular responses in *B. subtilis* cells growing in S7 minimal medium, at 37 °C with aeration. At early stages of transient toxin ζ expression (up to 15 min), the ζ UNAG-dependent ATPase decreases the ATP pool, and indirectly it alters the intracellular ATP/GTP ratios [37,38]. Such perturbation affects the concentration of the transcription initiation nucleoside triphosphates (ATP, GTP), and indirectly regulates RNA polymerase transcription initiation [39]. This mechanism, which is crucial for rapid adjustment of gene expression in response to environmental changes [40], alters the expression of ~2% of total *B. subtilis* genes (among those, it induces *relA* and *comGA* gene expression) [38]. However, an increased expression of SOS response genes or proteases that selectively induce cell-cycle arrest is not observed [38]. At 15–30 min times, the concerted action of *relA* and *comGA* increases the (p)ppGpp pool and indirectly the levels of the essential c-di-AMP second messenger [15,38]. Toxin ζ expression induces a cellular dormant state, but a reduced subpopulation, which shows no genetic modification, survives toxin action [10,15,38]. At 30–90 min, the synthesis of macromolecules (DNA, RNA, proteins) is inhibited and the membrane potential is impaired, with ~35% of total cells transiently stained with propidium iodide (an indicator of membrane compromised cells) [38]. At 90–120 min, the cell wall biosynthesis is reduced by ζ-mediated phosphorylation of a small UNAG fraction, leading to the accumulation of unreactive UNAG-3P [34,38]. In addition, (p)ppGpp mediates the inhibition of peptidoglycan metabolism [41]. Antitoxin ε expression blocks ATP binding by toxin ζ, thereby inhibiting its ATPase and phosphotransferase activities. Antitoxin ε reverses the toxic effect exerted by ζ in cell proliferation, and only ~10% of total cells remained stained with propidium iodide [38,42,43]. This is consistent with the fact that the toxin ζ action is reversible by nature even in distantly related bacteria, such as *E. coli* and *B. subtilis* cells [38,43]. Transient toxin ζ expression sensitizes cells to Amp treatment, and antitoxin ε expression reverses the ζ-induced dormant state, and the steady state condition is re-established [15,38,43]. The reversibility of ζ action upon longer periods of time in the presence of Amp is unknown. Therefore, we have re-examined the mode of action of toxin ζ, at or near physiological concentrations in *B. subtilis* cells growing in the minimal S7 medium, to understand how toxin ζ contributes to reduce Amp persistence.

In this study, we show that toxin ζ induces a dormant although still metabolically active state. The study of SPP1 in cells where toxin ζ is expressed reveals that ζ induces an active slow-growth rather than a growth-arrested condition and abortive infection. Amp addition at twice the minimum inhibitory concentration (2× MIC) halts *B. subtilis* proliferation, leading to a biphasic time-inactivation curve, which is maintained by a long period of time. Transient toxin ζ expression sensitizes cells to Amp action during a long time period. Antitoxin ε expression switches back ζ-induced dormancy, and promotes the exit of toxin-facilitated Amp dormants without observing a point of no return. Inactivation of *recA*, which suppresses the SOS response, marginally reduced the relative rate of antibiotic persistence. Transient ε expression reverses the negative effect of the toxin and indirectly awakes Amp dormants to levels observed in the rec^{+} control. This toxin-mediated sensitization of bacterial cells to Amp and the understanding of the conditions that inhibit antitoxin action will facilitate targeted engineering of their activity towards the development of anti-persistence agents.

2. Results and Discussion

2.1. Experimental Rationale

Resistance, tolerance, and persistence are independent solutions used by bacteria to survive an antibiotic action (Supplementary Figure S1) [1]. A bacterial population resistant to antimicrobial, which grows under drug pressure, uses mutation-associated defense mechanisms (such as antibiotic inactivation/modification, increased efflux, or target modification) to confer a resistant phenotype with or without a fitness cost (Figure S1) [19,44,45]. Persistence is a special case of tolerance, and often the two terms are interchangeable [45]. Tolerance is the general ability of a population to survive a longer antibiotic treatment without a significant change in the MIC (Figure S1, dotted line), whereas persistence represents a subpopulation of cells that can survive the antibiotic treatment (Figure S1, dotted vs. orange broken dotted line) [1].

The growth of susceptible bacteria is challenged by the addition of an antibiotic (e.g., Amp), but a subpopulation of stochastic variants, by undergoing a period of slow- or non-growth, escapes the antibiotic action, leading to a bimodal time-inactivation curve of cell persistence (Figure S1, orange broken dotted line) [1]. In *B. subtilis*, the mechanisms responsible for antibiotic persistence have not been delineated, and the fate of persistents after a prolonged exposure to the antibiotic before returning to permissive conditions is poorly understood. Here, Amp-induced cell-wall stress, by inducing *sasA* expression (one of the two small alarmone synthases), reduces the GTP pool and increases c-di-AMP and ppGpp levels [10,13,17]. This state might be transient and all cells might "awake" after removal of the antibiotic.

In previous studies, the number of toxin ζ and antitoxin ε molecules from a *S. pyogenes* plasmid-borne ε-ζ TA cassette was determined in the *B. subtilis* host grown in the minimal S7 medium [43,46]. An orthogonal control of ζ and ε was constructed by placing the ζ gene under the control of an isopropyl-β-D-thiogalactoside (IPTG) inducible promoter (*lacI*, P_{hsp}ζ, *spc* cassette), was ectopically integrated into the *B. subtilis amy* locus (Table 1) [38]. The antitoxin ε gene, which reads from a xylose (Xyl)-inducible promoter (*xylR*, P_{xylA}ε, *cat* cassette), was cloned into the middle-low-copy number pCB799 plasmid (~8 copies/cell) [38]. In *B. subtilis*, BG1125 pCB799 cells grown in S7 are prone to genome rearrangements, by transcriptional escape of the ζ gene from the P_{hsp} promoter. The Xyl concentration required to reduce genome rearrangements was estimated to be 0.005% [10,38]. BG1125 (pCB799) grows in the S7 medium with 0.005% Xyl at 37 °C with a doubling time of (58 ± 2 min) (Figure S2) [38].

Table 1. Bacterial strains used.

Strain	Relevant Genotype	Reference
BG214	*trpCE metA5 amyE1 ytsJ1 rsbV37 xre1 xkdA1* $att^{SP\beta}$, att^{ICEBs1}	Lab. Collection
BG1127	+ *lacI*, P_{hsp} [a], *spc*, [pCB799-borne *xylR*, P_{xylA}ε [b], *ermC*, *cat*]	[38]
BG1125	+ *lacI*, P_{hsp}ζ [a], *spc*, [pCB799-borne *xylR*, P_{xylA}ε [b], *ermC*, *cat*]	[38]
BG1125	+ *lacI*, P_{hsp}ζ [a], *spc*, [pCB1226-borne *xylR*, P_{xylA}ε [b], *cat*]	This work
BG1889	+ *lacI*, P_{hsp}ζ [a], *spc*, Δ*recA* (*recA:erm*) [pCB1226-borne *xylR*, P_{xylA}ε [b], *cat*]	This work

All *B. subtilis* strains are isogenic with the BG214 strain. [a] The P_{hsp} promoter is under the control of the LacI repressor. [b] The P_{xylA} promoter is under the control of the XylR repressor.

Then, the IPTG concentration necessary to express similar levels of toxin from the native system, and the Xyl concentration necessary to express sufficient ε_2 to inactivate the toxin ζ action was estimated [38]. Transient toxin ζ expression, at or near physiological concentrations, induces reversible cell dormancy. Here, a large cell fraction enters into a toxin-induced dormant state that cannot form colonies since they minimize their metabolic activity. By contrast, a small subpopulation escapes the toxin ζ action without diminishing their metabolic activity. When the inducer of toxin expression was removed, the cells exited the transient dormant state and formed colonies, leading to bimodal toxin survivors (Figure S1, blue broken dotted line) rather than toxin tolerants (Figure S1) [15]. These cells that escape the toxin ζ action show no genetic changes [38].

Transient toxin ζ expression sensitizes cells to the Amp treatment [10,15]. In the BG1125 [pCB799] strain, the presence of IPTG or Xyl does not affect the MIC for Amp [10]. The fate of the Amp persisters, which are sensitized by toxin ζ is unknown. At least two types of mechanisms can be envisioned. First, toxin-induced dormancy sensitizes cells to Amp action. Second, ζ-induced dormancy and stochastic slow growth cells trigger a viable but not-culturable state. In the latter condition, normal culturable cells stochastically lost their ability to grow on media, but remain viable [47]. If the first hypothesis is correct, antitoxin ε expression inactivates ζ, and indirectly reverses Amp dormants, as depicted in Figure S1 (red dotted curve). However, if the latter hypothesis is correct, transient ε_2 expression should reverse toxin-induced dormancy, but it should play no role on the exit from stochastic viable but the not-culturable state. These experiments performed here should contribute to understand the mechanism of genesis and exit of Amp persistence.

2.2. *Toxin ζ Induces a Slow-Growth Active State*

In γ-Proteobacteria, deletion of chromosomally-encoded type II TA loci neither confers a fitness benefit nor influences stress tolerance [48,49]. Similarly, deletion of *B. subtilis*-encoded type II TA systems has no effect on antibiotic persistence [12]. To gain insight on the molecular and physiological bases of toxin-induced dormancy, dormant cells were infected with a lytic bacteriophage (phage) and the infective cycle was followed.

Except TA systems that abort infection of a specific phage in selected hosts [50], and those phages that encode *bona fide* TA systems [23], it has been observed that overexpression of type II toxins contributes to abortive infection and causes exclusion of lytic phages from the bacterial population [26,51]. It has been also shown that in the arms race of a bacterium and its phage, a λ lytic variant was able to infect and kill survival bacteria, despite their dormancy state [52], and that T4 infection inactivates MazF toxin activity [53]. In contrast, it has been also shown that phage λ cannot overcome a toxin treat, since toxin overexpression, acting as a defense mechanism, protects the bacterial culture by increasing (p)ppGpp levels, and the alarmone might antagonize λ phage development [54]. Moreover, toxin RnlA overexpression cleaves phage T4 mRNAs to antagonize T4 amplification [55]. Therefore, phage infection of toxin-induced dormant cells can be a powerful tool to understand the physiology of toxin-induced dormancy and to address how the toxin ζ expression sensitizes cells to toxin ζ expression.

A set of isogenic *B. subtilis* strains, lacking prophages (SPβ and PBSX), conjugative genetic elements, CRISPR-Cas and restriction systems, and having a deficiency in the *rsbV* gene, whose function is crucial for the activation of RNA polymerase σ^B-mediated general stress response [56], were used to characterize the impact of toxin ζ expression on cell dormancy (Table 1). The BG1127 (pCB799) strain (in the presence or absence of IPTG, doubling time 59 ± 3 min) and BG1125 (pCB799) strain (in the absence of IPTG, doubling time 58 ± 2 min) grown in the minimal S7 medium supplemented with traces of Xyl (0.005%) at 37 °C with shaking had a similar doubling time (Figure S2). Both cell cultures reached a plateau at ~3.5×10^9 cells mL^{-1} (Figure S2). By contrast, if IPTG (2 mM) was added at $OD_{560} = 0.2$, the growth curve of the BG1125 (pCB799) strain was halted and after 720 min the OD_{560} was not significantly decreased (Figure 1 and Figure S2), but the plating efficiency was significantly reduced (Figure 2A) [38].

BG1125 (pCB799) cells were grown in the minimal S7 medium supplemented with 0.005% Xyl, to $OD_{560} = 0.2$ at 37 °C with shaking. At this time (-20 min), the culture was divided, 2 mM IPTG (time zero) was added to a half, and both cultures were incubated for 20 min to allow toxin-induced dormancy. The total number of cells were measured as colony-forming units (CFUs) by plating them on LB plates lacking IPTG (Figure 1A and Figure S3). At -5 min, both cell cultures, containing or not IPTG, were infected with phage SPP1 at a multiplicity of infection (*moi*) of 1 (to minimize the noise of free phages), and incubated for further 5 min. Then, free phages were removed by centrifugation, and this time was considered as 0 min in our experiment. At 0 min, the number of cells was measured by plating on LB agar plates and plaque-forming units (PFUs), to determine the number of infected centers, cells were measured by plating appropriate dilutions of the culture in a lawn of exponentially growing BG214 cells (indicator strain) onto LB-Mg (supplemented with 10 mM $MgCl_2$) plates (Figure 1A,B). In parallel, free phages from the supernatant of centrifugation were determined, to test whether centrifugation was sufficient to remove them or polyclonal anti-SPP1 antibodies were necessary to the inactive free phages. Since in the minus IPTG condition the number of total cells minus free phages was not significantly different from the number of infected centers, we have omitted the use of polyclonal anti-SPP1 antibodies.

After 20 min of toxin induction IPTG, CFUs were reduced by ~1000-fold at a time when compared to the parallel culture in the absence of IPTG (Figure 1A and Figure S3, black vs. grey bar). After infection with SPP1, the proportion of infected centers measured as PFUs was similar in the minus and plus IPTG conditions (Figure 1B, black and grey bars), suggesting that a large fraction of dormant cells were infected by SPP1. In the absence (grey bar) and presence of IPTG (black bar), the phage titer, which correlates to the cell population infected by SPP1, was similar and in good agreement with the expected for a one-hit kinetic. From the input phages, ~71% and ~67% infected the cells, respectively, defining the infected centers (Figure 1B) and ~30% were recovered from the supernatant after centrifugation. In parallel, the total number of cells at -20, 0, 60, and 120 min in the absence of phage SPP1 was also measured as CFUs by plating them on LB plates lacking IPTG (Figure S3).

The SPP1 infection cycle takes 35–45 min in cells growing in the minimal medium, thus the infected cells were incubated for 60 min, and then the relative number of progeny phage per infected cell (burst size) was deduced for the minus and plus IPTG conditions. At 60 min and in the presence of IPTG, cells remained in the dormant state (Figure 1A vs. Figure S3), suggesting that SPP1 does not encode (a) gene(s) whose function can neutralize toxin ζ action. At 60 min, the phage SPP1 titer from cultures where the toxin had been induced was similar to the number of infected cells at 0 min. However, in the absence of IPTG, the PFUs greatly increased. The SPP1 relative burst size (the ratio of the total number of SPP1 progeny to the number of input phage) from three independent experiments was quantified by counting cells and phages taking in to account the supernatant of the cultures (Figure 1A,B). The SPP1 burst size significantly increased (~80 SPP1 phages/infected cell). These results suggest that SPP1, upon infecting toxin-dormant cells neither enters in the amplification cycle nor promotes the exit of the dormant state.

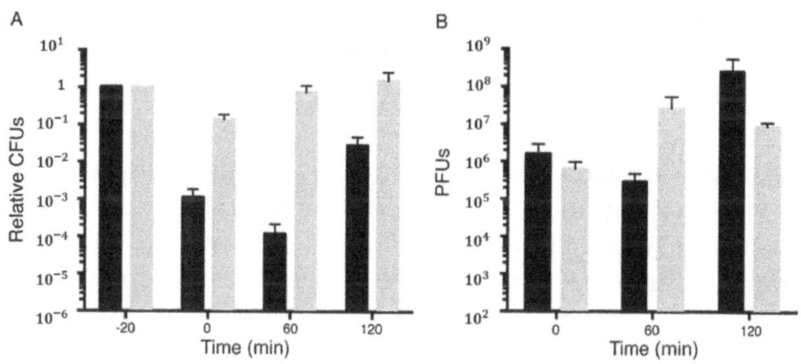

Figure 1. Toxin dormants infected with the lytic phage SPP1 are metabolically active. Survival rate of BG1125 (pCB799) cells infected by the SPP1 phage after ζ toxin induction. BG1125 (pCB799) cells were grown in the minimal S7 medium, containing traces of Xyl (0.005%) and 10 mM $MgCl_2$ to OD_{560} = 0.2 at 37 °C. At this time (-20 min) the culture was divided in two. To one half, IPTG (2 mM) was added (black bars) to induce ζ expression (-20 min) and incubated for 15 min. After 15 min, phage SPP1 at *moi* = 1 was added to both cultures and cells were incubated for further 5 min. At 0 min, cells were centrifuged to remove unadsorbed phages and resuspended in S7 medium, containing 0.005% Xyl and 10 mM $MgCl_2$. Colony-forming units (CFUs) were counted plating on a LB agar (time 0) and both infected cultures were incubated for further 60 min (60 min). At 60 min, 0.5% Xyl was added to induce antitoxin ε expression to neutralize toxin action, and both infected cultures were incubated for further 60 min (120 min). At 0, 60, and 120 min, samples were withdrawn and plated in LB agar plates lacking IPTG to count CFUs (**A**), or in a lawn of exponentially growing BG214 cells onto LB plates supplemented with 10 mM $MgCl_2$ to measure PFUs (**B**). Data are shown as mean ± standard error of the mean (SEM), from >3 independent experiments.

To test whether dormant cells antagonize or cannot support SPP1 amplification or if dormant cells can support SPP1 amplification upon toxin neutralization at 60 min post-infection, Xyl (0.5%) was added to induce antitoxin ε expression, and the cultures were incubated for further 60 min. At 120 min, the number of total cells and the SPP1 burst-size were quantified as described above. In the presence of IPTG and Xyl, the yield of SPP1 phage in cells expressing both ζ and ε genes (between 60 to 120 min) was significantly increased (~110 SPP1 phages/predicted infected cell), calculated from the PFU obtained and the number of infected centers at 60 min (Figure 1B). This suggests that only when infected cells exit toxin-induced dormancy, by the anti-toxin action, SPP1 was amplified. In the only Xyl condition (i.e., no toxin induced), the number of total cells increased (58 ± 2 min doubling time), but the phage titer did not, suggesting that these cells enter in a state that might not be further infected by the SPP1 phage.

These data altogether suggest that: (i) SPP1 infected toxin-induced dormant cells with a similar efficiency as normally growing cells (Figure 1B, time zero); (ii) ζ-induced dormancy does not impair phage absorption, but delays the SPP1 lytic cycle; (iii) toxin ζ-induced dormancy is a metabolically active state rather than a growth-arrested state leading to abortive infection; (iv) SPP1 does not undergo an abortive infection; and v) upon toxin ζ inactivation, by antitoxin ε expression, the exit of ζ-induced dormancy facilitates a viral infective cycle.

2.3. *Discrete Subpopulations of Toxin ζ Survivors and Amp Persisters*

The subpopulation of cells that survive an antibiotic treatment, are termed persisters [1,45]. The rate of killing by Amp is strictly proportional to the rate of bacterial growth, and Amp persistence is associated with a pre-existing phenotypic switching to a stochastic non-growing state (or leftover cells from high-stress stationary-phase inoculums) [57,58]. To analyze the interconnection between toxin survivors and Amp persisters and whether a prolonged growth arrest causes an irreversible cell cycle arrest (point of no return), we have used BG1125 (pCB799) cells (Table 1).

BG1125 (pCB799) cells were grown in the minimal S7 medium, supplemented with 0.005% Xyl, to $OD_{560} = 0.2$ at 37 °C with shaking, and then 2 mM IPTG (0 min) was added to induce toxin expression, and the reaction incubated for 900 min (Figure 2A). Toxin ζ expression quickly reduced the cell plating efficiency showing a biphasic time-inactivation curve (Figure 2A, blue empty rhombs), suggesting that toxin ζ expression reaches a steady-state in a short time and it remains for at least 900 min. It is likely that toxin ζ expression induces a slow-growth active state, but a rare fraction ($3–6 \times 10^{-5}$) exits this dormant state and survives, or a preexisting population of non-growing cells are insensitive to toxin action (type I-like persisters) (see [59]). Alternatively, expression of the ζ gene triggers the expression of any of the type I and/or type II TA modules present in the background, and the loss of cell viability and the observed surviving cells arise via the overlapping activation of multiple toxins, even with different activities.

To test whether the frequency of survivors may increase in response to the spontaneous induction of other chromosomally encoded toxins present in the background, IPTG was removed, by washing the culture with the pre-warmed S7 medium, containing 0.5% Xyl (to induce antitoxin ε expression), and cells were preincubated for 15 min before plating on LB plates containing 0.5% Xyl, but lacking IPTG. Antitoxin ε expression reversed ζ-induced dormancy, by significantly increasing cell viability nearly to full values almost similar to the time of IPTG addition (Figure 2A, blue filled rhombs). Similar results are observed when IPTG is not removed from the medium [15]. It is likely that: (i) Toxin ζ-induced dormancy rather than cell lysis, as judged by the non-significant variation in the $OD_{560} = 0.2$ during a 720-min interval (Figure S2); (ii) toxin ζ is not triggering a spontaneous induction of toxins from TA systems present in the background; and (iii) toxin ζ transiently halts DNA replication and triggers a phenotypic heterogeneity [38]. In addition, the expression of the antitoxin ε expression reversed ζ-induced growth arrest even after a prolonged dormancy (equivalent to ~15 mass doubling periods) (Figure 2A, blue filled rhombs).

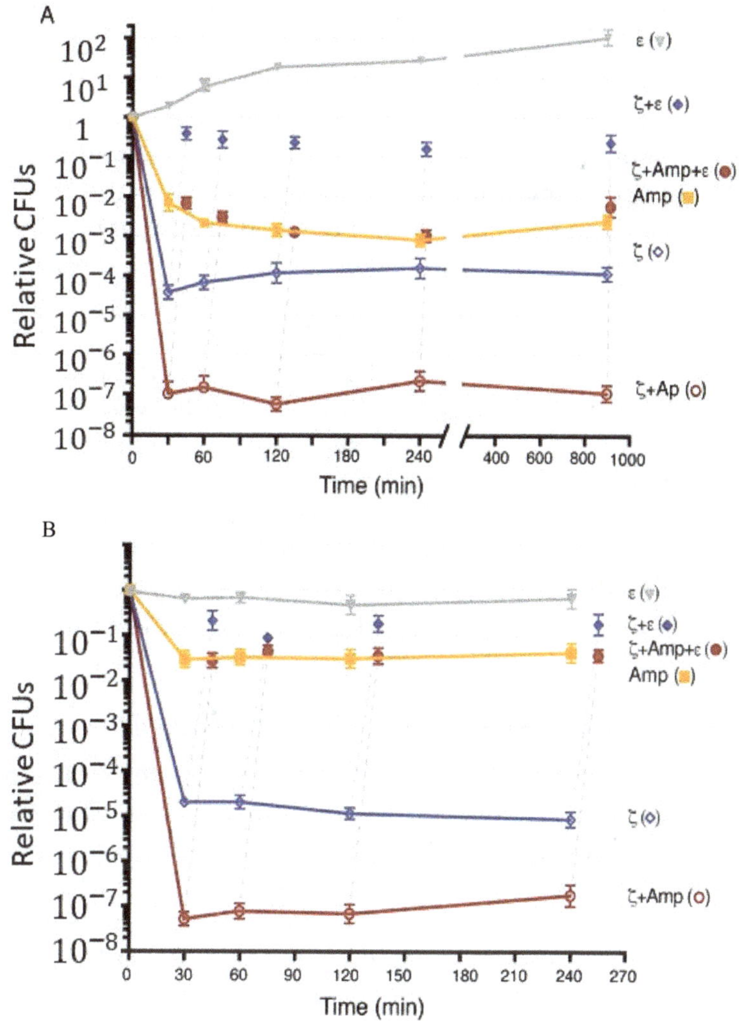

Figure 2. Toxin ζ induces reversible dormancy, and facilitates Amp dormancy. (**A**), BG1125 (pCB799) cells were grown in the S7 medium, containing traces of Xyl (0.005%) at 37 °C and the cultures were divided. Then, IPTG (2 mM) to induce ζ expression, Amp (at 2× MIC, 3 $\mu g\ mL^{-1}$), or both IPTG and Amp were added (0 min), and the cultures were incubated for 900 min. At various times, samples were withdrawn, centrifuged, and plated on LB agar plates to count ζ (empty blue rhomb), Amp (filled orange square), or both ζ and Amp survivals (empty purple circles). At 30, 60, 120, 240, and 900 min, aliquots were taken and 0.5% Xyl was added to induce antitoxin ε expression, and the cultures were incubated for 15 min before plating in LB agar plates containing Xyl, but lacking IPTG (filled blue rhomb) or both IPTG and Amp (filled purple circle). The vertical broken lines join the original point (ζ or ζ + Amp) with the reversed condition after ε expression. (**B**) An overnight culture of BG1125 (pCB799) cells grown in the S7 medium, containing traces of Xyl (0.005%) was normalized to ~1×10^{9} cells mL^{-1} (37 °C) and the cultures were divided. Xyl (0.5%) to induce ε expression as the control was added. IPTG to induce toxin, Amp, or both IPTG and Amp was added (0 min), and the cultures were incubated for 240 min. At various times, samples were withdrawn and plated in LB agar plates lacking IPTG (empty blue rhomb), Amp (filled orange square), or both IPTG and Amp (empty purple circles). At various times, 0.5% Xyl was added to induce antitoxin expression, and the cultures were incubated for 15 min before plating. Samples were withdrawn and plated in LB agar plates containing Xyl, but lacking IPTG (filled blue rhomb) or both IPTG and Amp (filled purple circle). The vertical broken lines join the original point (ζ or ζ + Amp) with the reversed condition after ε expression. Data are shown as mean ± standard error of the mean (SEM), from >4 independent experiments.

To address whether antibiotic persisters are produced due to a stochastic entrance into the stationary phase, the overnight culture was extensively diluted and BG1125 (pCB799) cells were grown in the minimal S7 medium, supplemented with 0.005% Xyl, to OD_{560} = 0.2 at 37 °C with shaking. It is expected that <0.3% of the total cells in the actively growing inoculum are old cells from the stationary-phase inoculum. Amp (at 2× MIC, 3 μg mL^{-1}) was added, and the culture was maintained up to 900 min. At indicated times, the cells were washed to remove the antibiotic, and plated in LB lacking Amp. The removal of Amp was sufficient to reveal that a large fraction of cells (>99% of total cells) cannot form colonies upon exposure to bacteriolytic Amp (Figure 2A, filled orange squares). As stated, the rate of Amp killing is proportional to the rate of bacterial growth [57], with >80% of total Amp-treated cells being stained with propidium iodide [10], which stains the membrane compromised cells. A small cell fraction (1–0.5 × 10^{-2} survivals), however, regained the ability to form colonies upon antibiotic removal (Figure 2A, filled orange squares). Similar results were observed when the overnight culture was less extensively diluted (~6% old cells), suggesting that old cells from the stationary-phase inoculum poorly contribute to the Amp persistence fraction.

Upon exposure to Amp for at least 900 min of incubation, cells survive following antibiotic removal (Figure 2A) rather than becoming metabolically inactive by entering a path of no return as described upon toxin MazF overexpression [60]. The culturable cells were still sensitive to Amp action, thus we have to believe that after prolonged incubation to the antibiotic there were no genetic changes [10].

2.4. *Transient Toxin ζ Expression and Amp Addition Reduce the Subpopulation of Persisters*

Transient toxin ζ expression sensitizes cells to the presence of different antibiotics [10,37]. At least two mechanisms can be envisioned: (i) Toxin dormants are sensitive to antibiotic action and stochastically enter in a viable but not-culturable state with resuscitation independent of antitoxin expression; or (ii) antibiotic persistence is the sum of stochastic and transient antibiotic-induced dormants with resuscitation dependent of antitoxin expression. To test the above hypotheses, BG1125 (pCB799) cells were grown in S7 medium, supplemented with 0.005% Xyl, to OD_{560} = 0.2 at 37 °C with shaking. Then, 2 mM IPTG and 3 μg mL^{-1} Amp were added, and at indicated times cells were washed and plated in the absence of both IPTG and Amp. A typical biphasic decline curve was observed (Figure 2A, purple empty circles). Amp and toxin action significantly decreased the proportion of survivals (1–2 × 10^{-7}) when compared to only ζ (3–6 × 10^{-5} survivors) or only Amp (1–0.5 × 10^{-2} persisters) (Figure 2A, purple empty circles vs. blue empty rhombs and orange filled squares). The persistent cells stochastically switch back to a growing state, upon plating in the absence of IPTG and Amp, with the fraction of persisters remaining nearly constant for ~900 min (Figure 2A, purple empty circles). It is likely that: (i) Amp persisters are in an active state that allows toxin expression; (ii) toxin ζ expression does not increase antibiotic persistence; (iii) toxin expression triggers a network of intracellular stress responses that significantly increases cell dormancy; and (iv) toxin-induced dormancy facilitates Amp dormants that resuscitate upon antitoxin expression.

2.5. *Transient Antitoxin ε Expression Reverses ζ-Facilitated Exit of Amp Dormants*

To test whether transient ζ expression enhances Amp efficacy with the cells entering in a viable but not-culturable state or if ζ expression induces dormancy of Amp persisters, BG1125 (pCB799) cells were grown in the minimal S7 medium, supplemented with 0.005% Xyl, to OD_{560} = 0.2 at 37 °C with shaking. Then, 2 mM IPTG and Amp (3 μg mL^{-1}) were added and the cells incubated up to 900 min (Figure 2A, purple empty circles). At indicated times, IPTG and Amp were removed by washing the culture with the pre-warmed S7 medium, containing 0.5% Xyl to induce antitoxin ε expression, and cells were preincubated (15 min) before plating (Figure 2A, purple filled circles). The antitoxin ε recovered cell proliferation to levels equivalent to that of Amp persisters, even after 900 min of toxin ζ and Amp action (Figure 2A, purple filled circles). These data altogether suggest that: (i) Toxin ζ expression and Amp addition induce prolonged dormancy (at least for 15 h) rather than programmed-cell death, as reported for the *E. coli* MazEF and RelBE locus (reviewed in [61]); (ii) ζ-induced dormancy is insensitive to Amp action; (iii) toxin-induced dormant cells awake and form colonies even after 15 h of toxin and Amp action, rather than triggering cell lysis or inducing a viable but not-culturable state (Figure 2A, purple filled circles); and (iv) the orthogonal control of ζ and ε_2 levels revealed that a TA system can be an important element in coping with Amp stress; and (v) toxin survivors and Amp persistence are not epistatic.

2.6. Toxin ζ Expression Induces Amp Dormancy in Non-Growing Cells

To re-evaluate whether Amp persisters are produced due to a stochastic entrance into the stationary phase or upon transient toxin ζ expression, and if toxin expression or Amp addition to stationary phase cells affects the persistence level and its subsequent exit, high-density slow- or non-growing stationary phase *B. subtilis* BG1125 (pCB799) cells, which grew overnight in the S7 medium supplemented with 0.005% Xyl, were used. The overnight culture was normalized to ~1×10^9 cells mL^{-1} with the pre-warmed S7 medium, supplemented with 0.005% Xyl and split. Amp (3 μg mL^{-1}), IPTG (2 mM), or both Amp and IPTG were added and the cells incubated up to 240 min. At the indicated times, IPTG, Amp, or both were removed from the medium and the culture was plated on LB plates. Since in the absence of both, Amp and IPTG, the number of cells did not significantly increase at 240 min (Figure 2B, filled grey triangles) and BG1125 (pCB799) cells grown in S7 supplemented with 0.005% Xyl usually had an usual lag-time of 30–45 min (Figure S2), we assume that the high-density cells remain in the stationary phase during the experimental period.

The MIC for Amp was estimated using low-density cells (1–3 × 10^6 cells mL^{-1} for 16 h, at 37 °C), and no correction by per-cell in the Amp concentration was performed. At a variable time, and upon removal of the antibiotic, a significant fraction could not form colonies, but a subpopulation of cells persisted to Amp action (3–5 × 10^{-2} survivals). As expected, the proportion of persisters was significantly higher than in exponentially growing cells (Figure 2A vs. Figure 2B, filled orange squares). This is consistent with the observation that the rate of killing of Amp is strictly proportional to the rate of bacterial growth [57]. Similar results were observed in the presence of 6 μg mL^{-1} Amp (data not shown). It is likely that the lower proportion of cell death might correlate with the less proportion of metabolically active cells in the stationary phase, but in the only Xyl condition the cells are not proliferating.

In the presence of IPTG, the subpopulation of toxin ζ (3–6 × 10^{-5} survivals) did not significantly increase during time and was similar in both, exponential growth and stationary phase (Figure 2A,B, empty blue rhombs). Similar results were observed when a low-lived ζ variant (ζY83C) is used [10], suggesting that when toxin ζ is expressed, it triggers dormancy in high-density stationary phase cells. When Amp and IPTG were added, a low proportion of high-density stationary phase cells survive toxin ζ and Amp action (4–7 × 10^{-8} survivals) (Figure 2B, empty purple circles).

To test whether the expression of ε antitoxin also reverses the effect of ζ-mediated dormancy in high-density non-growing cells, Xyl was added at the indicated times. The expression of ε antitoxin reversed the effect of the ζ toxin (3–5 × 10^{-1} survivals) and partially reversed toxin and Amp (4–5 × 10^{-2} survivals) (Figure 2B, filled blue rhombs and purple circles). It is likely that: (i) Toxin ζ induces a reversible dormant state in high-density, non-growing cells rather than entering in a stochastic viable but not-culturable state; (ii) there are two different subpopulations of Amp persisters: Stochastic (3–5 × 10^{-2} survivals) and toxin-induced (4–7 × 10^{-8} survivals); and iii) antitoxin expression reverses the apparent sensitization of toxin dormants to Amp action.

2.7. RecA Inactivation Does not Alter the Antitoxin Awake of Toxin and Amp Dormants

Toxin-induced dormants are metabolically active slow-growing cells, and this may enhance the development of heritable genetic changes, since many of the stress-response programs involved in the survival of persisters can also accelerate genome-wide mutagenesis [58]. In Proteobacteria, both stochastic and stress-induced slow-growth persisters have been shown to depend on the SOS responses [58,62]. Furthermore, the SOS response induces the expression of certain type I and II TA modules in a subpopulation of *E. coli* [5,63].

In Proteobacteria and Firmicutes, bactericidal antibiotics (e.g., Amp), regardless of the drug-target interaction, induce changes in cellular metabolism that promote the production of highly deleterious hydroxyl radicals, leading to cell killing [64]. Amp-induced hydroxyl radicals, may damage template bases and introduce single strand nicks. These nicks, through DNA replication, can be converted into one-ended double-strand breaks (DSBs), that cause a significant induction of the SOS response [64,65] and maintenance of Amp persistence in *E. coli* cells [66]. However, DNA damage caused by sublethal doses of a fluoroquinolone in *E. coli* promotes persistence, but not Amp [5]. The recombinase *recA*, which is a central player in the SOS response, contributes to the repair of the damaged template bases that escape specialized repair, and of the nicks that are converted in one-ended DSBs during DNA replication [67,68]. The contribution of the SOS response to the DSB repair, however, differs substantially between *E. coli* and *B. subtilis* cells. In *B. subtilis*, the inability to induce the SOS response does not compromise the DSB repair [69].

In *E. coli* cells, *recA* plays a critical role in the appearance of apoptotic phenotypes and in pushing the cell towards its death [65], and inactivation of *recA* potentiates killing by different antibiotics, Amp among them [64]. In *B. subtilis* cells, transient toxin ζ expression [38,43] or extended cell cycle arrest [70] does not induce the SOS response. To test whether *recA* inactivation sensitizes cells to toxin or Amp, a Δ*recA* mutation was mobilized onto the *B. subtilis* BG1125 (pCB1226) strain by SPP1-mediated generalized transduction, rendering the BG1889 (pCB1226) strain (Table 1). Plasmid pCB1226 is a derivative of pCB799, but the *erm*C gene was selectively inactivated so that the Δ*recA* mutation could be introduced in *B. subtilis* BG1125 (pCB1226) cells by Erm^R selection (Table 1).

BG1889 (pCB1226) cells were grown in S7 medium, supplemented with 0.005% Xyl, to $OD_{560} = 0.2$ at 37 °C with shaking. CFUs at this OD_{560} are 3–6 × 10^6 cells mL^{-1}, in agreement with previous reports showing that in *B. subtilis* only 10–20% of Δ*recA* cells can form colonies in plates in the absence of DNA damage [71]. Then, IPTG (2 mM), Amp (3 μg mL^{-1}), or both IPTG and Amp were added (0 min), and the culture was maintained for 900 min (Figure 3). In the absence of ITPG and presence of traces of Xyl (0.005%), the BG1889 (pCB1226) cells had a poor fitness, formed small-size colonies, and the plating efficiency was reduced ~100-fold after 900 min of incubation (Figure 3, filled grey triangles). This result indicates that a significant proportion of Δ*recA* cells may have a prolonged defect in DNA repair and chromosomal segregation, which reduces cell viability during prolonged growth.

The number of Amp persisters and toxin survivors during the first 120 min was lower ~2 × 10^{-3} and ~4 × 10^{-6}, respectively, but persistence to both Amp and toxin (~4 × 10^{-7} frequency of persisters) in the Δ*recA* context (Figure 3, orange filled squares and blue empty rhomb vs. grey filled triangles) was similar to that of the *rec*$^+$ control (see Figure 2A). The decrease of toxin survivors and Amp persisters was observed at 900 min (~7 × 10^{-9}) after correction of the poor viability of the BG1889 (pCB1226) strain in the absence of IPTG and Amp (Figure 3, purple empty circles vs. grey filled triangles). What is the significance of such large variations in persisters to both toxin and Amp action in the Δ*recA* context? First, to measure persisters, the culture was concentrated ~100-fold. Second, under this experimental condition, the relative persistence frequency is similar to the spontaneous mutation rate. Finally, *B. subtilis* has multiple forms of differentiation and development, thus we cannot rule out that any uncontrolled condition might affect the outcome.

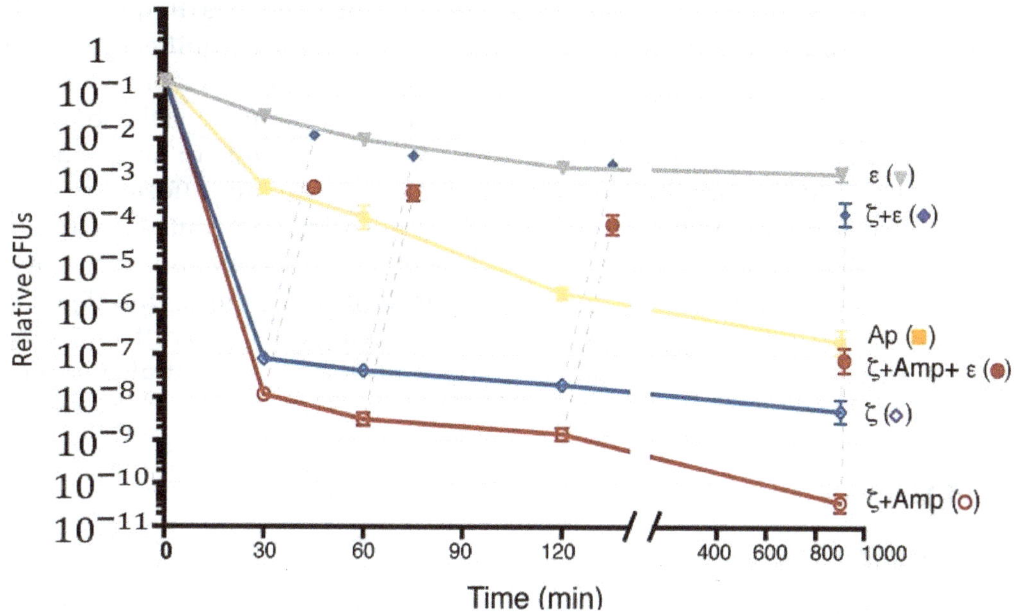

Figure 3. Inactivation of *recA* does not affect the frequency of ζ survivors and Amp persisters. BG1889 (pCB799) cells were grown in the minimal S7 medium, containing traces of xylose (Xyl; 0.005%) to OD_{560} = 0.2 (with CFUs of 1–3 × 10^6 cells mL^{-1}) at 37 °C. Then, the cultures were divided. To the indicated cultures, Xyl (0.5%) to induce ε expression as the control, IPTG (2 mM) to induce ζ expression, Amp (3 μg mL^{-1}) or both IPTG and Amp were added (0 min), and the cultures were incubated for 900 min. At various times, samples were withdrawn and plated in LB agar plates. At various times, aliquots were taken and 0.5% Xyl was added to induce antitoxin expression, and the cultures were incubated for 15 min before being plated in LB agar plates containing Xyl, but lacking IPTG (filled blue rhomb), or both IPTG and Amp (filled purple circle). Data are shown as mean ± standard error of the mean (SEM), from >4 independent experiments.

Transient antitoxin ε expression, by the addition of Xyl, reversed toxin action and the plating efficiency was increased to levels similar to the control strain in the absence of any treatment (Figure 3, blue filled rhombs vs. grey filled triangles). When Xyl was added to cells treated with both Amp and IPTG, the expression of ε antitoxin reversed the effect of ζ toxin and facilitated the exit of Amp dormants (Figure 3, purple filled circles). It is likely, therefore, that: (i) *RecA* inactivation, which impairs the SOS response, does not significantly increase the number of Amp or toxin survivors; (ii) *recA* inactivation reduces toxin or Amp persistence, suggesting that Amp-induced hydroxyl radicals accumulation is not significantly deleterious when compared to the *recA* control without Amp and IPTG (Figure 3, orange filled squares and blue empty rhombs vs. grey filled triangles); and (iii) *recA* does not induce an apoptotic phenotype upon Amp treatment, and *recA* inactivation does not sensitize cells to both toxin ζ and Amp action.

3. Conclusions

This study provides new insights that allow us to unravel the mechanisms underlying the mode of action of the toxin ζ dormancy and indirectly of toxin and Amp dormants. Transient toxin ζ expression, at or near physiological concentrations, induces a slow-growth dormant state, and a small fraction of cells (persisters) is insensitive to the toxin action in rec^+ and *recA* backgrounds. Toxin ζ-induced dormant cells can be infected with the lytic phage SPP1 with a similar efficiency as exponentially growing cells, but the amplification cycle is halted. Neutralization of the toxin, by inducing antitoxin expression, allows dormant cells to exit this state and this allows phage amplification. It is likely that toxin dormants are metabolically active rather than in a growth-arrested state.

Amp action kills ~99% of susceptible cells, but stochastic Amp persisters (frequency of 1–0.5 × 10^{-2}) can form colonies upon removal of the antibiotic. Toxin survivors are significantly rare (~3 × 10^{-5}), and in the presence of Amp the proportion of bacterial Amp and toxin survivors was further reduced (by 100–200-fold), suggesting a non-epistatic effect. To further evaluate the viable but non-culturable state hypothesis, we inactivated *recA*. In *E. coli* cells, *recA* contributes to apoptotic cell death in Amp treated cells [64], but the proportion of Amp persisters was not affected in the absence of *recA* [5]. In *B. subtilis* cells, toxin ζ and Amp persistence are not significantly affected in *recA* cells when compared to *rec*$^+$ control, and transient antitoxin ε expression is necessary and sufficient to switch off toxin-induced dormancy, and to awake the fraction of toxin-facilitated Amp-induced persisters both in the *rec*$^+$ and *recA* cells. We propose that in exponentially growing cells, two different subpopulations of Amp persisters are found, stochastic and toxin-induced dormancy, that also induce long term Amp dormancy.

4. Materials and Methods

4.1. Bacterial Strains and Plasmids

The bacterial strains and plasmids used in this study are listed in Table 1. All *B. subtilis* strains are isogenic with BG214. The strain BG1125 bearing *lacI*-P_{hsp} *wt*ζ and pCB799-borne *xylR*-P_{xylA} *wt*ε *ermC cat* cassette were previously reported (Table 1) [38]. The BG1885 strain was constructed in three steps. First, by site-directed mutagenesis (QuickChange Kit, Stratagene) using pCB799 as a template, the *ermC* gene was inactivated to render pCB1226. Second, the *lacI*-P_{hsp} *wt*ζ cassette was mobilized into BG214 cells bearing pCB1226-borne *xylR*-P_{xylA} *wt*ε *cat* cassette with selection for Cm^R. Third, the null *recA*::*Erm* gene was mobilized by chromosomal transformation into competent BG214 (pCB1226) cells with selection for Erm^R (Table 1). Integration by double crossover was analyzed by PCR.

Toxin ζ gene expression (transcribed from P_{hsp}) is regulated by IPTG (Calbiochem, Madrid, Spain) addition and the expression of ε gene (transcribed from P_{xylA}) is regulated by Xyl (Sigma-Aldrich, St. Louis, MO, USA) addition [38].

4.2. Growth Conditions

BG1125 (pCB799) and BG1125 (pCB1226) cells were grown to a mid-exponential phase (OD_{560} = 0.2) in S7 medium supplemented with methionine, tryptophan, and 0.005% Xyl at 37 °C with shaking [43]. Under this condition, cells grew in S7 medium, with a doubling time of 50–60 min. Transient toxin and/or antitoxin expression was induced by IPTG (2 mM) and/or Xyl (0.5%) addition. Before plating, cells were centrifuged and resuspended in the fresh LB medium to remove the inductor or the antibiotic, and dilutions were plated on LB agar plates. The survival rate was derived from the number of CFUs in a given condition relative to the CFU of the non-induced/non-antibiotic-treated control. All plates were incubated for 20 h at 37 °C.

The lytic SPP1 phage was used to test the toxin effect. BG1125 bearing *lacI*-P_{hsp} *wt*ζ and pCB799-borne *xylR*-P_{xylA} *wt*ε gene were grown to a mid-exponential phase at 37 °C in S7 medium, supplemented with methionine, tryptophan, 0.005% Xyl, and 10 mM $MgCl_2$. Toxin expression was induced by adding IPTG 2mM and after 15 min cells were infected with phage SPP1 at a multiplicity of infection (*moi*) of 1 in the presence or absence of IPTG. After 5 min, cells were centrifuged to remove unabsorbed phages and resuspended in S7 medium with 10 mM MgCl2 (0 min) and the culture was incubated for 60 or 120 min. The inducers were removed by washing, and the phage titer (PFUs) was calculated by plating the appropriate dilution deposited in a lawn of exponentially growing BG214 cells onto LB plates supplemented with 10 mM $MgCl_2$ (LB-Mg) to measure PFUs as described [72]. The plates were then incubated for 18–20 h at 37 °C.

The MIC for Amp (Sigma-Aldrich, St. Louis, MO, USA) was estimated by exposing 1–3 × 10^6 cells mL^{-1} (16 h, 37 °C) in the minimal S7 medium, with an increasing Amp concentration under shaking

(240 rpm). The minimal concentration that gave no growth overnight (1.5 μg mL^{-1}) was defined as MIC. The Amp concentration used was twice the MIC (2× MIC) or 3 μg mL^{-1}.

Author Contributions: M.M.-d.Á. and J.C.A. designed the experiments; M.M.-d.Á. and C.M. performed the experiments, J.C.A. coordinated the research; M.M.-d.Á. and J.C.A. interpreted the data; M.M.-d.Á., C.M. and J.C.A. drafted the manuscript; and J.C.A. wrote the manuscript. All authors have read and agreed to the published version of the manuscript.

Acknowledgments: We thank Silvia Ayora and Rubén Torres for critical reading and comments on the manuscript.

References

1. Balaban, N.Q.; Helaine, S.; Lewis, K.; Ackermann, M.; Aldridge, B.; Andersson, D.I.; Brynildsen, M.P.; Bumann, D.; Camilli, A.; Collins, J.J.; et al. Definitions and guidelines for research on antibiotic persistence. *Nat. Rev. Microbiol.* **2019**, *17*, 441–448. [CrossRef]
2. Hobby, G.L.; Meyer, K.; Chaffee, E. Observations on the mechanism of action of penicillin. *Proc. Soc. Exp. Biol. Med.* **1942**, *50*, 281–285. [CrossRef]
3. Bigger, J.W. Treatment of staphylococcal infections with penicillin. *Lancet* **1944**, *244*, 497–500. [CrossRef]
4. Johnson, P.J.; Levin, B.R. Pharmacodynamics, population dynamics, and the evolution of persistence in *Staphylococcus aureus*. *PLoS Genet.* **2013**, *9*, e1003123. [CrossRef]
5. Dörr, T.; Lewis, K.; Vulic, M. SOS response induces persistence to fluoroquinolones in *Escherichia coli*. *PLoS Genet.* **2009**, *5*, e1000760. [CrossRef]
6. Wu, Y.; Vulic, M.; Keren, I.; Lewis, K. Role of oxidative stress in persister tolerance. *Antimicrob. Agents Chemother.* **2012**, *56*, 4922–4926. [CrossRef]
7. Korch, S.B.; Henderson, T.A.; Hill, T.M. Characterization of the hipA7 allele of *Escherichia coli* and evidence that high persistence is governed by (p)ppGpp synthesis. *Mol. Microbiol.* **2003**, *50*, 1199–1213. [CrossRef]
8. Nguyen, D.; Joshi-Datar, A.; Lepine, F.; Bauerle, E.; Olakanmi, O.; Beer, K.; McKay, G.; Siehnel, R.; Schafhauser, J.; Wang, Y.; et al. Active starvation responses mediate antibiotic tolerance in biofilms and nutrient-limited bacteria. *Science* **2011**, *334*, 982–986. [CrossRef]
9. Amato, S.M.; Orman, M.A.; Brynildsen, M.P. Metabolic control of persister formation in *Escherichia coli*. *Mol. Cell* **2013**, *50*, 475–487. [CrossRef]
10. Tabone, M.; Lioy, V.S.; Ayora, S.; Machon, C.; Alonso, J.C. Role of toxin ζ and starvation responses in the sensitivity to antimicrobials. *PLoS ONE* **2014**, *9*, e86615. [CrossRef]
11. Conlon, B.P.; Rowe, S.E.; Gandt, A.B.; Nuxoll, A.S.; Donegan, N.P.; Zalis, E.A.; Clair, G.; Adkins, J.N.; Cheung, A.L.; Lewis, K. Persister formation in *Staphylococcus aureus* is associated with ATP depletion. *Nat. Microbiol.* **2016**, *1*, 16051. [CrossRef]
12. Fung, D.K.; Barra, J.T.; Schroeder, J.W.; Ying, D.; Wang, J.D. A shared alarmone-GTP switch underlies triggered and spontaneous persistence. *bioRxiv* **2020**. [CrossRef]
13. Nanamiya, H.; Kasai, K.; Nozawa, A.; Yun, C.S.; Narisawa, T.; Murakami, K.; Natori, Y.; Kawamura, F.; Tozawa, Y. Identification and functional analysis of novel (p)ppGpp synthetase genes in *Bacillus subtilis*. *Mol. Microbiol.* **2008**, *67*, 291–304. [CrossRef]
14. Kriel, A.; Brinsmade, S.R.; Tse, J.L.; Tehranchi, A.K.; Bittner, A.N.; Sonenshein, A.L.; Wang, J.D. GTP dysregulation in *Bacillus subtilis* cells lacking (p)ppGpp results in phenotypic amino acid auxotrophy and failure to adapt to nutrient downshift and regulate biosynthesis genes. *J. Bacteriol.* **2014**, *196*, 189–201. [CrossRef]
15. Moreno-Del Alamo, M.; Tabone, M.; Lioy, V.S.; Alonso, J.C. Toxin ζ triggers a survival response to cope with stress and persistence. *Front. Microbiol.* **2017**, *8*, 1130. [CrossRef]
16. Libby, E.A.; Reuveni, S.; Dworkin, J. Multisite phosphorylation drives phenotypic variation in (p)ppGpp synthetase-dependent antibiotic tolerance. *Nat. Commun.* **2019**, *10*, 5133. [CrossRef]
17. Eiamphungporn, W.; Helmann, J.D. The *Bacillus subtilis* σ^M regulon and its contribution to cell envelope stress responses. *Mol. Microbiol.* **2008**, *67*, 830–848. [CrossRef]

18. Gerdes, K.; Maisonneuve, E. Bacterial persistence and toxin-antitoxin loci. *Annu. Rev. Microbiol.* **2012**, *66*, 103–123. [CrossRef]
19. Page, R.; Peti, W. Toxin-antitoxin systems in bacterial growth arrest and persistence. *Nat. Chem. Biol.* **2016**, *12*, 208–214. [CrossRef]
20. Yamaguchi, Y.; Inouye, M. Regulation of growth and death in *Escherichia coli* by toxin-antitoxin systems. *Nat. Rev. Microbiol.* **2011**, *9*, 779–790. [CrossRef]
21. Diaz-Orejas, R.; Espinosa, M.; Yeo, C.C. The Importance of the expendable: Toxin-antitoxin genes in plasmids and chromosomes. *Front. Microbiol.* **2017**, *8*, 1479. [CrossRef]
22. Ogura, T.; Hiraga, S. Mini-F plasmid genes that couple host cell division to plasmid proliferation. *Proc. Natl. Acad. Sci. USA* **1983**, *80*, 4784–4788. [CrossRef]
23. Fraikin, N.; Goormaghtigh, F.; Van Melderen, L. Type II toxin-antitoxin systems: Evolution and revolutions. *J. Bacteriol.* **2020**, *202*, e00763-19. [CrossRef]
24. Alonso, J.C.; Balsa, D.; Cherny, I.; Christensen, S.K.; Espinosa, S.; Francuski, D.; Gazit, E.; Gerdes, K.; Hitchin, E.; Martín, M.T.; et al. *Bacterial Toxin-Antitoxin Systems as Targets for the Development of Novel Antibiotics*; ASM Press: Washington, DC, USA, 2007; pp. 313–329.
25. Moyed, H.S.; Bertrand, K.P. HipA, a newly recognized gene of *Escherichia coli* K-12 that affects frequency of persistence after inhibition of murein synthesis. *J. Bacteriol.* **1983**, *155*, 768–775. [CrossRef]
26. Bokinsky, G.; Baidoo, E.E.; Akella, S.; Burd, H.; Weaver, D.; Alonso-Gutierrez, J.; Garcia-Martin, H.; Lee, T.S.; Keasling, J.D. HipA-triggered growth arrest and β-Lactam tolerance in *Escherichia coli* are mediated by RelA-dependent ppGpp synthesis. *J. Bacteriol.* **2013**, *195*, 3173–3182. [CrossRef]
27. Germain, E.; Castro-Roa, D.; Zenkin, N.; Gerdes, K. Molecular mechanism of bacterial persistence by HipA. *Mol. Cell* **2013**, *52*, 248–254. [CrossRef]
28. Kaspy, I.; Rotem, E.; Weiss, N.; Ronin, I.; Balaban, N.Q.; Glaser, G. HipA-mediated antibiotic persistence via phosphorylation of the glutamyl-tRNA-synthetase. *Nat. Commun.* **2013**, *4*, 3001. [CrossRef]
29. Vogwill, T.; Comfort, A.C.; Furio, V.; MacLean, R.C. Persistence and resistance as complementary bacterial adaptations to antibiotics. *J. Evol. Biol.* **2016**, *29*, 1223–1233. [CrossRef]
30. Ramisetty, B.C.; Ghosh, D.; Roy Chowdhury, M.; Santhosh, R.S. what is the link between stringent response, endoribonuclease encoding type II toxin-antitoxin systems and persistence? *Front. Microbiol.* **2016**, *7*, 1882. [CrossRef]
31. Goormaghtigh, F.; Fraikin, N.; Putrins, M.; Hallaert, T.; Hauryliuk, V.; Garcia-Pino, A.; Sjodin, A.; Kasvandik, S.; Udekwu, K.; Tenson, T.; et al. Reassessing the role of type II toxin-antitoxin systems in formation of Escherichia coli type II persister cells. *MBio* **2018**, *9*, e00640-18. [CrossRef]
32. Khoo, S.K.; Loll, B.; Chan, W.T.; Shoeman, R.L.; Ngoo, L.; Yeo, C.C.; Meinhart, A. Molecular and structural characterization of the PezAT chromosomal toxin-antitoxin system of the human pathogen *Streptococcus pneumoniae*. *J. Biol. Chem.* **2007**, *282*, 19606–19618. [CrossRef]
33. Meinhart, A.; Alonso, J.C.; Strater, N.; Saenger, W. Crystal structure of the plasmid maintenance system ε/ζ: Functional mechanism of toxin ζ and inactivation by $\varepsilon_2\zeta_2$ complex formation. *Proc. Natl. Acad. Sci. USA* **2003**, *100*, 1661–1666. [CrossRef]
34. Mutschler, H.; Gebhardt, M.; Shoeman, R.L.; Meinhart, A. A novel mechanism of programmed cell death in bacteria by toxin-antitoxin systems corrupts peptidoglycan synthesis. *PLoS Biol.* **2011**, *9*, e1001033. [CrossRef]
35. Meinhart, A.; Alings, C.; Strater, N.; Camacho, A.G.; Alonso, J.C.; Saenger, W. Crystallization and preliminary X-ray diffraction studies of the $\varepsilon\zeta$ addiction system encoded by *Streptococcus pyogenes* plasmid pSM19035. *Acta Cryst. D Biol. Cryst.* **2001**, *57*, 745–747. [CrossRef]
36. Moreno-Del Alamo, M.; Tabone, M.; Munoz-Martinez, J.; Valverde, J.R.; Alonso, J.C. Toxin ζ reduces the ATP and modulates the uridine diphosphate-*N*-acetylglucosamine pool. *Toxins* **2019**, *11*, 29. [CrossRef]
37. Tabone, M.; Ayora, S.; Alonso, J.C. Toxin ζ reversible induces dormancy and reduces the UDP-N-acetylglucosamine pool as one of the protective responses to cope with stress. *Toxins* **2014**, *6*, 2787–2803. [CrossRef]
38. Lioy, V.S.; Machon, C.; Tabone, M.; Gonzalez-Pastor, J.E.; Daugelavicius, R.; Ayora, S.; Alonso, J.C. The ζ toxin induces a set of protective responses and dormancy. *PLoS ONE* **2012**, *7*, e30282. [CrossRef]
39. Krasny, L.; Gourse, R.L. An alternative strategy for bacterial ribosome synthesis: *Bacillus subtilis* rRNA transcription regulation. *EMBO J.* **2004**, *23*, 4473–4483. [CrossRef]

40. Kriel, A.; Bittner, A.N.; Kim, S.H.; Liu, K.; Tehranchi, A.K.; Zou, W.Y.; Rendon, S.; Chen, R.; Tu, B.P.; Wang, J.D. Direct regulation of GTP homeostasis by (p)ppGpp: A critical component of viability and stress resistance. *Mol. Cell* **2012**, *48*, 231–241. [CrossRef]
41. Eymann, C.; Homuth, G.; Scharf, C.; Hecker, M. *Bacillus subtilis* functional genomics: Global characterization of the stringent response by proteome and transcriptome analysis. *J. Bacteriol.* **2002**, *184*, 2500–2520. [CrossRef]
42. Lioy, V.S.; Rey, O.; Balsa, D.; Pellicer, T.; Alonso, J.C. A toxin-antitoxin module as a target for antimicrobial development. *Plasmid* **2010**, *63*, 31–39. [CrossRef]
43. Lioy, V.S.; Martin, M.T.; Camacho, A.G.; Lurz, R.; Antelmann, H.; Hecker, M.; Hitchin, E.; Ridge, Y.; Wells, J.M.; Alonso, J.C. pSM19035-encoded ζ toxin induces stasis followed by death in a subpopulation of cells. *Microbiology* **2006**, *152*, 2365–2379. [CrossRef]
44. McKeegan, K.S.; Borges-Walmsley, M.I.; Walmsley, A.R. Microbial and viral drug resistance mechanisms. *Trends Microbiol.* **2002**, *10*, S8–S14. [CrossRef]
45. Brauner, A.; Fridman, O.; Gefen, O.; Balaban, N.Q. Distinguishing between resistance, tolerance and persistence to antibiotic treatment. *Nat. Rev. Microbiol.* **2016**, *14*, 320–330. [CrossRef]
46. Camacho, A.G.; Misselwitz, R.; Behlke, J.; Ayora, S.; Welfle, K.; Meinhart, A.; Lara, B.; Saenger, W.; Welfle, H.; Alonso, J.C. *In vitro* and *in vivo* stability of the $\varepsilon_2\zeta_2$ protein complex of the broad host-range *Streptococcus pyogenes* pSM19035 addiction system. *Biol. Chem.* **2002**, *383*, 1701–1713. [CrossRef]
47. Ayrapetyan, M.; Williams, T.C.; Oliver, J.D. Bridging the gap between viable but non-culturable and antibiotic persistent bacteria. *Trends Microbiol.* **2015**, *23*, 7–13. [CrossRef]
48. LeRoux, M.; Culviner, P.H.; Liu, Y.J.; Littlehale, M.L.; Laub, M.T. Stress can induce transcription of toxin-antitoxin systems without activating toxin. *Mol. Cell* **2020**, *79*, 280–292e288. [CrossRef]
49. Rosendahl, S.; Tammana, H.; Brauera, A.; Remma, M.; Hõraka, R. *Pseudomonas putida* chromosomal toxin-antitoxin systems carry neither clear fitness benefits nor big costs. *bioRxiv* **2020**. [CrossRef]
50. Dy, R.L.; Przybilski, R.; Semeijn, K.; Salmond, G.P.; Fineran, P.C. A widespread bacteriophage abortive infection system functions through a Type IV toxin-antitoxin mechanism. *Nucleic Acids Res.* **2014**, *42*, 4590–4605. [CrossRef]
51. Hazan, R.; Engelberg-Kulka, H. *Escherichia coli mazEF*-mediated cell death as a defense mechanism that inhibits the spread of phage P1. *Mol. Genet. Genom.* **2004**, *272*, 227–234. [CrossRef]
52. Pearl, S.; Gabay, C.; Kishony, R.; Oppenheim, A.; Balaban, N.Q. Nongenetic individuality in the host-phage interaction. *PLoS Biol.* **2008**, *6*, e120. [CrossRef]
53. Alawneh, A.M.; Qi, D.; Yonesaki, T.; Otsuka, Y. An ADP-ribosyltransferase Alt of bacteriophage T4 negatively regulates the *Escherichia coli* MazF toxin of a toxin-antitoxin module. *Mol. Microbiol.* **2016**, *99*, 188–198. [CrossRef]
54. Lyzen, R.; Kochanowska, M.; Wegrzyn, G.; Szalewska-Palasz, A. Transcription from bacteriophage lambda P_R promoter is regulated independently and antagonistically by DksA and ppGpp. *Nucleic Acids Res.* **2009**, *37*, 6655–6664. [CrossRef]
55. Koga, M.; Otsuka, Y.; Lemire, S.; Yonesaki, T. *Escherichia coli rnlA* and *rnlB* compose a novel toxin-antitoxin system. *Genetics* **2011**, *187*, 123–130. [CrossRef]
56. Pane-Farre, J.; Quin, M.B.; Lewis, R.J.; Marles-Wright, J. Structure and function of the stressosome signalling hub. *Subcell Biochem.* **2017**, *83*, 1–41. [CrossRef]
57. Tuomanen, E.; Cozens, R.; Tosch, W.; Zak, O.; Tomasz, A. The rate of killing of *Escherichia coli* by β-lactam antibiotics is strictly proportional to the rate of bacterial growth. *J. Gen. Microbiol.* **1986**, *132*, 1297–1304. [CrossRef]
58. Cohen, N.R.; Lobritz, M.A.; Collins, J.J. Microbial persistence and the road to drug resistance. *Cell Host Microbe* **2013**, *13*, 632–642. [CrossRef]
59. Balaban, N.Q.; Merrin, J.; Chait, R.; Kowalik, L.; Leibler, S. Bacterial persistence as a phenotypic switch. *Science* **2004**, *305*, 1622–1625. [CrossRef]
60. Amitai, S.; Yassin, Y.; Engelberg-Kulka, H. MazF-mediated cell death in *Escherichia coli*: A point of no return. *J. Bacteriol.* **2004**, *186*, 8295–8300. [CrossRef]
61. Franch, T.; Gerdes, K. Programmed cell death in bacteria: Translational repression by mRNA end-pairing. *Mol. Microbiol.* **1996**, *21*, 1049–1060. [CrossRef]

62. Levin, B.R.; Rozen, D.E. Non-inherited antibiotic resistance. *Nat. Rev. Microbiol.* **2006**, *4*, 556–562. [CrossRef]
63. Dörr, T.; Vulic, M.; Lewis, K. Ciprofloxacin causes persister formation by inducing the TisB toxin in *Escherichia coli. PLoS Biol.* **2010**, *8*, e1000317. [CrossRef]
64. Kohanski, M.A.; Dwyer, D.J.; Hayete, B.; Lawrence, C.A.; Collins, J.J. A common mechanism of cellular death induced by bactericidal antibiotics. *Cell* **2007**, *130*, 797–810. [CrossRef]
65. Dwyer, D.J.; Camacho, D.M.; Kohanski, M.A.; Callura, J.M.; Collins, J.J. Antibiotic-induced bacterial cell death exhibits physiological and biochemical hallmarks of apoptosis. *Mol. Cell* **2012**, *46*, 561–572. [CrossRef]
66. Cui, P.; Niu, H.; Shi, W.; Zhang, S.; Zhang, W.; Zhang, Y. Identification of genes involved in bacteriostatic antibiotic-induced persister formation. *Front. Microbiol.* **2018**, *9*, 413. [CrossRef]
67. Ayora, S.; Carrasco, B.; Cardenas, P.P.; Cesar, C.E.; Canas, C.; Yadav, T.; Marchisone, C.; Alonso, J.C. Double-strand break repair in bacteria: A view from *Bacillus subtilis*. *FEMS Microbiol. Rev.* **2011**, *35*, 1055–1081. [CrossRef]
68. Kowalczykowski, S.C. An overview of the molecular mechanisms of recombinational DNA repair. *Cold Spring Harb. Perspect. Biol.* **2015**, *7*, a016410. [CrossRef]
69. Simmons, L.A.; Goranov, A.I.; Kobayashi, H.; Davies, B.W.; Yuan, D.S.; Grossman, A.D.; Walker, G.C. Comparison of responses to double-strand breaks between *Escherichia coli* and *Bacillus subtilis* reveals different requirements for SOS induction. *J. Bacteriol.* **2009**, *191*, 1152–1161. [CrossRef]
70. Arjes, H.A.; Kriel, A.; Sorto, N.A.; Shaw, J.T.; Wang, J.D.; Levin, P.A. Failsafe mechanisms couple division and DNA replication in bacteria. *Curr. Biol.* **2014**, *24*, 2149–2155. [CrossRef]
71. Sanchez, H.; Kidane, D.; Reed, P.; Curtis, F.A.; Cozar, M.C.; Graumann, P.L.; Sharples, G.J.; Alonso, J.C. The RuvAB branch migration translocase and RecU Holliday junction resolvase are required for double-stranded DNA break repair in *Bacillus subtilis*. *Genetics* **2005**, *171*, 873–883. [CrossRef]
72. Valero-Rello, A.; Lopez-Sanz, M.; Quevedo-Olmos, A.; Sorokin, A.; Ayora, S. Molecular mechanisms that contribute to horizontal transfer of plasmids by the bacteriophage SPP1. *Front. Microbiol.* **2017**, *8*, 1816. [CrossRef]

4

Impact of Dendrimer Terminal Group Chemistry on Blockage of the Anthrax Toxin Channel: A Single Molecule Study

Goli Yamini, Nnanya Kalu and Ekaterina M. Nestorovich *

Department of Biology, The Catholic University of America, Washington DC, WA 20064, USA; 00yamini@cua.edu (G.Y.); 56kalu@cua.edu (N.K.)
* Correspondence: nestorovich@cua.edu.

Abstract: Nearly all the cationic molecules tested so far have been shown to reversibly block K^+ current through the cation-selective PA_{63} channels of anthrax toxin in a wide nM–mM range of effective concentrations. A significant increase in channel-blocking activity of the cationic compounds was achieved when multiple copies of positively charged ligands were covalently linked to multivalent scaffolds, such as cyclodextrins and dendrimers. Even though multivalent binding can be strong when the individual bonds are relatively weak, for drug discovery purposes we often strive to design multivalent compounds with high individual functional group affinity toward the respective binding site on a multivalent target. Keeping this requirement in mind, here we perform a single-channel/single-molecule study to investigate kinetic parameters of anthrax toxin PA_{63} channel blockage by second-generation (G2) poly(amido amine) (PAMAM) dendrimers functionalized with different surface ligands, including G2-NH_2, G2-OH, G2-succinamate, and G2-COONa. We found that the previously reported difference in IC_{50} values of the G2-OH/PA_{63} and G2-NH_2/PA_{63} binding was determined by both on- and off-rates of the reversible dendrimer/channel binding reaction. In 1 M KCl, we observed a decrease of about three folds in k_{on} and a decrease of only about ten times in t_{res} with G2-OH compared to G2-NH_2. At the same time for both blockers, k_{on} and t_{res} increased dramatically with transmembrane voltage increase. PAMAM dendrimers functionalized with negatively charged succinamate, but not carboxyl surface groups, still had some residual activity in inhibiting the anthrax toxin channels. At 100 mV, the on-rate of the G2-succinamate binding was comparable with that of G2-OH but showed weaker voltage dependence when compared to G2-OH and G2-NH_2. The residence time of G2-succinamate in the channel exhibited opposite voltage dependence compared to G2-OH and G2-NH_2, increasing with the *cis*-negative voltage increase. We also describe kinetics of the PA_{63} ion current modulation by two different types of the "imperfect" PAMAM dendrimers, the mixed-surface G2 75% OH 25% NH_2 dendrimer and G3-NH_2 dendron. At low voltages, both "imperfect" dendrimers show similar rate constants but significantly weaker voltage sensitivity when compared with the intact G2-NH_2 PAMAM dendrimer.

Keywords: multivalency; planar lipid bilayer technique; *Bacillus anthracis*; protective antigen; pore blockage

---

1. Introduction

Many bacterial exotoxins oligomerize during invasion to form ion-conductive channels or pores in the host cell or organelle membranes. This oligomeric centrosymmetric organization represents an ideal multivalent receptor target to explore a variety of multivalent channel-blocking ligands with a controlled number of preassembled functional groups (reviewed in [1]). In the past decade, two classes of multivalent compounds, cationic cyclodextrins (CDs) and dendrimers have been reported to directly block the channel-forming B components of the AB type anthrax, C2, iota, and CDT toxins [2–5], PA_{63}, C2IIa, Ib, and CDTb, respectively. The potency of these multivalent blockers compares well with the most effective blockers of the classical channels of electrophysiology (Table 3 in ref. [6]) and exceeds activities of the small-molecule cationic ligands [7–11]. A key advantage of cyclodextrins (reviewed in [12]) is their arrangement into rigid 6-, 7-, and 8-fold centrosymmetric structures with controlled number and position of potential attachment sites and the ability to form water-soluble "host-guest" inclusion complexes with poorly soluble small molecules and macromolecule fragments. The poly(amido amine) (PAMAM) dendrimers (reviewed in [13]) are repeatedly branched polymers with all bonds forming amidoamine branches emanating from a central alkyldiamine core, where each consecutive growth step represents a new dendrimer "generation" with an increased diameter and doubled number of reactive surface functional groups. PAMAM dendrimers form monodispersed, starburst-shaped polymers that are synthesized in generations with a growing but well-controlled number of attachment sites (Figure S1). While many details of the biophysical mechanisms of cyclodextrin interaction with the channel-forming components of anthrax, C2, and iota AB type toxins are known [14,15], the physical forces involved in the dendrimer/channel binding reaction require further analysis. In this small-scale study written for the special issue of Toxins (Basel) on "Novel Pharmacological Inhibitors for Bacterial Protein Toxins", we perform a single channel investigation of the PAMAM dendrimer/PA_{63} channel binding reaction focusing on two specific aspects of the dendrimer-induced channel blockage. Firstly, we perform a single-channel analysis of the kinetic parameters of the dendrimer/PA_{63} binding reaction using generation 2 amino-terminated (G2-NH_2), hydroxyl-terminated (G2-OH), succinamate-terminated (G2-SA), and carboxyl-terminated (G2-COONa) PAMAM dendrimers (Figure 1A and Figure S1). The amino-terminated PAMAM dendrimers of different generations were recently reported to effectively block the PA_{63} channel lumen in multichannel experiments under near-physiological conditions [5]. This effect was explained by the direct electrostatic interaction of the positively-charged terminal amino-groups on the PAMAM dendrimers with the negatively charged lumen of PA_{63}. However, second and third generation (G2 and G3) PAMAM dendrimers functionalized with surface hydroxyl groups were reported to have an avidity decrease of only 20 and 9 times compared to the G2 and G3 amino-terminated dendrimers, respectively. This difference is comparable with the variation observed between different PAMAM dendrimer generations. Moreover, even PAMAM dendrimers terminated with negatively charged carboxylic and succinamic surface groups showed some residual PA_{63} binding. When fine-tuned, some of these dendrimers may offer the advantage of being effective channel inhibitors with decreased cytotoxicity. In fact, while all dendrimers are less toxic than linear polymers [16,17], the cationic amino-terminated PAMAM dendrimers have been reported to display concentration- and generation-dependent cytotoxicity, and are therefore less biocompatible compared to their neutral and negatively charged analogues [18,19]. Secondly, we analyze the kinetic parameters of the blockage of a single PA_{63} channel by G2 75% OH 25% NH_2 PAMAM dendrimer and G3-NH_2 dendron (Figure 1B). It has been demonstrated that a favorable therapeutic window for dendrimers can also be achieved by either partial surface modification, aiming to lower amino group density, or by degradation of the dendrimers to "imperfect" dendrimers or fractured dendron-like branches [20]. The channel-blocking activity of these two types of the "imperfect" dendrimers was previously investigated on a multichannel level [5]. The mixed-surface G2 75% OH 25% NH_2 PAMAM dendrimer that, on average, had only four surface positive charges was about 17 times less active (IC_{50} = 122 nM vs. 7 nM) than the 16+ charged G2-NH_2 dendrimer, and its activity was comparable (IC_{50} = 122 nM vs. 128 nM) with the 4+ charged G0 dendrimer. The structurally incomplete 4+ charged G1 dendron was about 26 times more effective (IC_{50} = 4.9 nM) than the 4+ charged G0 dendrimer (IC_{50} = 128 nM). The 8+ charged G2 dendron and G1 dendrimer had similar inhibitory activity. The five different commercially available generation 2 dendrimers and one generation 3 dendron, investigated in this study, were chosen with the purpose of specifically focusing on blocker terminal group chemistry and dendrimer flexibility. The number of terminal amino, hydroxyl, succinamate, carboxyl, or mixed OH/NH_2 surface groups was fixed and equal to 16.

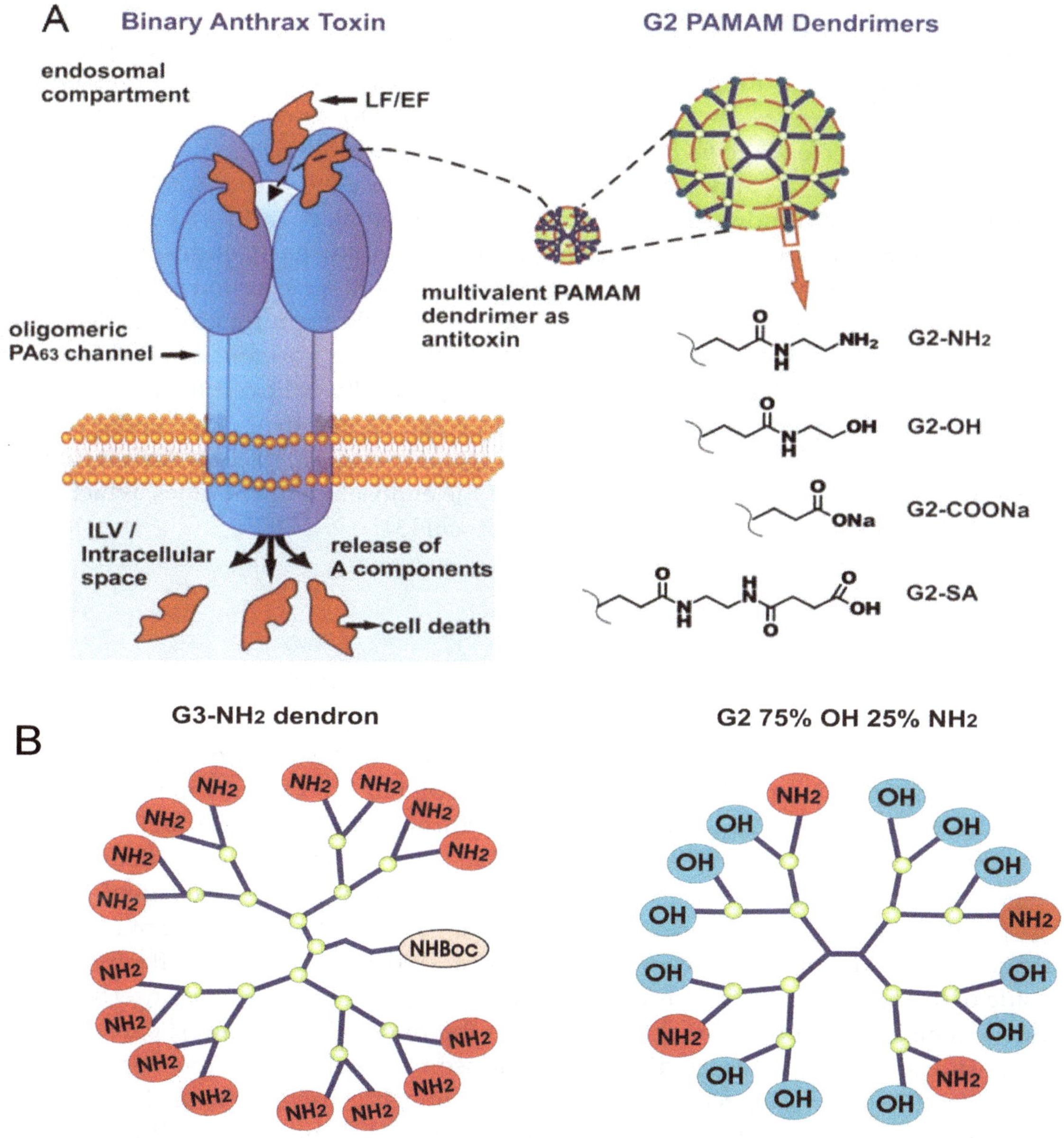

Figure 1. Inhibition of anthrax toxin by poly(amido amine) (PAMAM) dendrimers. (**A**) Oligomeric PA_{63} channel, produced by *B. anthracis* (left), is responsible for translocation of lethal factor (LF) and edema factor (EF) into the host cell. The cartoon is a simplified illustration of the early/late endosomal stages of the LF and EF host cell transport. Second generation (G2) PAMAM dendrimers (right) tested in this study. All shown G2 dendrimers have 16 terminal groups that could be charged positively (G2-NH_2), negatively (G2-COONa, G2-SA) or are neutral (G2-OH); (**B**) Schematic representation of the mixed-surface G3-NH_2 dendron (left) and G2 75% OH 25% NH_2 dendrimer (right). Note that in contrast to all other dendrimers, G2 75% OH 25% NH_2 is not monodisperse and contains 75% of terminal OH groups and 25% of terminal NH_2 groups on average.

In this study, we investigated the oligomeric channel-forming B component of the anthrax toxin, protective antigen (PA), as a multivalent target for the multivalent dendrimer binding. Traditionally,

PA, because of its main role in anthrax toxin uptake, has been one of the key targets for small molecule and multivalent antitoxin development [21]. The AB type anthrax toxin is composed of three individually nontoxic proteins. The two A components, lethal factor (LF) and edema factor (EF), are the intracellularly active enzymes. LF is a Zn-metalloprotease that cleaves MAP kinase kinases [22,23] and Nlrp1 [24]. EF is a Ca^{2+} and calmodulin-activated adenylyl cyclase [25,26]. Protective antigen (PA), named this way for its ability to elicit protective antibodies (the property utilized in the anthrax vaccines), is a receptor for LF/EF binding, which mediates their translocation. The anthrax toxin intracellular delivery occurs in several stages. After binding to its cellular CMG2 and/or TEM8 receptors and proteolytic cleavage to PA_{63}, by extracellular furin, PA oligomerizes on the host cell surface to form heptameric [27] and/or oligomeric ring-shaped pre-pores [28,29] creating three [30] or four [28] LF and/or EF binding sites. After receptor-mediated endocytosis [31], the anthrax toxin AB complexes are delivered to the acidic environment of the early endosome. There, the PA oligomers undergo substantial conformational changes leading to their insertion in endosomal limiting, and possibly in intraluminal vesicle membranes [32], eventually forming an extended 180-Å long "flower-on-a-stem" cation-selective [33] channel with a 75 Å long bud and a 105 Å long stem and radius varying from 16 Å to 3.5 Å [34]. This channel is generally believed to work as an effective translocase that unfolds and allows for translocation of LF and EF into the cytosol under a pH gradient across the late endosomal limiting membrane ($pH_{endosome} < pH_{cytosol}$) [35,36]. An alternative model suggests that the anthrax toxin catalyzes the rupture of the endosomal membranes, which leads to the consequent delivery of the toxin complexes into the cytosol [37].

2. Results

2.1. Two Modes of G2-NH_2 PAMAM Dendrimer Inhibition of PA_{63} Channel

Figure 2 illustrates the bimodal effect of G2-NH_2 PAMAM *cis*-solution addition on the ionic current through a single PA_{63} channel incorporated into planar lipid bilayer membranes. To obtain reliable statistics on G2-NH_2/PA_{63} interaction, we performed the single channel measurements in 1 M KCl. A decrease in salt concentration led to a dramatic increase in the blocker lifetimes, suggesting the involvement of the long-range Coulomb interactions. Quantitative analysis of the process at lower, e.g., physiological salt concentrations, proved impossible over the course of our experiments. Previously, the PAMAM dendrimers were reported to be ~100–900 times more effective when added to the *cis*-side of the membrane, which is also the side of PA_{63} addition [5]. PA_{63} insertion was shown to be almost exclusively unidirectional [2,9,33], with the bud, LF/EF binding part of the channel, facing the *cis*-side solution (corresponding to the endosome interior), and the stem part facing the *trans*-side solution (corresponding to the cytosol or ILV interior). The single channel current recordings show that, in a manner similar to the cationic β-cyclodextrin [14] and G1-NH_2 PAMAM dendrimer blockers [5], the G2-NH_2 inhibitive action is bimodal (Figure 2A). Firstly, G2-NH_2 addition generates complete but reversible blockages of ion current through a single channel (marked with two blue ovals, Figure 2A). Frequency of these events increases in a concentration-dependent manner and is a strong function of the applied transmembrane voltage (Figure 2B). Note: *cis*-positive sign of the applied transmembrane voltages corresponds to the inside-positive voltage gradient across endosomal limiting membranes. Secondly, G2-NH_2 addition led to a dramatic increase in the voltage-dependent gating of PA_{63} channels (Figure S2), seen as prolonged closing events (marked with red ovals, Figure 2A, middle and right). Higher concentrations of G2-NH_2 and higher voltages compared to the ones reported earlier ($K_D = (7.2 \pm 4.7) \times 10^{-9}$ M at $V = 20$ mV) in the multichannel systems were needed because of the increased supporting electrolyte concentrations (1 M vs. 0.1 M) that, by electrostatically screening charges on both the blocker and the channel, weakened blocker binding.

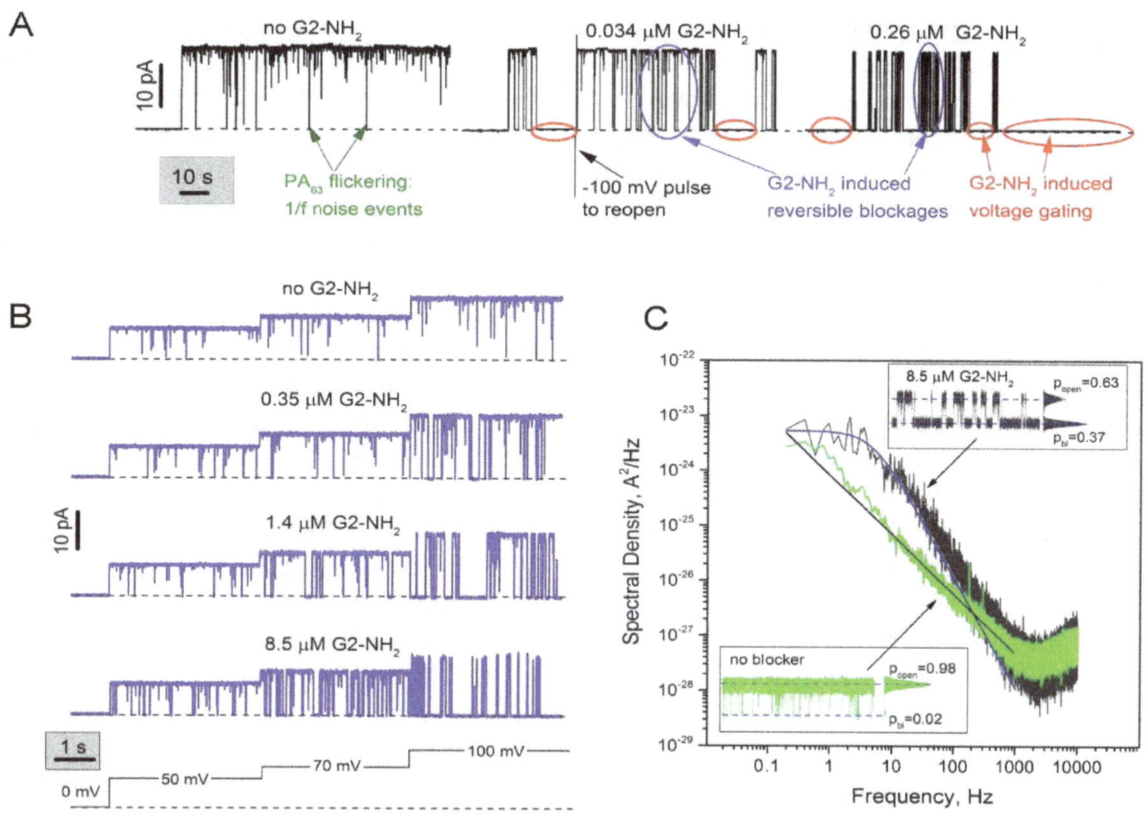

Figure 2. Modulation of a single PA_{63} channel current by G2-NH_2 PAMAM dendrimer. (**A**) Two modes of G2-NH_2 action on a single PA_{63} channel. In the absence of G2-NH_2 (left), PA_{63} remains in an open state. Fast flickering between the open and closed states (the so-called $1/f$ noise) is mostly but not completely removed by averaging over a time interval of 50 ms. In the presence of two different G2-NH_2 concentrations (middle and right), both blocker-induced reversible blockage and prolonged voltage gating events are seen. Recordings were taken at 100 mV applied voltage; (**B**) First mode of action. In the absence of the blocker (top) the ion movement is determined by the geometry and surface properties of the PA_{63} pore. In the presence of G2-NH_2 in the *cis* compartment of the bilayer chamber (three following rows), the channel gets reversibly blocked. At higher concentrations of G2-NH_2 (bottom) the blockages, which are seen as downward spikes, are more frequent. The probability of finding PA_{63} in the blocked state increases with *cis*-positive transmembrane voltage increase (50, 70, and 100 mV are shown). Current tracks were averaged over a time interval of 2 ms; (**C**) Power spectral densities of the G2-NH_2–induced PA_{63} current fluctuations (spectrum in grey) can be fitted by a single Lorentzian at frequencies of <1000 Hz in contrast to $1/f$ noise in the absence of G2-NH_2 (spectrum in green). A clear deviation from a single Lorentzian dependence at f > 100 Hz could to some degree be explained by the $1/f$ noise and the partial dendrimer degradation and breakage with formation of the imperfect cationic substrates, capable of blocking PA_{63} channels with shorter lifetimes (mass spectra and NMR characterization is given in Figures S5 and S6). Inserts: 1-s current tracks averaged over a time interval of 0.2 ms for blocker-free (bottom left) and 8.5 μM G2-NH_2 (upper right) recordings. All-point histograms are shown to the right of the recordings; p_{open} and p_{bl} denote the probability of the PA_{63} channel being in conductive and non-conductive states, respectively. Applied voltage was 100 mV; dashed lines represent zero current levels.

The fast reversible current fluctuation induced by G2-NH_2 in the parts of current tracks with excluded voltage-dependent gating can be described as a two-state memoryless Markov process, where both the residence time in the blocked state and the channel lifetime in the unblocked state (the time between blockages) are described by exponential distributions. This is demonstrated by the Lorentzian shape of the power spectral density of G2-NH_2-induced current fluctuations at f < 1000 Hz (Figure 2C, spectrum in grey fitted by the smooth blue solid line through the experimental curve). This relatively straightforward kinetic analysis was to a certain extent complicated by a number of factors, namely the two types of complex non-Markovian channel gating described in detail previously [14,15]. The first type of gating is induced by the applied voltage that brings the PA_{63} channel into a nonconductive state, which seems to be characteristic for β-barrel channels in general [38]. This voltage dependent gating was especially prominent at *cis*-side negative voltages [33]; thus applied voltages as low as − (10–20) mV led to the prolonged channel closures. The fact that the β-barrel PA_{63} channel tends to stay closed when positioned under non-physiological inside-negative voltage gradients, adds fuel to the little rusty but still very interesting debate about the significance [39] and mechanism [40] of voltage gating for the unconventional channel [41] function. Note that while some researchers show that the voltage-dependent β-barrel channel closure represents nothing more than an artifact of bilayer lipid experiments [42], others report clear evidence of physiological significance of voltage gating in β-barrel channels [43,44]. In one way or another, this circumstance has largely limited our ability to collect reliable statistics on channel/blocker binding reaction at negative and high positive voltages, especially because in many cases the dendrimer addition has significantly enhanced the voltage sensitivity of the channel. The second type of PA_{63} non-Markov gating is the so-called voltage-independent fast flickering $1/f$ noise between the open and completely closed states that was earlier described as a universal intrinsic property of the pore-forming components of AB type toxins, PA_{63}, C2IIa, and Ib, both at the single [2,3,14,15] and multi-channel [8,9] level. The current noise power spectrum of the non-modified PA_{63} channel contains a $1/f$-like voltage-independent [14,15] component (Figure 2C, see the spectrum in green and the corresponding current track (left insert) shown at 0.2 ms time resolution). Even though the $1/f$ flickering is not among the immediate points of interests of the current publication, the universality of the $1/f$ flickering and the fact that the F427A mutant of PA_{63}, which lacks the ϕ-clamp [7] and therefore A-component translocation functionality, was devoid of the $1/f$ noise behavior, deserves to be studied more closely. Within the limits of this study, we had to be very careful to uncouple the $1/f$ fast-flickering events and the dendrimer-induced reversible blockages, especially under conditions where the closed time distributions of these events partially overlap. For example, in Figure 2A (left) we show several relatively long $1/f$ flickering events that are still seen even at low, 50 ms time resolution. To quantify kinetic parameters of the G2-NH_2-induced reversible blockages, we primarily used current noise spectral analysis (Figure 2C) instead of the direct counting of open and closed event durations. The direct counting approach does not allow us to distinguish between open times of the $1/f$ noise closures and the dendrimer-induced reversible blockages, because a combined open time distribution for these two processes was single-exponential, which may be explained by the fact that there is only one open state of the channel. The average lifetime of G2-NH_2 in the channel pore (t_{res}) and average time between blockages (t_{on}) were found correspondingly as $t_{res} = {}^{1}/_{2\pi f_c(1-p_{bl})}$ and $t_{on} = {}^{1}/_{2\pi f_c p_{bl}}$, where f_c is the corner frequency of Lorentzian and p_{bl} is the probability of finding PA_{63} in the blocked state [45]. Ideally, the probability of the channels being in the blocked state could be directly determined as $p_{bl} = \frac{I_0 - I_{ave}}{I_0}$, where I_0 is the ion current through the completely open channel, and I_{ave} is the average ion channel current modified by the blocker. However here, to account for the $1/f$ fast flickering, the equation was corrected, assuming independence of these two processes as follows: $p_{bl} = \frac{I_{ave}^{free} - I_{ave}}{I_{ave}^{free}}$, where I_{ave}^{free} is the average current through the PA_{63} channel measured in blocker-free solutions [14]. Note: to determine I_{ave}^{free} and I_{ave}, the prolonged voltage gating closures (both intrinsic and dendrimer-induced) were excluded from the open and closed states probability analysis.

The second mode of dendrimer-induced current inhibition was hard to describe quantitatively because the blocker-induced channel closures often appeared to be irreversible, lasting for minutes or longer. To reopen the channel, we either had to apply 0 mV or to reverse the voltage sign (shown in Figure 2A, middle) which did not allow us to collect reliable statistics on kinetic parameters of the second mode of channel blockage. Moreover, as described above, the voltage-induced closures were also recorded in the absence of blocker, and voltage-sensitivity and probability of finding a channel in the closed state varied from channel to channel. However qualitatively, this process evinced all the key characteristics of the voltage-induced gating of β-barrel channels, such as strong voltage dependence, prolonged closures (minutes), and difficulties in reopening channels even when voltage was reduced to zero. Paradoxically, channel reopening was often possible with abrupt second-long pulses of high voltages of opposite sign (marked in Figure 2A, middle track), that, in turn, also induced voltage dependent closures if applied longer.

2.2. Role of PAMAM Dendrimer Surface Chemistry in PA_{63} Blockage

The common mechanism of inhibiting cation-selective PA_{63} involves blockage of the channel's lumen by positively charged molecules [2,3,5,7,9,10,14,33,46,47]. However, on a multichannel level in 0.1 M KCl, it has been demonstrated that G2 and G3 PAMAM dendrimers functionalized with surface hydroxyl groups (G2-OH and G3-OH) inhibited PA_{63} channels in a concentration-dependent manner [5]. The IC_{50} values were about 20 (Table 1) and 9 times lower compared to those of G2-NH_2 and G3-NH_2, respectively. Typical multichannel titration curve analyses are given in Figure S3. Interestingly, even in the presence of PAMAM dendrimer functionalized with negatively charged carboxyl surface group (G1-COONa, but named G0.5 by the manufacturer), some small current decrease was recorded, though compound activity was not high enough to reach 50% inhibitory concentrations. In order to understand how the PA_{63} channel selects among the PAMAM dendrimer blockers, here we investigated kinetic parameters of the binding reaction directly comparing G2-NH_2, G2-OH, G2-SA, and G2-COONa PAMAM dendrimers binding to PA_{63} on a single channel level (Figure 3 and Figure S4). As before, the single channel measurements are performed in 1 M KCl (Figure 3).

Table 1. Inhibition of PA_{63} channel ion current by intact G2 PAMAM dendrimers and G3 PAMAM dendron expressed as experimental IC_{50} values.

	PA_{63}/PAMAM Dendrimer Binding Reaction, IC_{50}					
	G2-NH_2	G2-OH	G2-SA	G2-COONa	G2 75% OH 25% NH_2	G3-NH_2 Dendron
0.1 M KCl	7.2 ± 4.7 nM	142 ± 36 nM	879 ± 50 μM	>14 mM	122 ± 35 nM	16.4 ± 4.0 nM
1 M KCl	5.1 ± 2.6 mM	>30 mM	1.7 ± 0.2 mM	not determined	7.7 ± 0.2 mM	7.8 ± 1.0 mM

All data were calculated as means from at least two separate multichannel experiments; the errors are standard deviations. 0.1 M and 1 M KCl solutions at pH 6 were buffered by 5 mM MES. Recordings were taken at 20 mV applied voltage, which was *cis*-side positive. Note: because G2-COONa activity in 0.1 M KCl was too low to reliably detect IC_{50}, we did not attempt to perform experiments in 1 M KCl bathing solutions, where charges on both the channel and the dendrimer were to a certain extent screened by the bathing solution counterions.

Figure 3A gives four representative recordings of a single PA_{63} channel current modified by different concentrations of G2-NH_2, G2-OH, G2-SA (three lower rows) compared to dendrimer-free solution (top). When added to the *cis* compartment solutions, all dendrimers reversibly blocked the channel, with *cis*-positive transmembrane voltages significantly enhancing the blockage with G2-NH_2 and G2-OH, but not G2-SA (Figure 3B,C). Two modes of G2-NH_2-induced current blockage are evident at a very low sub-μM blocker concentration at 100 mV applied voltage. In the presence of 0.35 μM of G2-NH_2, probability of finding the channel in the closed state increases to 35% (second row) compared with 0.3% in the dendrimer-free solution (first row). G2-OH dendrimer addition (third row) causes similar complete but reversible channel closures (first mode of blockage), however more than 2000 times higher blocker concentrations were required to increase probability of the closed state to 20%. At voltages ≥100 mV, we also detected the second mode of channel closure (voltage gating events), however this effect was significantly less pronounced compared to the one observed for G2-NH_2. At comparable sub-mM concentrations and 100 mV, G2-SA, modified with negatively charged surface groups, (fourth row) shows only weak interaction with probability of closed state equal to 2%. We could not detect any concentration-dependent channel fluctuations upon addition of

G2-COONa to the *cis* compartment of the bilayer chamber. Instead, high concentration of G2-COONa (3 mM) caused membrane instability and, eventually, breakage (Figure S4).

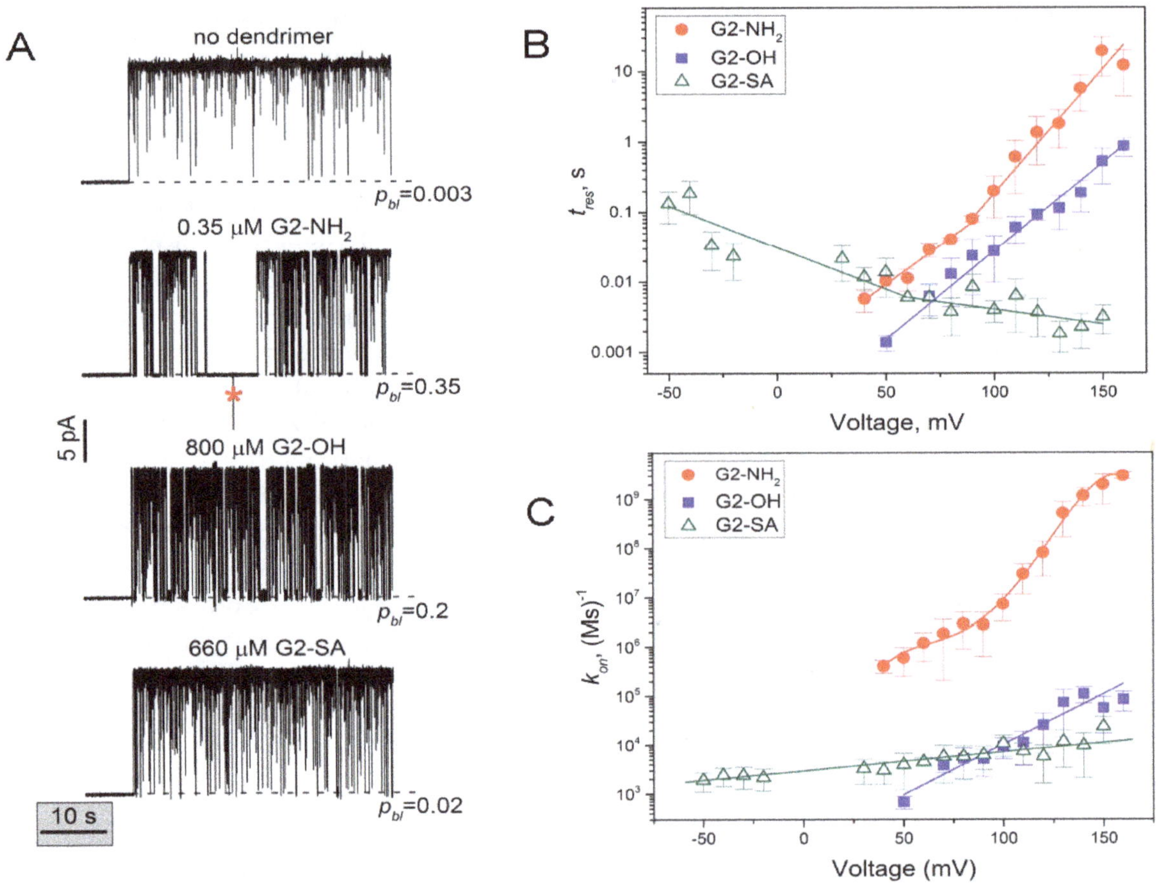

Figure 3. Influence of the PAMAM dendrimer surface chemistry on the PA_{63} single channel current inhibition. (**A**) PA_{63} channel current tracks in the absence (top) and presence of G2-NH_2, G2-OH, and G2-SA PAMAM dendrimers in the *cis* side of the bilayer chamber (three lower rows). In the dendrimer-free solution, the channel remains open with only 0.3% probability of being in the closed state (p_{bl}) due to intrinsic $1/f$ noise flickering (mostly filtered). Addition of G2-NH_2 (second row) induces two modes of PA_{63} current blockage, fast reversible fluctuations and longer voltage gating type of closures (marked by "*"). At 0.35 μM G2-NH_2, the probability of finding the channel in the closed state increases to 35%. The addition of 800 μM G2-OH (third row) causes fast reversible blockages, whereas the voltage gating events are less pronounced; the probability of finding PA_{63} in the closed state is 20%. G2-SA addition (bottom row) causes reversible blockages at similar sub-mM concentrations, with a 2% probability of finding PA_{63} in the closed state. Applied voltage was 100 mV; current tracks were averaged over a time interval of 10 ms. The dashed lines represent zero current levels; (**B**) Residence times of dendrimer binding reaction plotted as functions of transmembrane voltage. While residence time of G2-OH (filled squares) increases exponentially with voltage (solid line through the data points), residence times of G2-NH_2 (filled circles) and G2-SA (open triangles) show more complex progression, increasing and decreasing with the voltage increase respectively. The data were split into two voltage intervals and fitted with two separate exponents (solid lines) (**C**) On-rate of dendrimer blockage defined as the inverse of the average open channel life time as a function of voltage for G2-NH_2 (filled circles), G2-OH (filled squares), and G2-SA (open triangles) blockers. Note the voltage sensitivity of k_{on} for all three dendrimers and the significantly higher absolute values of the on-rate constant in the case of G2-NH_2.

2.3. The Rate Constants of Dendrimer's First Mode of Binding Reaction are Voltage Dependent

For all three blockers, the on- and off-rate constants of the PA_{63}/dendrimer binding reaction varied with the applied voltage (Figure 3B,C). The shown data were obtained with power spectral analysis of the reversible current fluctuations, using the fitting by single Lorentzian spectra as described above (Figure 2C). For voltages ≥130 mV, when strong blocker-induced voltage gating (second type of blockage) prevented us from collecting long current recordings suitable for the power spectral analysis, the data were analyzed by averaging over direct measurements of blocked and open event durations. The on-rate was calculated as $k_{on} = 1/t_{on}c_{PAMAM}$, where t_{on} is time between successful blockages, and c_{PAMAM} is dendrimer bulk concentration. While in the case of G2-OH, the dendrimer with the positively charged tertiary amine interior and neutral hydroxyl surface groups, t_{res} was shown to increase exponentially (linear fit in the semi-logarithmic scale in Figure 3B), in the case of the "highly" cationic G2-NH_2 binding, we observed non-exponential voltage dependence of t_{res} (fitted with two linear dependences in Figure 3B). The binding time in the presence of G2-SA, the dendrimer that is made of the positively charged PAMAM core and 16 negatively charged succinamate surface groups, was shown to have inverse voltage dependence, with the t_{res} binding time decreasing with *cis*-positive voltage decrease. Figure 3C shows the on-rate dependence of the PAMAM dendrimer/binding reaction on the applied voltage. Interestingly, for all three dendrimers we observed a voltage-dependent increase in the binding reaction rate constants (Figure 3C), which indicates that high voltages make PAMAM dendrimer capture by the PA_{63} easier. At the same time, k_{on}, and, therefore the number of individual blockage events, was drastically higher for G2-NH_2 compared with both G2-OH and G2-SA.

2.4. PA_{63} Blockage by Imperfect Cationic PAMAM Dendrimers

Activity of PAMAM dendrimers was previously shown to increase dramatically when they were degraded at the amide linkage, to a heterodisperse population of dendrimers of different molecular weights [20]. These less sterically constrained and hence, more flexible, "imperfect" dendrimers were reported to show significantly enhanced transfection activity compared to the intact dendrimers. To test if a similar effect could be achieved with the pore blockage, we previously investigated the channel blocking activity of two different types of "imperfect" PAMAM dendrimers. First, we tested a mixed-surface G2 75% OH 25% NH_2 charge-dispersed PAMAM dendrimer, where the proportion of the positively charged surface amino groups was only 25% on average. In 0.1 M KCl, this dendrimer was about 17 times less active against PA_{63} compared to G2-NH_2. A notable increase in the blocker's activity was achieved with the structurally incomplete 8+ charged G2 dendron, which was about 26 times more active, compared to the 8+ charged G1-NH_2 dendrimer [5]. Here we investigate the G2 75% OH 25% NH_2 dendrimer activity and also use the 16+ charged G3-NH_2 dendron to directly compare their activity with the 16+ charged G2-NH_2 dendrimer on a single-channel level (Figure 4). In a manner similar to G2-NH_2 intact dendrimers, the inhibitive action of the G2 75% OH 25% NH_2 dendrimer and G3-NH_2 dendron was bimodal. Both the short reversible blockage events and the prolonged closures were detected after the addition of sub-μM concentrations of either of these two imperfect dendrimers to the *cis*-compartment solutions (Figure 4A). While the effective concentrations were comparable to those used for G2-NH_2 (and significantly lower compared to G2-OH and G2-SA), the voltage sensitivity of the binding reaction on-rates was substantially different (Figure 4B,C).

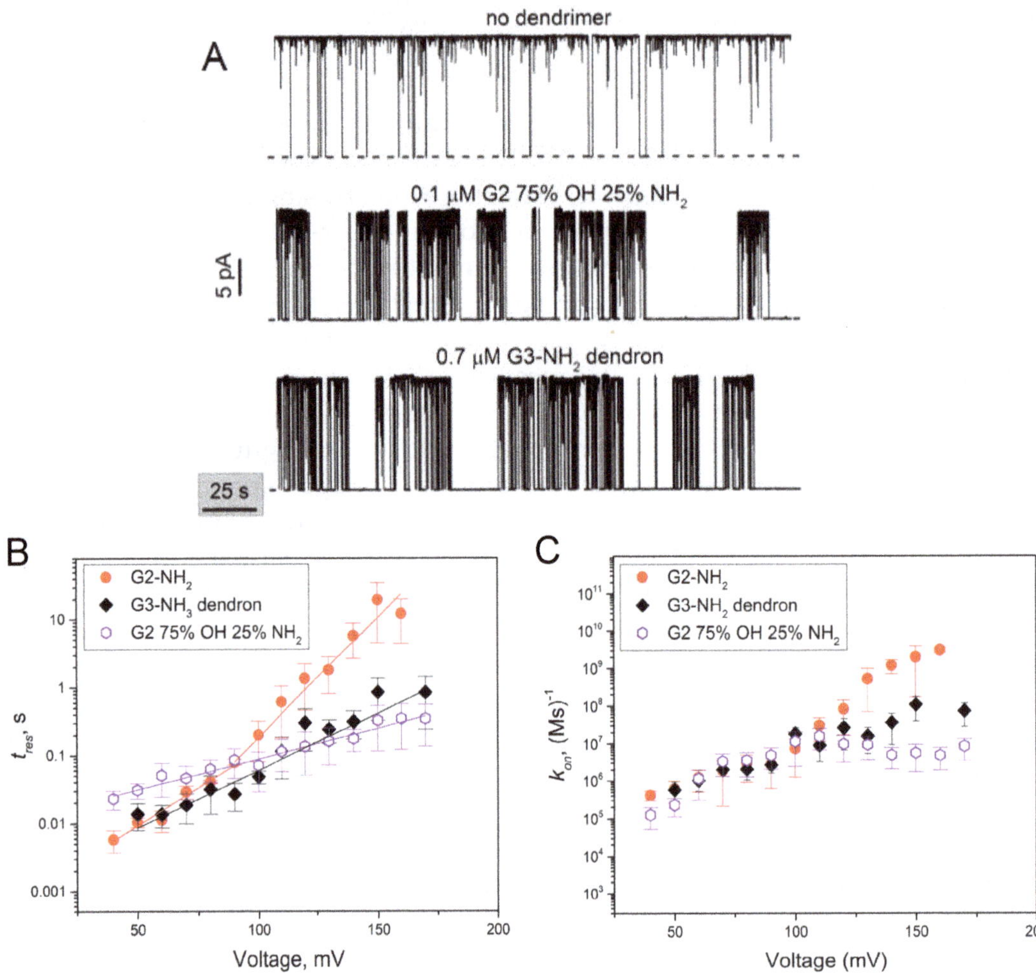

Figure 4. Modulation of a single PA_{63} channel current by imperfect G2-75% OH 25% NH_2 dendrimer and G3-NH_2 dendron. (**A**) The sub-μM addition of G2 75% OH 25% NH_2 dendrimer (middle) and G3-NH_2 dendron (bottom) to the *cis*-side of the bilayer chamber results in PA_{63} current inhibition. Two modes of the blocker action are clearly seen. In the absence of a blocker (top), the channel is open; the intrinsic $1/f$ flickering events are mostly filtered by averaging over a time interval of 50 ms. Applied voltage was 100 mV, measurements were performed in 1 M KCl. The dashed lines represent zero current levels; (**B**) Residence time of imperfect G2 75% OH 25% NH_2 dendrimer (open hexagons) and G3 dendron (filled diamonds) binding increases exponentially with voltage increase. Residence time for G2-NH_2 dendrimer (filled circles) is as discussed in Figure 3B; (**C**) On-rate constant as a function of voltage increases and then shows saturation at $V > 100$ mV in the case of G2 75% OH 25% NH_2 dendrimer and G3 dendron, but not with G2-NH_2 where it continues increasing.

3. Discussion

3.1. Two Modes of G2-PAMAM Dendrimer Inhibition of PA_{63} Channel

In this paper, we used the single channel planar lipid bilayer technique to present evidence that second generation PAMAM dendrimers inhibit the PA_{63} channel of anthrax toxin by ion current blockage. For all tested G2 PAMAM dendrimers and the G3-NH_2 PAMAM dendron, the channel current inhibition was bimodal. The first mode of the ion current inhibition was observed in the form of complete (100% of total channel conductance) but reversible ion current fluctuations that were described by a two-state Markov process, with one state being an open, dendrimer-free state and second state being a blocked, dendrimer-bound state. Interestingly, not only the three tested cationic blockers (intact G2-NH_2, mixed surface 75% OH 25% NH_2 dendrimers, and G3-NH_2 dendron) but also G2-OH and G2-SA dendrimers (functionalized, respectively, with neutral and negatively charged terminal groups) reversibly blocked the K^+ current through PA_{63}, apparently physically entering the channel's permeation pathway. Figure 5 summarizes the equilibrium dissociation constants for all five blockers that were calculated using the blocker/channel binding reaction kinetic constants shown in Figures 3 and 4 as $K_D = \frac{k_{off}}{k_{on}}$, where $k_{off} = \frac{1}{t_{res}}$. It can be seen (Figure 5) that G2-NH_2 activity significantly exceeds (lower K_D values) that of G2-OH and G2-SA, which is mostly determined by the dramatically higher on-rates of the G2-OH/PA_{63} and G2-SA/PA_{63} binding reactions rather than by the reactions' off-rates. Thus, under 50 mV of applied voltage, t_{res} of G2-OH was about 10 times lower, and t_{res} of G2-SA was comparable with that of G2-NH_2 (Figure 3B). The on-rates were about three-fold and two-and-a-half-fold higher for G2-NH_2 compared with G2-SA and G2-OH, respectively (Figure 3C). In contrast, the on-rate of the binding reaction was nearly identical when G2 75% OH 25% NH_2 dendrimer and G3-NH_2 dendron were investigated and compared with G2-NH_2 dendrimer at 50 mV (Figure 4C).

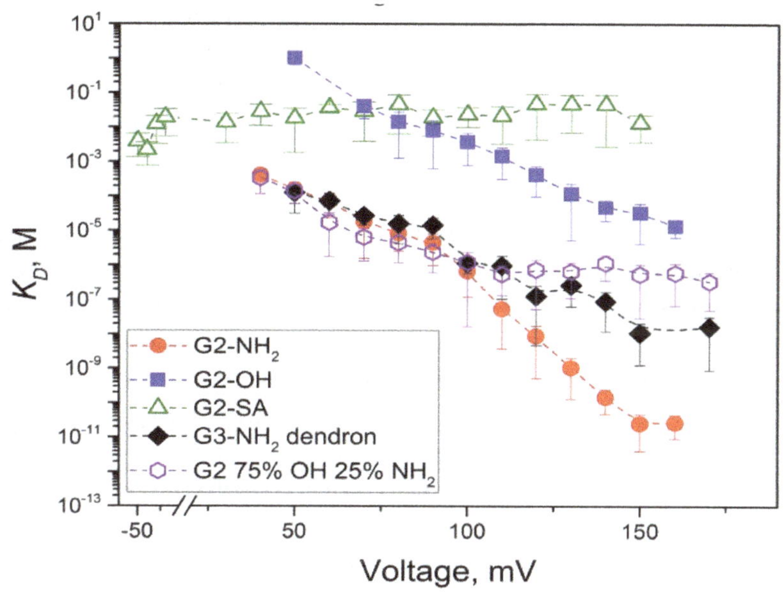

Figure 5. Equilibrium dissociation constants of the PAMAM dendrimer binding reaction to a single PA_{63} pore plotted as a function of applied transmembrane voltage. Note that only first mode of blockage type events were considered. K_D values significantly decrease with voltage (stronger binding) for all the shown blockers, except G2-SA. The G2-OH dendrimer K_D values are significantly higher (less effective binding) compared to those for G2-NH_2, G2 75% OH 25% NH_2 dendrimers and the G3-NH_2 dendron. The experiments were performed in 1 M KCl at pH 6.

In addition to the reversible two-state Markov process blockage events, all five blockers caused long voltage-dependent PA_{63} closures that, in many aspects, resembled the typical characteristics of the classical voltage gating of β-barrel ion channels [42] (Figure S2). The quasi-irreversible character of the voltage gating allowed us to perform only qualitative analysis of this type of blocker-induced closure. In a manner similar to the reversible Markov PA_{63} blockage process, the voltage gating type of closures appeared to be stronger when the dendrimers were decorated with positively-charged terminal groups and was very weak in the case of G2-SA (see e.g., Figure 3A where no voltage dependent closures are seen in the presence of 660 μL of G2-SA over ~50 s recording). Previously, we described the two similar modes of the blocker-induced channel inhibition with the 7-fold symmetrical β-cyclodextrin inhibitors [14,15]. Because the molecular mechanism, possibly a universal one, of the voltage gating observed for many functionally distinct β-barrel channels in planar lipid bilayers has yet to be clarified [40] and its physiological relevance has been called in question [39], we cannot make a strong judgement on the biological importance of the second mode of the dendrimer-induced PA_{63} current inhibition. However, we want to emphasize that when we attempted to study the dendrimer and cyclodextrin binding reaction single-channel kinetics at low, close to physiological salt concentrations (e.g., in 0.1 M solutions), the second type of the current inhibition was very strong. This led to prolonged and complete channel closures at very low, (20–50) mV applied voltages, precluding us from performing kinetic investigation of the PA_{63}/multivalent blocker binding reaction. For the same reason, we had difficulties studying the first type of the blockages at voltages >(150–180) mV, because the quasi-irreversible voltage gating events were observed at extremely low blocker concentrations (0.01–0.1 nM) when the number of the first type of reversible blockage events was too low compared to the natural $1/f$ current fluctuations of this very complex channel. Previously we reported a reasonably good linear correlation ($R = 0.84$) between activity of the PA_{63} channel cyclodextrin inhibitors in RAW 264.7 cells and in the multichannel reconstitution assays [48], allowing us to suggest that the second voltage gating type of closures is physiologically relevant. Moreover, the strong asymmetry that we and others observed in non-modified PA_{63} voltage gating with the voltage sign (the channel tends to close even at very low *cis*-negative voltages) [33,49], suggests that the voltage gating could be an internal tool for PA_{63} to stay closed when being occasionally inserted into off-target bilayers.

Note that the K_D data in Figure 5, when approximated to 20 mV, are higher compared with the equilibrium IC_{50} values measured at 20 mV (Table 1). All the blockers appear to be more effective when used at the multichannel level. The explanation of this apparent discrepancy is simple and identical to the one previously suggested for the PA_{63}/7+β-CD binding reaction [14]. The IC_{50} values measured at the multichannel level contain both the fast reversible dendrimer-induced blockage events (first mode of dendrimer action) and the prolonged voltage gating closures (second mode of action), whereas the voltage gating events were intentionally excluded from binding reaction kinetics analysis at the single channel level. As discussed before, we believe that the second mode of PA_{63} current inhibition is related to the voltage gating of β-barrel channels, well-known to any electrophysiologist who works with these channels. It appears that β-CD and dendrimer addition significantly lowers the so-called "critical voltage" needed for the channel gating. Because we currently lack understanding of the voltage gating mechanism, we cannot comment on the physiological relevance of this second mode of current blockage (frequently a more intense one) in anthrax toxin inhibition. Note that the β-CD and dendrimer blockers were previously shown to effectively protect cells [2,3] and animals [50] against the anthrax toxin.

3.2. *Voltage Dependence of the Reversible Dendrimer/PA_{63} Interaction*

With all previously tested β-cyclodextrin PA_{63} blockers, nearly exponential (linear in semi logarithmic scale) voltage dependence of t_{res} was reported with almost identical voltage sensitivity between two tested 7+ charged β-CD blockers, whereas the k_{on} values were only weakly voltage-dependent [14,15]. Moreover, it was the off-rate and not the on-rate that mainly determined the earlier reported difference in 7+β-CD potency. Details of the mechanism of the PAMAM dendrimer-induced PA_{63} current blockage are different. Thus in the case of G2-OH, the dendrimer with a positively charged tertiary amine interior and neutral hydroxyl surface groups, t_{res} was shown to increase exponentially with voltage (linear fit in the semi-logarithmic scale in Figure 3B, filled squares) with the slope of the logarithm of the lifetimes versus voltage dependence $\mathrm{dlg}t_{res}/\mathrm{d}V = (25 \pm 0.6) \times 10^{-3}\ (\mathrm{mV})^{-1}$. In contrast, with the "highly" cationic (16 terminal amino groups) G2-NH_2, we observed a non-exponential t_{res} increase with voltage. The t_{res} voltage dependence data were broken into two intervals and each was fitted with a single exponent (shown as two linear dependences in the semi-logarithmic scale in Figure 3B, filled circles). The slopes of the logarithm of the lifetimes versus voltage dependence for low and high voltages were: $\mathrm{dlg}t_{res}/\mathrm{d}V = (22 \pm 2) \times 10^{-3}\ (\mathrm{mV})^{-1}$ at low *cis*-positive voltages and $\mathrm{dlg}t_{res}/\mathrm{d}V = (35 \pm 2) \times 10^{-3}\ (\mathrm{mV})^{-1}$ at high *cis*-positive voltages. The binding time in the presence of G2-SA, the dendrimer that is made of the positively charged

PAMAM core and 16 negatively charged succinamate surface groups, was shown to have inverse voltage dependence, with the t_{res} decreasing with *cis*-positive voltage decrease. The G2-SA residence time voltage dependence was also approximated with two separate single exponential dependences (Figure 3B, open triangles), one at lower voltages ($\mathrm{dlg}t_{res}/\mathrm{d}V = -(12 \pm 1) \times 10^{-3}\ (\mathrm{mV})^{-1}$) and another at higher voltages ($\mathrm{dlg}t_{res}/\mathrm{d}V = -(4.2 \pm 0.9) \times 10^{-3}\ (\mathrm{mV})^{-1}$). It appears that the high *cis*-positive and high *cis*-negative applied voltages increase voltage sensitivity of the $PA_{63}/G2\text{-}NH_2$ and PA_{63}/G2-SA binding reactions, respectively. Thus, the t_{res}/voltage dependence data could be used to determine the so-called effective "gating charge" [51], a parameter characterizing sensitivity of the blockage reaction to voltage [52], as $\delta z = k_B T \frac{\mathrm{d}2.303\mathrm{lg}t_{res}}{\mathrm{d}V}$. Here, δ is the dimensionless "apparent electrical distance" to the blocking site, z is the blocker charge, V is the applied voltage, and k_B and T have their usual meaning as the Boltzmann constant and absolute temperature (in degrees Kelvin), respectively. The G2-OH binding could be characterized by $\delta z = 1.50 \pm 0.03$ elementary charges, the G2-NH_2 binding by $\delta z = 1.3 \pm 0.2$ (low voltage) and $\delta z = 2.10 \pm 0.07$ (high voltage) elementary charges, and G2-SA binding by $\delta z = -0.72 \pm 0.08$ (low voltage) and $\delta z = -0.25 \pm 0.05$ (high voltage) elementary charges. The twisted residence time voltage dependence observed with G2-NH_2 and G2-SA but not with G2-OH could tentatively be explained by structural reorientation of the terminal amino and succinamate groups or solvation structure (shell water and counterions) reorganization under the applied electrostatic field. Interestingly, k_{on} increases as a function of voltage (Figure 3C) showing that high voltages facilitate dendrimer delivery to the binding site, in contrast to the very weak voltage dependence of the binding reaction on-rate reported earlier for 7+β-CDs but in a manner similar to that reported for alpha-synuclein/α-hemolysin binding reaction [53]. Surprisingly, we also observed a slight increase in the on-rate with *cis*-positive voltage increase even for G2-SA dendrimer, which carries 16 negatively-charged terminal groups.

We have also investigated the role of the so-called "imperfect" dendrimers in the PA_{63} blockage dynamics using G2 75% OH 25% NH_2 dendrimer which, on average, has only four terminal positive charges and G3-NH_2 dendron decorated with 16 positively-charged terminal groups. The residence times of the blockers inside the channel changed exponentially with voltage showing more shallow slopes compared with those earlier discussed for G2-NH_2: $\mathrm{dlg}t_{res}/\mathrm{d}V = (16.8 \pm 1.5) \times 10^{-3}\ (\mathrm{mV})^{-1}$ for the G3-NH_2 dendron and $\mathrm{dlg}t_{res}/\mathrm{d}V = (9.1 \pm 0.3) \times 10^{-3}\ (\mathrm{mV})^{-1}$ for G2 75% OH 25% NH_2 dendrimer (Figure 4B). These two systems can be characterized by the effective "gating charge" $\delta z = 0.92 \pm 0.16$ (G2-NH_2 dendron) and $\delta z = 0.53 \pm 0.03$ (G2 75% OH 25% NH_2) elementary charges showing weaker residence time voltage sensitivity compared to G2-NH_2. At the same time at voltages <80 mV, G3-NH_2 dendron lifetime is comparable and G2 75% OH 25% NH_2 lifetime is even higher than that for G2-NH_2 (Figure 4C). At $V < 120$ mV the binding reaction on-rates are comparable for all three blockers, however at $V > 120$, t_{on} shows only weak voltage dependence for both the imperfect dendrimers.

4. Conclusions

Since their discovery in 1985 by Tomalia [54,55], PAMAM dendrimers have been the subject of thorough theoretical and experimental investigation in soft-matter physics, not only due to their seemingly high commercial potential but also because of their unique "ultrasoft colloid" properties bridging the gap between polymers and hard spheres [56]. PAMAM dendrimer conformational flexibility in solutions and the effect of solvent and pH on their structure, swelling, charge, counterion distribution, degree of protonation, and deformability have been addressed in a significant number of publications [57–71]. At the same time, single channel studies investigating specifics of the dendrimer dynamics in ion channel confinement are limited. In 2000, a rapid nuclear pore sizing patch-clamp method based on the calibrated fluorescently-labeled amino-terminated dendrimers was described [72]. In 2007, sulfhydryl-reactive poly(amido amine) G2, G3, and G5 dendrimers of second, third and fifth generations decorated with a mixed surface of terminal hydroxyl and amine groups were designed to interact with α-hemolysin channels that contained engineered cysteine residues [73], with the ultimate goal of modifying the stochastic-sensing properties of α-hemolysin upon addition of the charged and dense dendrimers into its lumen. The dendrimers acted as both an ion-selectivity filter and a molecular sieve, regulating the passage of small- and macromolecules. In 2013, polypropylenimine dotriaconta-amine G3 and G4 dendrimers were tested against the *E. coli* E69 pore-forming Wza K30 capsular polysaccharide transporter; however, no detectable inhibitory activity was reported [74]. In 2014, PAMAM dendrimers were reported to effectively block the ion-channel forming components of the anthrax and C2 toxins using planar lipid bilayer measurements and cell assays [5]. More recently, α-hemolysin/PAMAM dendrimer system was used to investigate the molecular process of ion channel confinement and its effect on dendrimer conformation using single channel measurements and molecular dynamics simulations [75]. The authors have shown that the electrophoretic migration of the polycationic dendrimers into a confined space is determined by the generation-dependent compressibility of the dendrimers rather than by their diameter. The ion channel nanoscale confinement had also reduced the PAMAM dendrimer protonation. Just recently, fully atomistic molecular dynamics simulations were used to investigate pH-dependent blockage (the authors call it "gating") of the cytolysin A pore by PAMAM dendrimers [76]. The protonated dendrimers were able to adapt a more extended conformation, effectively blocking about 91% of the channel current, whereas the non-protonated dendrimers were more compact, which created some void space for water and ion passage and about 31% reduction in current.

In this study, we investigated the effect of the PAMAM dendrimer surface chemistry and structural integrity on their ability to enter and block the ion (and probably lethal and edema factor) translocation pathway of the anthrax toxin channel. Considering the electrostatic nature of the cationic blocker interaction with the strongly cation-selective PA_{63} pore, it comes as no surprise that the residence time of the dendrimer/channel binding reaction turned out to be dependent on both blocker chemistry and transmembrane voltage. One of the unforeseen findings made in this study is the increase in the dendrimer capture rate (the binding reaction on-rate) with the *trans*-negative voltage increase and the strong on-rate dependence on dendrimer surface chemistry. Indeed, the on-rate, which is proportional to the number of effective binding events, is traditionally believed to be determined by a correlation between the particle size and the channel entry area (or squared radius). The applied electrical field that falls primarily across the channel (traditionally, across the membrane) is not expected to significantly influence the capture kinetics at distances greater than 3 Å away (1 Debye length in 1 M KCl) from the channel entrance. Thus, previously reported strong voltage sensitivity of a particle capture for the tubulin/VDAC [77] and alpha-synuclein/α-hemolysin [53] binding reactions was explained by the tubulin and alpha-synuclein interaction with the bilayer lipid membranes as an essential first step before the channel lumen binding. This explanation is clearly not appropriate for dendrimer/PA_{63} binding because of its compact nature. The size of the blocker (29 Å) and extremely elongated structure of the PA_{63} channel [34] means that it may extend to a distance greater than 100 Å above the membrane surface. Nevertheless, PAMAM dendrimers were previously shown to interact with the bilayer lipid membranes [78,79], and this therefore leaves room for the possible membrane-binding related effects on the ion channel conductance. For example, the blocker-induced PA_{63} voltage gating (the second mode of action), shown to depend on both the multivalent blocker [5] addition and the bilayer lipid composition [14], may originate from the dendrimer/membrane interaction process. At the same time, with the exception of G2-COONa (Figure S4), we did not observe any significant membrane instability upon PAMAM dendrimer addition to the membrane bathing solutions.

5. Materials and Methods

5.1. Reagents

PA_{63} was purchased from List Biological Laboratories, Inc., (Campbell, CA, USA). The following chemical reagents were used: KCl, MES, KOH, and HCl (Sigma-Aldrich, St. Louis, MO, USA), "purum" hexadecane (Fluka, Buchs, Switzerland), diphytanoylphosphatidylcholine, (DPhPC, Avanti Polar lipids,

Inc., Alabaster, AL, USA), pentane (Burdick and Jackson, Muskegon, MI, USA), and agarose (Bethesda Research Laboratory, Gaithersburg, MD, USA). MQ water was used to prepare solutions. Primary amine (generation 2) and hydroxyl (generation 2) PAMAM dendrimers, commercially available at Dendritech, Inc., (Midland, MI, USA) as w/w H_2O solutions, were a kind gift from Dr. Sergey Bezrukov. G3 primary amino dendrons, mixed-surface 75% OH 25% G2-NH_2 dendrimers, G2 carboxylate-Na terminated PAMAM dendrimers and G2 succinamic acid terminated PAMAM dendrimers were purchased from Dendritech, Inc. (Midland, MI, USA) as w/w H_2O solutions. Note that G2-COONa dendrimers were named G1.5 by the manufacturer because instead of only the terminal amino groups (like in the case of G2-NH_2 vs. G2-OH substitution), all –NH–CH_2–CH_2–NH_2 end groups were replaced with the COONa substituents, shortening the terminal chain of each branch by about a half. Dendritech, Inc has provided us with the analytical measurements on their PAMAM dendrimer products which are given in Figures S5 and S6. However even though mass spectrometry is often considered as the main method to characterize the presence and nature of defects in the dendrimer structure, it was demonstrated that "dendrimer purity needs to be interpreted with care and may be misleading in the sense that falsely negative results are obtained" [80].

5.2. Channel Reconstitution into Planar Lipid Bilayers

To form solvent-free planar lipid bilayers with the lipid monolayer opposition technique [81], we used a 5 mg/mL stock solution of diphytanoylphosphatidylcholine (DPhPC) in pentane. Bilayer lipid membranes were formed on a 60-μm-diameter aperture in the 15-μm-thick Teflon film that separated the electrolyte chamber into two compartments, as described in detail elsewhere [14]. The 0.1 and 1 M aqueous solutions of KCl were buffered at pH 6 (5 mM MES) at room temperature (23 ± 0.5 °C). Single channels were formed by adding 0.5 to 1 μL of 20 $\mu g \cdot mL^{-1}$ solution of PA_{63} to the 1.5 mL aqueous phase in the *cis*-half of the bilayer chamber. Under this protocol, PA_{63} channel insertions were always directional, as judged by channel conductance asymmetry in the applied transmembrane voltage. Multichannel experiments were performed in 0.1 M and 1 M KCl solutions, buffered at pH 6 by 5 mM MES, at 20 mV applied voltage; $\sim 1-2$ μL of 1 $mg \cdot mL^{-1}$ stock PA_{63} solution was added to the *cis*-side of the chamber. The electrical potential difference across the lipid bilayer was applied with a pair of Ag-AgCl electrodes in 2 M KCl, 1.5% agarose bridges. In all experiments, the PAMAM dendrimers were added to the *cis*-compartment of a bilayer chamber, which was the side of PA_{63} addition. The *cis* compartment is believed to correspond to the endosome-facing "flower" side of the channel. Single-channel measurements were performed at −50 to +180 mV. Multichannel experiments were performed at 20 mV. The applied potential is defined as positive if it is higher on the side of protein addition (*cis*-side). Conductance measurements were done using an Axopatch 200B amplifier (Molecular Devices, LLC., Sunnyvale, CA, USA) in the voltage clamp mode. Signals were filtered by a low-pass 8-pole Butterworth filter (Model 9002, Frequency Devices, Inc., Haverhill, MA, USA) at 15 Hz for multichannel and 15 kHz for single channel systems and sampled with a frequency of 50 Hz and 50 kHz for multichannel and single channel experiments respectively. Amplitude, lifetime, and fluctuation analysis was performed with ClampFit 10.2 (Molecular Devices, Sunnyvale, CA, USA) and OriginPro 8.5 (OriginLab, Northampton, MA, USA) software, as well as with software developed in-house.

5.3. Reproducibility of the Experiments and Statistics

All multichannel planar lipid measurements, performed to obtain the IC_{50} data shown in Table 1, were repeated at least two times. Values are given as the means ± standard deviations. The single-channel statistical analysis with G2-NH_2, G2-OH, G2-SA dendrimers and G3 dendron was performed with the power spectral analysis of hundreds of dendrimer-induced current fluctuation events. At voltages >150 mV, because of the strong intrinsic and blocker-induced voltage gating, collection of a significant number of the reversible binding events proved to be difficult. Therefore the on- and off-rates were determined by averaging over lifetimes of several dozen events. Because of the wide distribution of lifetimes in the case of non-homogeneous G2 75% NH_2 25% OH dendrimer, the recordings were analyzed by averaging over the event lifetimes at all voltages. The standard deviation for t_{res} ($SD_{t_{res}}$) and k_{on} ($SD_{k_{on}}$) were then determined by averaging over 2–6 recordings taken independently from different PA_{63} channel reconstitution experiments. The standard deviations for K_D values were calculated from SDs determined for the on- and off-rates as $SD_{K_D} = \frac{1}{t_{res}k_{on}}\sqrt{\left(\frac{SD_{t_{res}}}{t_{res}}\right)^2 + \left(\frac{SD_{k_{on}}}{k_{on}}\right)^2}$.

Supplementary Materials:
Figure S1, Chemical structures of the PAMAM dendrimers used in this study. (A) Cationic PAMAM dendrimers G2-NH_2, with 16 positively charged terminal groups (left), G2-OH, with positively-charged PAMAM core and neutral OH terminal groups (right). (B) G2 PAMAM dendrimers with negatively charged succinamate (left) and carboxyl (right) terminal groups, G2-SA and G2-COONa respectively. (C) Imperfect G2 PAMAM dendrimers G2 75% OH 25% NH_2, with 12 neutral OH and 4 positively charged NH_2 terminal groups on average (left), and G3-NH_2 dendron, with a fractured more flexible structure and 16 positively charged terminal groups (right). Similar to the Figure 1B color coding, terminal primary amines are colored in red; core tertiary amines are colored in green; terminal hydroxyl groups are colored in blue. The images were created using chemical drawing software ChemDoodle 8.1.0, iChemLabs, LLC. Note that in contrast to all other dendrimers, G2 75% OH 25% NH_2 is not monodisperse and contains 75% of terminal OH groups and 25% of terminal NH_2 groups on average. Figure S2, Second mode of G2-NH_2-induced modulation of a single PA_{63} channel current. At the relatively low applied voltages (70 and 80 mV), PA_{63} mostly remains in an open state in the blocker-free solutions (left, two upper rows). Fast flickering between the open and closed states (the so-called $1/f$ noise) is mostly removed by averaging over a time interval of 100 ms. At 90 mV (left, low row), several pronounced voltage gating events are seen; $p_{bl} = 0.12$. In the presence of 0.35 μM of G2-NH_2 (right), the voltage gating of the channel is significantly increased. Multiple fast current blockages (first mode of dendrimer-induced current inhibition) are observed but they are partially filtered over a time interval of 100 ms. Figure S3, Influence of the PAMAM dendrimer terminal group chemistry on the PA_{63} channel inhibition studied on a multichannel level. (A) A typical dendrimer-induced PA_{63} inhibition curve (shown for G2-OH dendrimer). G2-OH additions are marked with the downward arrows; total bulk dendrimer concentration is indicated. The dashed line represents zero current level; (B) Typical multichannel titration curves of the PA_{63} channel inhibition by G2-NH_2, G2-OH, and G2-SA dendrimers. The dashed line represents 50% of the original current level. The recordings were taken in 0.1 M KCl solutions at pH 6 under 20 mV applied voltage. Figure S4, Effect of G2-COONa *cis*-side addition on a single PA_{63} channel. Both in blocker-free solution and in presence of 3 mM G2-COONa, PA_{63} mostly remains in an open state. The $1/f$ events are to a large extent removed by averaging over a time interval of 10 ms. G2-COONa addition causes lipid bilayer instability (upward events) and, eventually, breakage. Recordings were taken in 1 M KCl solutions at pH 6 and 100 mV applied voltage. Figure S5, MALDI-TOF mass spectra of G2-NH_2 (A); G2-OH (B); G2-SA (C); G2-COONa (D) dendrimers and G3-NH_2 dendron (E). The data were provided by Dendritech, Inc. (Midland, MI, USA). Figure S6, Characterization of generation 2 PAMAM dendrimers and generation 3 PAMAM-NH_2 dendron by 13 C NMR. The data were provided by Dendritech, Inc. (Midland, MI, USA), (A) 13 C NMR of G2-NH_2. (B) 13 C NMR of G2-OH; (C) 13 C NMR of G2-SA; (D) 13 C NMR of G2-COONa; (E) 13 C NMR G3-NH_2 dendron.

Acknowledgments: The project was financially supported by National Institute of Allergy and Infectious Diseases of the National Institutes of Health under award number 1R15AI099897-01A1 and by The Catholic University startup funds (to Ekaterina Nestorovich). Many thanks to Sergey M. Bezrukov and Edward L. Mertz (both NICHD, NIH) for fruitful discussion.

Author Contributions: Goli Yamini and Ekaterina M. Nestorovich designed and performed experiments, analyzed the data, and wrote the manuscript. Nnanya Kalu performed experiments and contributed to the manuscript proofreading.

References

1. Yamini, G.; Nestorovich, E.M. Multivalent inhibitors of channel-forming bacterial toxins. *Curr. Top. Microbiol. Immunol.* **2016**. [CrossRef]
2. Karginov, V.A.; Nestorovich, E.M.; Moayeri, M.; Leppla, S.H.; Bezrukov, S.M. Blocking anthrax lethal toxin at the protective antigen channel by using structure-inspired drug design. *Proc. Natl. Acad. Sci. USA* **2005**, *102*, 15075–15080. [CrossRef] [PubMed]
3. Nestorovich, E.M.; Karginov, V.A.; Popoff, M.R.; Bezrukov, S.M.; Barth, H. Tailored ß-cyclodextrin blocks the translocation pores of binary exotoxins from *C. botulinum* and *C. perfringens* and protects cells from intoxication. *PLoS ONE* **2011** *6*. [CrossRef] [PubMed]

4. Roeder, M.; Nestorovich, E.M.; Karginov, V.A.; Schwan, C.; Aktories, K.; Barth, H. Tailored cyclodextrin pore blocker protects mammalian cells from clostridium difficile binary toxin CDT. *Toxins (Basel)* **2014**, *6*, 2097–2114. [CrossRef] [PubMed]
5. Forstner, P.; Bayer, F.; Kalu, N.; Felsen, S.; Fortsch, C.; Aloufi, A.; Ng, D.Y.; Weil, T.; Nestorovich, E.M.; Barth, H. Cationic PAMAM dendrimers as pore-blocking binary toxin inhibitors. *Biomacromolecules* **2014**, *15*, 2461–2474. [CrossRef] [PubMed]
6. Nestorovich, E.M.; Bezrukov, S.M. Obstructing toxin pathways by targeted pore blockage. *Chem. Rev.* **2012**, *112*, 6388–6430. [CrossRef] [PubMed]
7. Krantz, B.A.; Melnyk, R.A.; Zhang, S.; Juris, S.J.; Lacy, D.B.; Wu, Z.; Finkelstein, A.; Collier, R.J. A phenylalanine clamp catalyzes protein translocation through the anthrax toxin pore. *Science* **2005**, *309*, 777–781. [CrossRef] [PubMed]
8. Bachmeyer, C.; Orlik, F.; Barth, H.; Aktories, K.; Benz, R. Mechanism of C2-toxin inhibition by fluphenazine and related compounds: Investigation of their binding kinetics to the C2II-channel using the current noise analysis. *J. Mol. Biol.* **2003**, *333*, 527–540. [CrossRef] [PubMed]
9. Orlik, F.; Schiffler, B.; Benz, R. Anthrax toxin protective antigen: Inhibition of channel function by chloroquine and related compounds and study of binding kinetics using the current noise analysis. *Biophys. J.* **2005**, *88*, 1715–1724. [CrossRef] [PubMed]
10. Beitzinger, C.; Bronnhuber, A.; Duscha, K.; Riedl, Z.; Huber-Lang, M.; Benz, R.; Hajós, G.; Barth, H. Designed azolopyridinium salts block protective antigen pores in vitro and protect cells from anthrax toxin. *PLoS ONE* **2013**, *8*. [CrossRef] [PubMed]
11. Kronhardt, A.; Beitzinger, C.; Barth, H.; Benz, R. Chloroquine Analog Interaction with C2- and Iota-Toxin in Vitro and in Living Cells. *Toxins (Basel)* **2016**, *8*. [CrossRef] [PubMed]
12. Crini, G. Review: A history of cyclodextrins. *Chem. Rev.* **2014**, *114*, 10940–10975. [CrossRef] [PubMed]
13. Wu, L.P.; Ficker, M.; Christensen, J.B.; Trohopoulos, P.N.; Moghimi, S.M. Dendrimers in medicine: Therapeutic concepts and pharmaceutical challenges. *Bioconjug. Chem.* **2015**, *26*, 1198–1211. [CrossRef] [PubMed]
14. Nestorovich, E.M.; Karginov, V.A.; Berezhkovskii, A.M.; Bezrukov, S.M. Blockage of anthrax PA63 pore by a multicharged high-affinity toxin inhibitor. *Biophys. J.* **2010**, *99*, 134–143. [CrossRef] [PubMed]
15. Bezrukov, S.M.; Liu, X.; Karginov, V.A.; Wein, A.N.; Leppla, S.H.; Popoff, M.R.; Barth, H.; Nestorovich, E.M. Interactions of high-affinity cationic blockers with the translocation pores of *B. anthracis*, *C. botulinum*, and *C. perfringens* binary toxins. *Biophys. J.* **2012**, *103*, 1208–1217. [CrossRef] [PubMed]
16. Lee, C.C.; MacKay, J.A.; Frechet, J.M.; Szoka, F.C. Designing dendrimers for biological applications. *Nat. Biotechnol.* **2005**, *23*, 1517–1526. [CrossRef] [PubMed]
17. Svenson, S.; Tomalia, D.A. Dendrimers in biomedical applications-reflections on the field. *Adv. Drug Deliv. Rev.* **2005**, *57*, 2106–2129. [CrossRef] [PubMed]
18. Duncan, R.; Izzo, L. Dendrimer biocompatibility and toxicity. *Adv. Drug Deliv. Rev.* **2005**, *57*, 2215–2237. [CrossRef] [PubMed]
19. Hong, S.; Bielinska, A.U.; Mecke, A.; Keszler, B.; Beals, J.L.; Shi, X.; Balogh, L.; Orr, B.G.; Baker, J.R., Jr.; Banaszak Holl, M.M. Interaction of poly(amidoamine) dendrimers with supported lipid bilayers and cells: Hole formation and the relation to transport. *Bioconjug. Chem.* **2004**, *15*, 774–782. [CrossRef] [PubMed]
20. Tang, M.X.; Redemann, C.T.; Szoka, F.C., Jr. In vitro gene delivery by degraded polyamidoamine dendrimers. *Bioconjug. Chem.* **1996**, *7*, 703–714. [CrossRef] [PubMed]
21. Nestorovich, E.M.; Bezrukov, S.M. Designing inhibitors of anthrax toxin. *Expert Opin. Drug Discov.* **2014**, *9*, 299–318. [CrossRef] [PubMed]
22. Duesbery, N.S.; Webb, C.P.; Leppla, S.H.; Gordon, V.M.; Klimpel, K.R.; Copeland, T.D.; Ahn, N.G.; Oskarsson, M.K.; Fukasawa, K.; Paull, K.D.; et al. Proteolytic inactivation of MAP-kinase-kinase by anthrax lethal factor. *Science* **1998**, *280*, 734–737. [CrossRef] [PubMed]
23. Vitale, G.; Bernardi, L.; Napolitani, G.; Mock, M.; Montecucco, C. Susceptibility of mitogen-activated protein kinase kinase family members to proteolysis by anthrax lethal factor. *Biochem. J.* **2000**, *352*, 739–745. [CrossRef] [PubMed]
24. Levinsohn, J.L.; Newman, Z.L.; Hellmich, K.A.; Fattah, R.; Getz, M.A.; Liu, S.; Sastalla, I.; Leppla, S.H.; Moayeri, M. Anthrax lethal factor cleavage of Nlrp1 is required for activation of the inflammasome. *PLoS Pathog.* **2012**, *8*. [CrossRef] [PubMed]

25. Leppla, S.H. Anthrax toxin edema factor: A bacterial adenylate cyclase that increases cyclic AMP concentrations of eukaryotic cells. *Proc. Natl. Acad. Sci. USA* **1982**, *79*, 3162–3166. [CrossRef] [PubMed]
26. Leppla, S.H. *Bacillus anthracis* calmodulin-dependent adenylate cyclase: Chemical and enzymatic properties and interactions with eucaryotic cells. *Adv. Cycl. Nucleotide Protein Phosphorylation Res.* **1984**, *17*, 189–198.
27. Petosa, C.; Collier, R.J.; Klimpel, K.R.; Leppla, S.H.; Liddington, R.C. Crystal structure of the anthrax toxin protective antigen. *Nature* **1997**, *385*, 833–838. [CrossRef] [PubMed]
28. Kintzer, A.F.; Thoren, K.L.; Sterling, H.J.; Dong, K.C.; Feld, G.K.; Tang, I.I.; Williams, E.R.; Berger, J.M.; Krantz, B.A. The protective antigen component of anthrax toxin forms functional octameric complexes. *J. Mol. Biol.* **2009**, *392*, 614–629. [CrossRef] [PubMed]
29. Kintzer, A.F.; Sterling, H.J.; Tang, I.I.; Williams, E.R.; Krantz, B.A. Anthrax toxin receptor drives protective antigen oligomerization and stabilizes the heptameric and octameric oligomer by a similar mechanism. *PLoS ONE* **2010**, *5*. [CrossRef] [PubMed]
30. Mogridge, J.; Cunningham, K.; Collier, R.J. Stoichiometry of anthrax toxin complexes. *Biochemistry* **2002**, *41*, 1079–1082. [CrossRef] [PubMed]
31. Pilpa, R.M.; Bayrhuber, M.; Marlett, J.M.; Riek, R.; Young, J.A. A receptor-based switch that regulates anthrax toxin pore formation. *PLoS Pathog.* **2011**, *7*. [CrossRef] [PubMed]
32. Abrami, L.; Brandi, L.; Moayeri, M.; Brown, M.J.; Krantz, B.A.; Leppla, S.H.; van der Goot, F.G. Hijacking multivesicular bodies enables long-term and exosome-mediated long-distance action of anthrax toxin. *Cell. Rep.* **2013**, *5*, 986–996. [CrossRef] [PubMed]
33. Blaustein, R.O.; Koehler, T.M.; Collier, R.J.; Finkelstein, A. Anthrax toxin: Channel-forming activity of protective antigen in planar phospholipid bilayers. *Proc. Natl. Acad. Sci. USA* **1989**, *86*, 2209–2213. [CrossRef] [PubMed]
34. Jiang, J.; Pentelute, B.L.; Collier, R.J.; Zhou, Z.H. Atomic structure of anthrax protective antigen pore elucidates toxin translocation. *Nature* **2015**, *521*, 545–549. [CrossRef] [PubMed]
35. Zhang, S.; Udho, E.; Wu, Z.; Collier, R.J.; Finkelstein, A. Protein translocation through anthrax toxin channels formed in planar lipid bilayers. *Biophys. J.* **2004**, *87*, 3842–3849. [CrossRef] [PubMed]
36. Zhang, S.; Finkelstein, A.; Collier, R.J. Evidence that translocation of anthrax toxin's lethal factor is initiated by entry of its N terminus into the protective antigen channel. *Proc. Natl. Acad. Sci. USA* **2004**, *101*, 16756–16761. [CrossRef] [PubMed]
37. Nablo, B.J.; Panchal, R.G.; Bavari, S.; Nguyen, T.L.; Gussio, R.; Ribot, W.; Friedlander, A.; Chabot, D.; Reiner, J.E.; Robertson, J.W.; et al. Anthrax toxin-induced rupture of artificial lipid bilayer membranes. *J. Chem. Phys.* **2013**, *139*. [CrossRef] [PubMed]
38. Rappaport, S.M.; Teijido, O.; Hoogerheide, D.P.; Rostovtseva, T.K.; Berezhkovskii, A.M.; Bezrukov, S.M. Conductance hysteresis in the voltage-dependent anion channel. *Eur. Biophys. J.* **2015**, *44*, 465–472. [CrossRef] [PubMed]
39. Sen, K.; Hellman, J.; Nikaido, H. Porin channels in intact cells of Escherichia coli are not affected by Donnan potentials across the outer membrane. *J. Biol. Chem.* **1988**, *263*, 1182–1187. [PubMed]
40. Robertson, K.M.; Tieleman, D.P. Molecular basis of voltage gating of OmpF porin. *Biochem. Cell Biol.* **2002**, *80*, 517–523. [CrossRef] [PubMed]
41. Delcour, A.H. *Electrophysiology of Unconventional Channels and Pores*; Springer: Cham, Switzerland, 2015.
42. Bainbridge, G.; Gokce, I.; Lakey, J.H. Voltage gating is a fundamental feature of porin and toxin beta-barrel membrane channels. *FEBS Lett.* **1998**, *431*, 305–308. [CrossRef]
43. Teijido, O.; Rappaport, S.M.; Chamberlin, A.; Noskov, S.Y.; Aguilella, V.M.; Rostovtseva, T.K.; Bezrukov, S.M. Acidification asymmetrically affects voltage-dependent anion channel implicating the involvement of salt bridges. *J. Biol. Chem.* **2014**, *289*, 23670–23682. [CrossRef] [PubMed]
44. Rostovtseva, T.K.; Tan, W.; Colombini, M. On the role of VDAC in apoptosis: Fact and fiction. *J. Bioenerg. Biomembr.* **2005**, *37*, 129–142. [CrossRef] [PubMed]
45. Kullman, L.; Winterhalter, M.; Bezrukov, S.M. Transport of maltodextrins through maltoporin: A single-channel study. *Biophys. J.* **2002**, *82*, 803–812. [CrossRef]
46. Blaustein, R.O.; Finkelstein, A. Diffusion limitation in the block by symmetric tetraalkylammonium ions of anthrax toxin channels in planar phospholipid bilayer membranes. *J. Gen. Physiol.* **1990**, *96*, 943–957. [CrossRef] [PubMed]

47. Blaustein, R.O.; Finkelstein, A. Voltage-dependent block of anthrax toxin channels in planar phospholipid bilayer membranes by symmetric tetraalkylammonium ions. Effects on macroscopic conductance. *J. Gen. Physiol.* **1990**, *96*, 905–919. [CrossRef] [PubMed]
48. Karginov, V.A.; Nestorovich, E.M.; Yohannes, A.; Robinson, T.M.; Fahmi, N.E.; Schmidtmann, F.; Hecht, S.M.; Bezrukov, S.M. Search for cyclodextrin-based inhibitors of anthrax toxins: Synthesis, structural features, and relative activities. *Antimicrob. Agents Chemother.* **2006**, *50*, 3740–3753. [CrossRef] [PubMed]
49. Anderson, D.S.; Blaustein, R.O. Preventing voltage-dependent gating of anthrax toxin channels using engineered disulfides. *J. Gen. Physiol.* **2008**, *132*, 351–360. [CrossRef] [PubMed]
50. Moayeri, M.; Robinson, T.M.; Leppla, S.H.; Karginov, V.A. In vivo efficacy of beta-cyclodextrin derivatives against anthrax lethal toxin. *Antimicrob. Agents Chemother.* **2008**, *52*, 2239–2241. [CrossRef] [PubMed]
51. French, R.J.; Shoukimas, J.J. An ion's view of the potassium channel. The structure of the permeation pathway as sensed by a variety of blocking ions. *J. Gen. Physiol.* **1985**, *85*, 669–698. [CrossRef] [PubMed]
52. Gurnev, P.A.; Queralt-Martin, M.; Aguilella, V.M.; Rostovtseva, T.K.; Bezrukov, S.M. Probing tubulin-blocked state of VDAC by varying membrane surface charge. *Biophys. J.* **2012**, *102*, 2070–2076. [CrossRef] [PubMed]
53. Gurnev, P.A.; Yap, T.L.; Pfefferkorn, C.M.; Rostovtseva, T.K.; Berezhkovskii, A.M.; Lee, J.C.; Parsegian, V.A.; Bezrukov, S.M. Alpha-synuclein lipid-dependent membrane binding and translocation through the alpha-hemolysin channel. *Biophys. J.* **2014**, *106*, 556–565. [CrossRef] [PubMed]
54. Tomalia, D.A.; Frechet, J.M.J. Discovery of dendrimers and dendritic polymers: A brief historical perspective. *J. Polym. Sci. A Polym. Chem.* **2002**, *40*, 2719–2728. [CrossRef]
55. Tomalia, D.A.; Baker, H.; Dewald, J.; Hall, M.; Kallos, G.; Martin, S.; Roeck, J.; Ryder, J.; Smith, P. A new class of polymers: Starburst-dendritic macromolecules. *Polym. J.* **1985**, *17*, 117–132. [CrossRef]
56. Ballauff, M.; Likos, C.N. Dendrimers in solution: Insight from theory and simulation. *Angew. Chem. Int. Ed. Engl.* **2004**, *43*, 2998–3020. [CrossRef] [PubMed]
57. Huissmann, S.; Likos, C.N.; Blaak, R. Conformations of high-generation dendritic polyelectrolytes. *J. Mater. Chem.* **2010**, *20*, 10486–10494. [CrossRef]
58. Maiti, P.K.; Lin, S.T.; Cagin, T.; Goddard, W.A. Effect of Solvent and pH on the Structure of PAMAM Dendrimers. *Macromolecules* **2005**, *38*, 979–991. [CrossRef]
59. Huissmann, S.; Wynveen, A.; Likos, C.N.; Blaak, R. The effects of pH, salt and bond stiffness on charged dendrimers. *J. Phys. Condens Matter* **2010**, *22*, 232101. [CrossRef] [PubMed]
60. Liu, Y.; Bryantsev, V.S.; Diallo, M.S.; Goddard, W.A., 3rd. PAMAM dendrimers undergo pH responsive conformational changes without swelling. *J. Am. Chem. Soc.* **2009**, *131*, 2798–2799. [CrossRef] [PubMed]
61. Garcia-Fernandez, E.; Paulo, P.M. Deswelling and electrolyte dissipation in free diffusion of charged PAMAM dendrimers. *J. Phys. Chem. Lett.* **2014**, *5*, 1472–1478. [CrossRef] [PubMed]
62. Maiti, P.K.; Bagchi, B. Diffusion of flexible, charged, nanoscopic molecules in solution: Size and pH dependence for PAMAM dendrimer. *J. Chem. Phys.* **2009**, *131*, 214901. [CrossRef] [PubMed]
63. Maiti, P.K.; Messina, R. Counterion Distribution and ζ-Potential in PAMAM Dendrimer. *Macromolecules* **2008**, *41*, 5002–5006. [CrossRef]
64. Mecke, A.; Lee, I.; Baker, J.R., Jr.; Holl, M.M.; Orr, B.G. Deformability of poly(amidoamine) dendrimers. *Eur. Phys. J. E. Soft Matter* **2004**, *14*, 7–16. [CrossRef] [PubMed]
65. Liu, Y.; Porcar, L.; Hong, K.; Shew, C.Y.; Li, X.; Liu, E.; et al. Effect of counterion valence on the pH responsiveness of polyamidoamine dendrimer structure. *J. Chem. Phys.* **2010**, *132*. [CrossRef] [PubMed]
66. Porcar, L.; Liu, Y.; Verduzco, R.; Hong, K.; Butler, P.D.; Magid, L.J.; Smith, G.S.; Chen, W. Structural investigation of PAMAM dendrimers in aqueous solutions using small-angle neutron scattering: Effect of generation. *J. Phys. Chem. B* **2008**, *112*, 14772–14778. [CrossRef] [PubMed]
67. Wu, B.; Kerkeni, B.; Egami, T.; Do, C.; Liu, Y.; Wang, Y.; Porcar, L.; Hong, K.; Smith, S.C.; Liu, E.L.; et al. Structured water in polyelectrolyte dendrimers: Understanding small angle neutron scattering results through atomistic simulation. *J. Chem. Phys.* **2012**, *136*, 144901. [CrossRef] [PubMed]
68. Cakara, D.; Kleimann, J.; Borkovec, M. Microscopic protonation equilibria of poly(amidoamine) dendrimers from macroscopic titrations. *Macromolecules* **2003**, *36*, 4201–4207. [CrossRef]
69. Huang, Q.R.; Dubin, P.L.; Moorefield, C.N.; Newkome, G.R. Counterion binding on charged spheres: Effect of pH and ionic strength on the mobility of carboxyl-terminated dendrimers. *J. Phys. Chem. B* **2000**, *104*, 898–904. [CrossRef]

70. Dobrovolskaia, M.A.; Patri, A.K.; Simak, J.; Hall, J.B.; Semberova, J.; de Paoli Lacerda, S.H.; McNeil, S.E. Nanoparticle size and surface charge determine effects of PAMAM dendrimers on human platelets in vitro. *Mol. Pharm.* **2012**, *9*, 382–393. [CrossRef] [PubMed]
71. Böhme, U.; Klenge, A.; Hänel, B.; Scheler, U. Counterion condensation and effective charge of PAMAM dendrimers. *Polymers* **2011**, *3*, 812–819. [CrossRef]
72. Bustamante, J.O.; Michelette, E.R.; Geibel, J.P.; Hanover, J.A.; McDonnell, T.J.; Dean, D.A. Dendrimer-assisted patch-clamp sizing of nuclear pores. *Pflugers Arch.* **2000**, *439*, 829–837. [CrossRef] [PubMed]
73. Martin, H.; Kinns, H.; Mitchell, N.; Astier, Y.; Madathil, R.; Howorka, S. Nanoscale protein pores modified with PAMAM dendrimers. *J. Am. Chem. Soc.* **2007**, *129*, 9640–9649. [CrossRef] [PubMed]
74. Kong, L.; Harrington, L.; Li, Q.; Cheley, S.; Davis, B.G.; Bayley, H. Single-molecule interrogation of a bacterial sugar transporter allows the discovery of an extracellular inhibitor. *Nat. Chem.* **2013**, *5*, 651–659. [CrossRef] [PubMed]
75. Ficici, E.; Andricioaei, I.; Howorka, S. Dendrimers in nanoscale confinement: The interplay between conformational change and nanopore entrance. *Nano Lett.* **2015**, *15*, 4822–4828. [CrossRef] [PubMed]
76. Mandal, T.; Kanchi, S.; Ayappa, K.G.; Maiti, P.K. pH controlled gating of toxic protein pores by dendrimers. *Nanoscale* **2016**, *8*, 13045–13058. [CrossRef] [PubMed]
77. Rostovtseva, T.K.; Gurnev, P.A.; Chen, M.Y.; Bezrukov, S.M. Membrane lipid composition regulates tubulin interaction with mitochondrial voltage-dependent anion channel. *J. Biol. Chem.* **2012**, *287*, 29589–29598. [CrossRef] [PubMed]
78. Klajnert, B.; Epand, R.M. PAMAM dendrimers and model membranes: Differential scanning calorimetry studies. *Int. J. Pharm.* **2005**, *305*, 154–166. [CrossRef] [PubMed]
79. Mecke, A.; Uppuluri, S.; Sassanella, T.M.; Lee, D.K.; Ramamoorthy, A.; Baker, J.R., Jr.; Orr, B.G.; Banaszak Holl, M.M. Direct observation of lipid bilayer disruption by poly(amidoamine) dendrimers. *Chem. Phys. Lipids* **2004**, *132*, 3–14. [CrossRef] [PubMed]
80. Baytekin, B.; Werner, N.; Luppertz, F.; Engeser, M.; Brüggemann, J.; Bitter, S.; Henkel, R.; Felder, T.; Schalley, C.A. How useful is mass spectrometry for the characterization of dendrimers? "Fake defects" in the ESI and MALDI mass spectra of dendritic compounds. *Int. J. Mass Spectrom.* **2006**, *249–250*, 138–148. [CrossRef]
81. Montal, M.; Mueller, P. Formation of bimolecular membranes from lipid monolayers and a study of their electrical properties. *Proc. Natl. Acad. Sci. USA* **1972**, *69*, 3561–3566. [CrossRef] [PubMed]

5

The Toxin-Antitoxin Systems of the Opportunistic Pathogen *Stenotrophomonas maltophilia* of Environmental and Clinical Origin

Laurita Klimkaitė, Julija Armalytė *, Jūratė Skerniškytė and Edita Sužiedėlienė *

Institute of Biosciences, Life Sciences Center, Vilnius University, LT-1025 Vilnius, Lithuania; laurita.klimkaite@gmail.com (L.K.); jurate.skerniskyte@gmc.vu.lt (J.S.)

* Correspondence: julija.armalyte@gf.vu.lt (J.A.); edita.suziedeliene@cr.vu.lt (E.S.)

Abstract: *Stenotrophomonas maltophilia* is a ubiquitous environmental bacterium that has recently emerged as a multidrug-resistant opportunistic pathogen causing bloodstream, respiratory, and urinary tract infections. The connection between the commensal environmental *S. maltophilia* and the opportunistic pathogen strains is still under investigation. Bacterial toxin–antitoxin (TA) systems have been previously associated with pathogenic traits, such as biofilm formation and resistance to antibiotics, which are important in clinical settings. The same species of the bacterium can possess various sets of TAs, possibly influencing their overall stress response. While the TA systems of other important opportunistic pathogens have been researched, nothing is known about the TA systems of *S. maltophilia*. Here, we report the identification and characterization of *S. maltophilia* type II TA systems and their prevalence in the isolates of clinical and environmental origins. We found 49 putative TA systems by bioinformatic analysis in *S. maltophilia* genomes. Despite their even spread in sequenced *S. maltophilia* genomes, we observed that *relBE*, *hicAB*, and previously undescribed COG3832-ArsR operons were present solely in clinical *S. maltophilia* isolates collected in Lithuania, while *hipBA* was more frequent in the environmental ones. The kill-rescue experiments in *Escherichia coli* proved *higBA*, *hicAB*, and *relBE* systems to be functional TA modules. Together with different TA profiles, the clinical *S. maltophilia* isolates exhibited stronger biofilm formation, increased antibiotic, and serum resistance compared to environmental isolates. Such tendencies suggest that certain TA systems could be used as indicators of virulence traits.

Keywords: *Stenotrophomonas maltophilia*; opportunistic pathogen; clinical origin; environmental origin; toxin-antitoxin system; biofilm; antibiotic resistance

Key Contribution: Previously undescribed *Stenotrophomonas maltophilia* type II TA systems were identified and their abundance in clinical and environmental *S. maltophilia* isolates evaluated. A difference in TA content and virulence-related characteristics between *S. maltophilia* isolates of clinical and environmental origins was observed.

---

1. Introduction

Stenotrophomonas maltophilia is a bacterium of class *Gammaproteobacteria*, ubiquitously found in the natural environment (soil, plants, animals, foods, or aqueous habitats) [1]. Recently, *S. maltophilia* came into attention as a globally emerging multidrug-resistant opportunistic pathogen causing nosocomial and community-acquired infections to immunocompromised patients [2]. The mortality rate of *S. maltophilia* infection can exceed 35% [3], with the World Health Organization (WHO) listing *S. maltophilia* as one of the leading multidrug-resistant pathogens [4]. The most known *S. maltophilia* features contributing to its successful adaptation in clinical environments include intrinsic antibiotic resistance mechanisms and its ability to form biofilms [5], although other less characterized factors (synthesis of extracellular enzymes, bacterial motility, and quorum sensing) are also known to influence its pathogenesis [6]. *S. maltophilia* is also known as a highly diverse species, since numerous studies have revealed genotypic diversity between clinical *S. maltophilia* isolates [7–9], even when analyzing samples from the same hospital [10,11]. Moreover, no significant genotypic or phenotypic differences between clinical and environment *S. maltophilia* isolates could be identified [12]. There is still an open question whether all the members of this species can become virulent and is there a specific marker to distinguish clinical and environmental *S. maltophilia* isolates.

The ability of pathogens to adapt to the clinical environment and to develop resistance to treatment can often be attributed to their ability to quickly exchange genetic information [13]. Gaining antibiotic resistance genes or virulence factors increases the chances of survival and gives rise to nosocomial opportunistic pathogens [14]. The ability of the bacterial species to mobilize their DNA could, therefore, be considered as a virulence factor in itself.

Toxin–antitoxin (TA) systems are abundant genetic elements mostly found in prokaryotes to encode a toxic protein, which inhibits cell growth and an antitoxin molecule (protein or RNA), which protects cells from toxin effects [15]. Human pathogenic bacteria have been shown to possess multiple sets of TAs, the type II TA systems, consisting of toxin and antitoxin proteins, being the most represented module and encoded by both plasmids and chromosomes [16]. Once the bacteria acquires DNA encoding the TA system, it becomes difficult to lose it due to the toxin's higher stability and eventual suppression of the cell growth in case of loss of the genes [17]. Apart from being addictive, other functions have been proposed for the TA systems, such as a role in virulent phenotype and alleviating bacterial survival under stress conditions [18]. Moreover, the connection between TA systems and virulence traits, such as biofilm formation [19,20] and host colonization [18], was demonstrated. Interestingly, the same species can have various sets of TA systems [21], indicating their ability to be transferred horizontally. Analysis of the prevalence of TA systems in the genomes of 12 dangerous epidemic bacteria has shown that pathogens encoded a higher number of TA systems and reduced genomes compared to their closest non-pathogenic relatives [22]. This suggests that TAs might participate in the manifestation of the evolved pathogenic phenotype. The pathogenic *Mycobacterium tuberculosis* represents the prominent example, which genome encodes a remarkably high number of TA systems compared with nonpathogenic *Mycobacterium* strains [23]. Altogether, a possible TA systems role in bacteria pathogenicity is possible.

In this study, we aimed to identify and characterize previously undescribed *S. maltophilia* type II TA systems, as well as to evaluate and compare the abundance of these TA modules in clinical and environmental *S. maltophilia* isolates. Additionally, we accessed virulence-related features of *S. maltophilia*, including antibiotic and serum resistance, biofilm, pellicle, and capsule formation in order to find out if the correlation between the presence of TA systems and virulence-associated traits is apparent in *Stenotrophomonas*. We observed a difference in TA content and virulence-related characteristics between *S. maltophilia* isolates of clinical and environmental origins.

2. Results

2.1. Detection of *S. maltophilia* TA Systems in Genomes

To date, there are several tools available to detect TA systems in bacterial genomes: TADB 2.0 database tool TA-finder [24], RASTA-Bacteria [25], and TASmania [26]. However, only TA-finder was useful in our case, since RASTA-Bacteria limited the size of the analyzed DNA fragments, while TASmania relied on the Ensembl database, which at the time of this search did not have *S. maltophilia* genomes. On the date of our analysis, 21 fully assembled genomes were available to analyze with TA-finder. If the TA pair contained the same predicted domains, they were considered homologous; if they matched with a protein of a different domain, they were considered a distinct TA pair (e.g., RelE-Xre and RelE-RHH were considered as two distinct TAs). The presence of each predicted TA system was checked against all the genomes using BLAST. The final results for 49 detected TA systems are presented in Figure 1. Interestingly, some of the TA system genes were not identified in the genomes by TA-finder, but their presence was later detected by BLAST. After additional analysis, we found that the gene sequences in those genomes contained mutations that disrupted the reading frame, which could be due to sequencing errors or genuine mutations. The possible pseudogenes belonging to TAs are indicated in Figure 1.

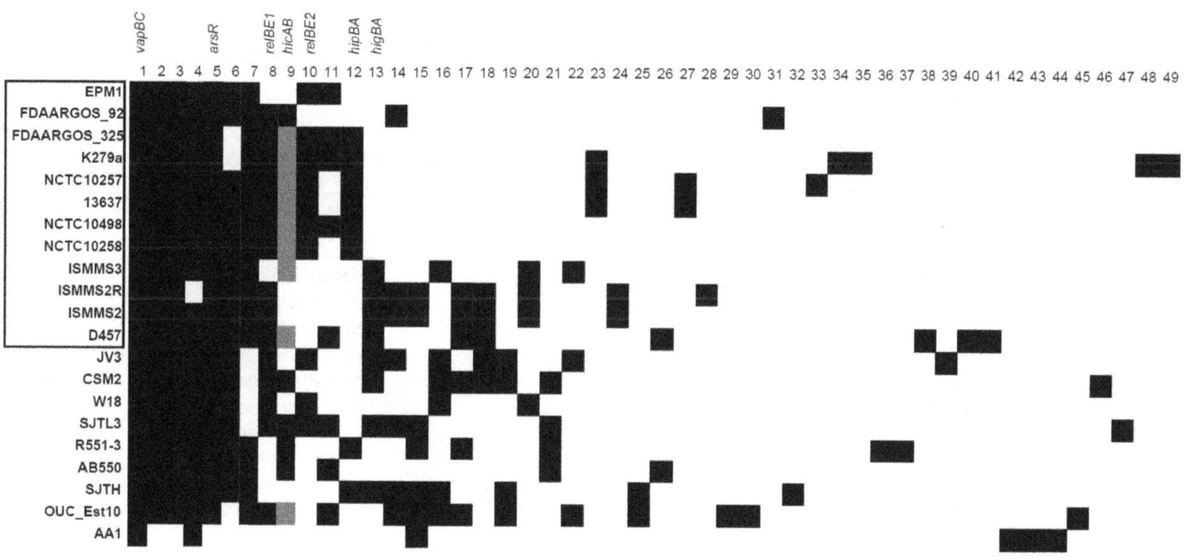

Figure 1. The predicted toxin–antitoxin (TA) systems in *S. maltophilia* genomes. The genomes with names bracketed in black rectangle were isolated from clinical sources. Black squares indicate the presence of a TA system, whereas grey squares indicate the presence of a pseudogene version. The information about the presented TA systems (1–49) is described in Table S1. TA systems analyzed in this work are indicated above.

As can be seen from Figure 1, there were no clear TA distribution patterns characteristic to annotated genomes of *S. maltophilia* from clinical or environmental origins. The average number of predicted TA systems per genome did not significantly differ between the *S. maltophilia* of both origins (Figure 2a).

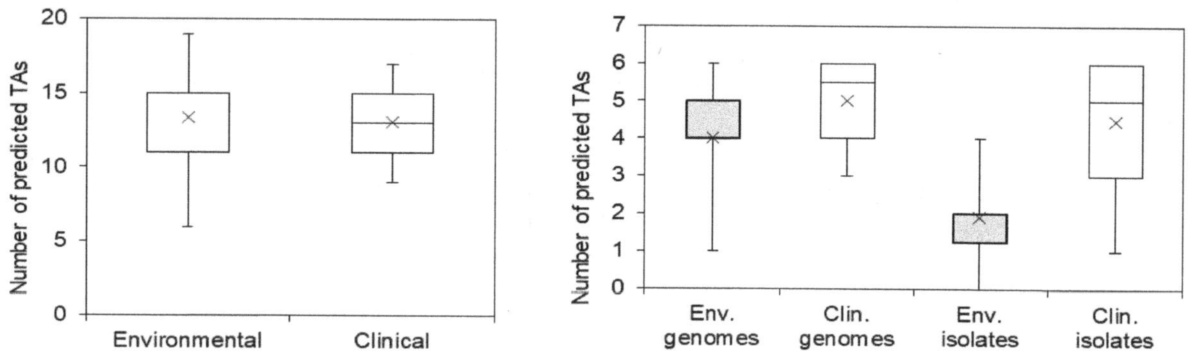

(a) (b)

Figure 2. The prevalence of predicted TA systems in sequenced *S. maltophilia* genomes and *Stenotrophomonas* spp. isolates. (**a**) The distribution of total predicted TA systems in annotated *S. maltophilia* genomes; (**b**) the distribution of selected most prevalent TA systems in annotated genomes of *S. maltophilia* and in isolates of clinical and environmental origin. Boxes indicate upper and lower quartiles, whiskers indicate minimum and maximum values, and crosses indicate mean values.

Not all gene pairs predicted as TA systems are active TA modules [23,27,28]. Therefore, we aimed to check if the prevailing predicted TA systems of *S. maltophilia* were functional. For functional analysis, the TA systems that were present in at least 1/3 of the analyzed genomes were chosen. To increase the likelihood of the selected operons acting as TA systems, we chose TAs where both genes were no longer than 1 kb [25] (with the exception of *hipBA* system, where the toxin gene is known to be larger) [29]. The distance between the toxin and antitoxin genes was set to be 50 bp or less, which increased the possibility that the genes belonged to an operon and were expressed from a single transcript. The genes, which were clearly annotated for a function and not related to TAs (e.g., antibiotic resistance genes, membrane transporters), were excluded from further analysis. The final set of seven TAs, selected for functional characterization, is presented in Table 1.

Table 1. The characteristics of TA systems prevalent in annotated genomes of *S. maltophilia*.

Predicted TA Family	Protein Domains (T-A) [1]	Found in Genomes ($n = 21$)	Conservativity of Operon (%) [2]	Prevalence in Other Bacteria [3]	Remarks
vapBC	PIN-AbrB	21	84	*Stenotrophomonas* spp., *Xantomonas* spp.	
-	COG3832- ArsR	20	90	*Stenotrophomonas* spp. and other genera	
relBE	RelE-Xre	15	73	*Stenotrophomonas* spp.	T before A
relBE	RelE-RHH	10	92	*Stenotrophomonas* spp.	
relBE (higBA)	HigB-Xre	8	78	*Stenotrophomonas* spp., *Pseudomonas* spp.	T before A
hicAB	HicA-HicB	14	79	*Stenotrophomonas* spp.	T before A
hipBA	HipA-Xre	8	70	*Stenotrophomonas* spp. (T common to *Burkholderia* spp.)	T is 440 a.a.

[1] Predicted domains (domain-like); [2] in *Stenotrophomonas* spp.; [3] the prevalence of selected TA systems was identified by BLAST analysis against domain *Bacteria*; T: toxin, A: antitoxin.

The majority of the selected systems belonged to well-known type II TA families. There were three TA systems belonging to the *relBE* family: *relBE1* (containing RelE-Xre domains), *relBE2* (RelE-RHH domains), and *higBA* (HigB-Xre domains). However, one of the most common predicted TAs in *S. maltophilia*, containing COG3832-ArsR domains, was the least described. It was predicted as a TA system bioinformatically more than 10 years ago [30], yet only once was it analyzed thoroughly [23].

The predicted TA systems were conservative not only in *S. maltophilia* but in *Stenotrophomonas* spp. with high sequence conservativity (>70%) (Table 1). Therefore, we decided to test the presence and spread of these TAs in both *S. maltophilia* and *Stenotrophomonas* spp. isolates from our collection.

2.2. The Prevalence of TA Systems in Stenotrophomonas spp. of Clinical and Environmental Origin

To detect selected TA systems, a set of primers was designed after aligning all available sequences for a TA system in a *Stenotrophomonas* genus and identifying the most conserved regions (Table S2). Next, the collection of *Stenotrophomonas* spp. of clinical ($n = 21$) and environmental ($n = 14$) origins (Figure 3) was screened for the presence of TAs. All isolates of clinical origin were attributed to *S. maltophilia* species, while six of the environmental isolates belonged to *S. maltophilia*. The rest of the environmental isolates were attributed to either *S. rhizophila* or *Stenotrophomonas* sp.

All of the examined TA systems were detected in at least four isolates (Figure 3). *vapC* and *hipBA* modules were the most prevalent (31 and 21 isolates, respectively), and were evenly spread in both clinical and environmental bacteria. However, *hipBA* was more common in isolates of environmental origin. Three TAs (*relBE1*, *relBE2*, and COG3832-ArsR) were prevalent in isolates of clinical origin yet were absent in environmental isolates. *hicAB* system was detected in the majority of clinical isolates and in only two isolates of environmental origin. Surprisingly, we observed a clear difference in TA profiles between *Stenotrophomonas* isolates of clinical and environmental origins, contradicting a similar distribution of these TAs in the sequenced genomes. We found that certain TAs were more likely to be present in the clinical isolates of *S. maltophilia*. Moreover, clinical isolates of *S. maltophilia* had more TA systems per isolate, compared to the environmental isolates (Figure 2b).

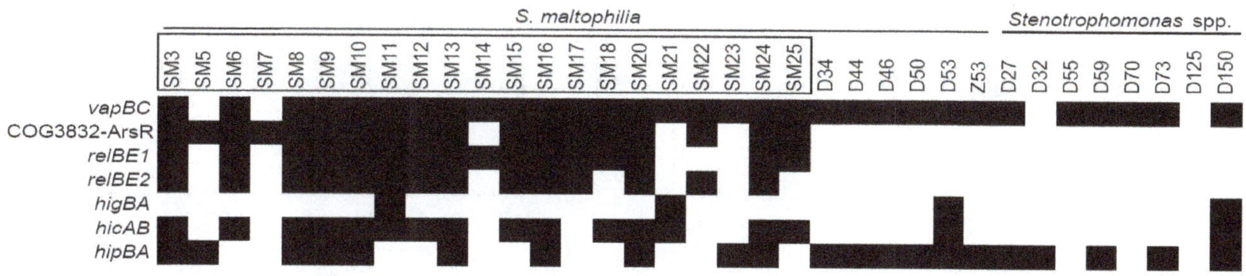

Figure 3. The prevalence of predicted TA systems in *Stenotrophomonas* spp. of clinical and environmental origins. The isolates with their names bracketed in black rectangles were derived from clinical sources. The presence of TAs, indicated as black boxes, was detected by PCR as described in the Materials and Methods section.

2.3. Functionality of *S. maltophilia* TA Systems

To test if the predicted TA systems were functional, we cloned toxin and antitoxin genes into separate inducible vectors to perform kill-rescue experiments in *E. coli*, as described in Materials and methods. The predicted toxin and antitoxin expression was induced with arabinose, and IPTG, respectively.

Only *S. maltophilia relBE1*, *relBE2* and *higBA* TA systems, belonging to *relBE* family, were functional in kill-rescue experiments in *E. coli*, as evident from the growth inhibition upon induction of the toxin and growth restoration, when antitoxin was concomitantly induced (Figure 4a–c). The predicted toxins of *hipBA*, *vapBC*, *hicAB*, and COG3832-ArsR TA systems did not display any toxicity in the kill-rescue assay (Figure 4d–f).

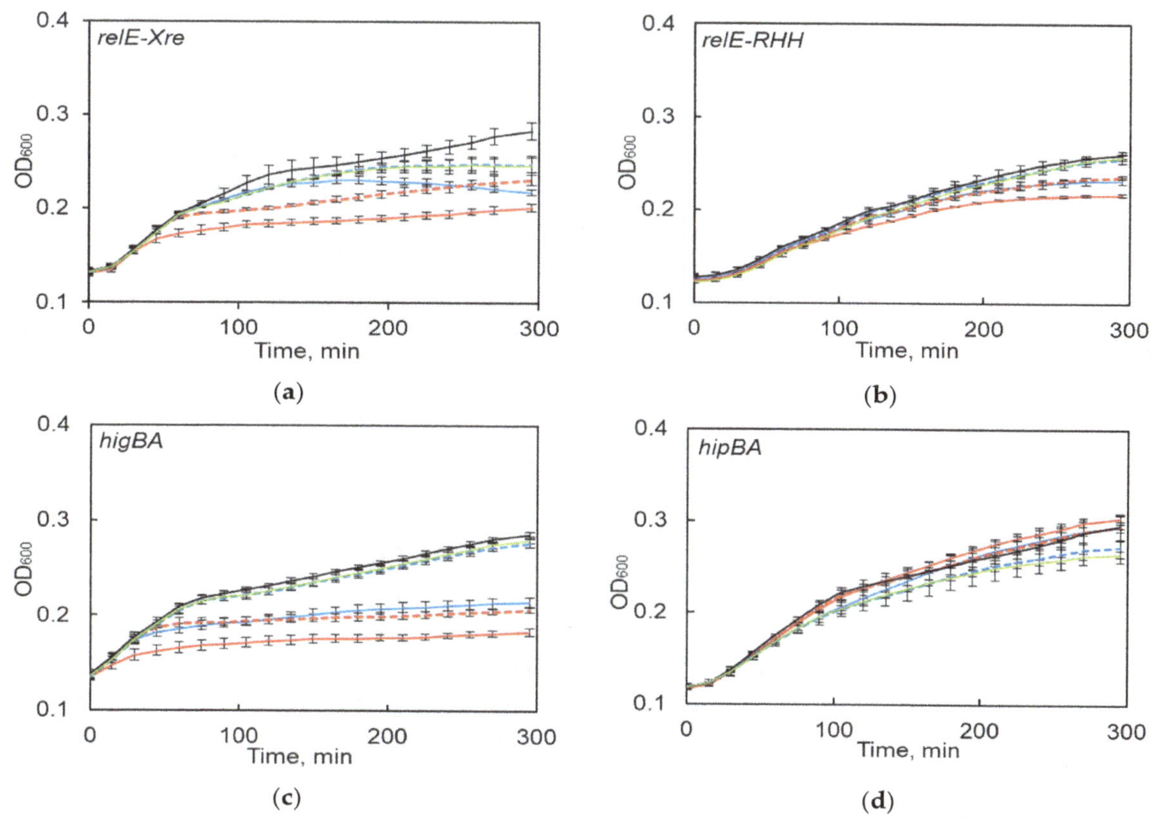

Figure 4. *Cont.*

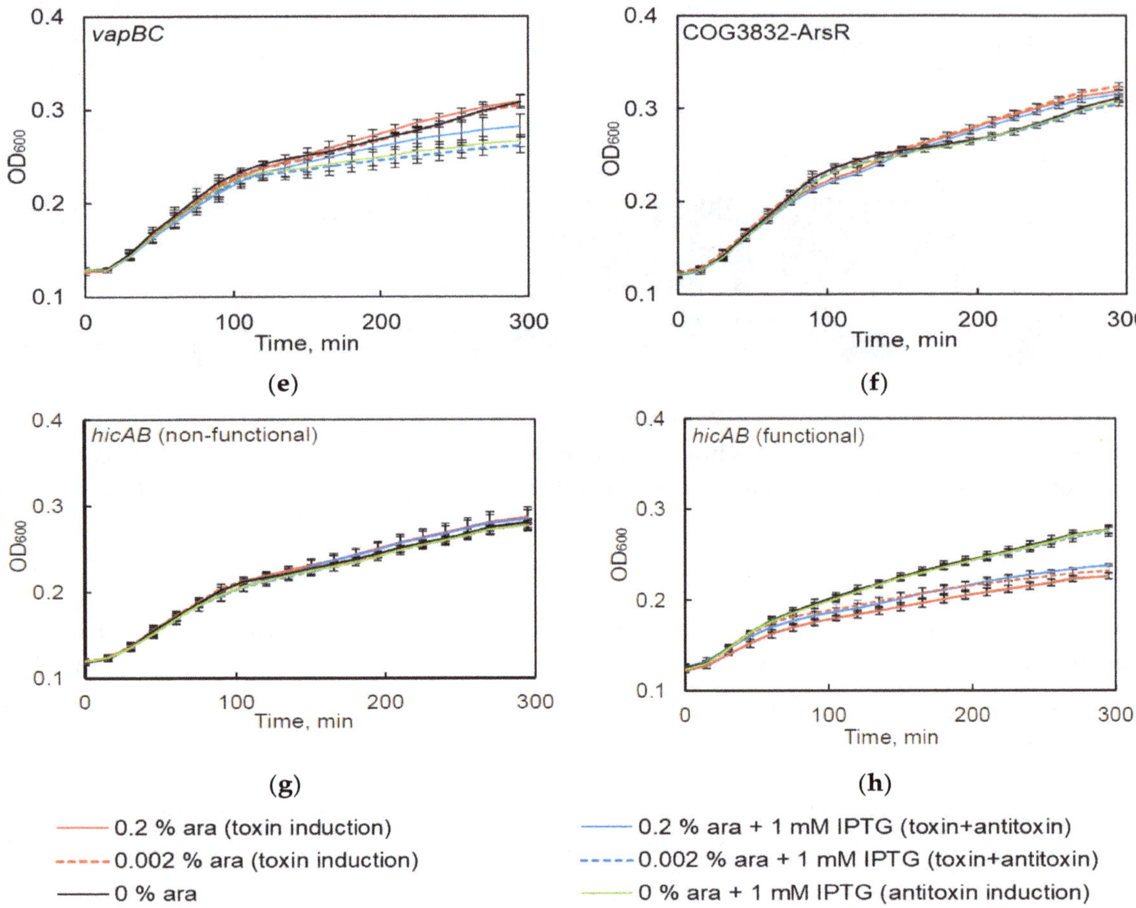

Figure 4. The functionality of predicted *S. maltophilia* TA systems as tested by the kill-rescue assay in *E. coli.* (**a**–**c**) the growth inhibition and restoration of functional TA systems; (**d**–**f**) the kill-rescue assays for the predicted TA that appear to be non-functional in *E. coli* system; (**g**,**h**) representative growth curves for a functional and a non-functional version of *hicAB*. The toxin and antitoxin genes were cloned into separate inducible plasmids, the toxin was induced by adding arabinose (ara), and its activity counteracted by inducing antitoxins with IPTG. The bacterial growth was measured as OD_{600}, as described in the Materials and Methods section. The experiments were repeated at least three times; error bars indicate standard deviation.

We previously observed that one of the tested TAs (i.e., *hicAB*) was often present in the sequenced *S. maltophilia* genomes as a pseudogene (Figure 1). Therefore, when we first observed *hicAB* to be non-functional in the kill-rescue assay (Figure 4g), we decided to clone it from several other *S. maltophilia* isolates and to use the kill-rescue assay. Out of eight cloned versions of *hicAB*, three were non-functional, and five systems displayed a distinct kill-rescue phenotype. The growth of *E. coli* after the induction of a representative functional *S. maltophilia hicAB* is presented in Figure 4h. Sequencing of the cloned non-functional versions of *hicAB* revealed the same frameshift mutation in the toxin gene.

According to our results, *S. maltophilia* COG3832-ArsR, *hipBA*, and *vapBC* predicted TA systems were non-functional in *E. coli*. To find out, if this was due to the differences of the host or TAs, which were indeed non-functional, we created an expression system for *S. maltophilia* by using a broad-host range plasmid pBAD1 with an arabinose controlled promoter and cloned the predicted toxins. *S. maltophilia relE* toxin of the *relE-Xre* TA system was proven to be functional in *E. coli* and was thus used as a control. Further, *relE* toxin showed toxicity upon induction (not shown), while COG3832-ArsR, *hipBA*, and *vapC* toxins did not show any toxicity in *S. maltophilia* under the same induction conditions (not shown).

2.4. Characterization of *Stenotrophomonas* spp. Isolates

Despite the similar distribution of predicted TAs observed in the *S. maltophilia* sequenced genomes of clinical and environmental origins (Figure 1), the most prevalent TA modules in bacterial isolates showed characteristic profiles regarding their origin, with clear prevalence for some TAs in clinical isolates (Figure 3). Since the majority of the clinical *S. maltophilia* isolates were obtained in the same hospital, we wanted to find out if there was a clonality between the analyzed isolates. The genotyping of *Stenotrophomonas* spp. isolates was undertaken by a random amplified polymorphic DNA (RAPD) assay with two sets of random primers as described in the Materials and Methods section. Each isolate significantly differed and was not grouped into clusters according to their origin (Figure 5). This thereby confirmed that the *S. maltophilia* of clinical origin were sporadic cases.

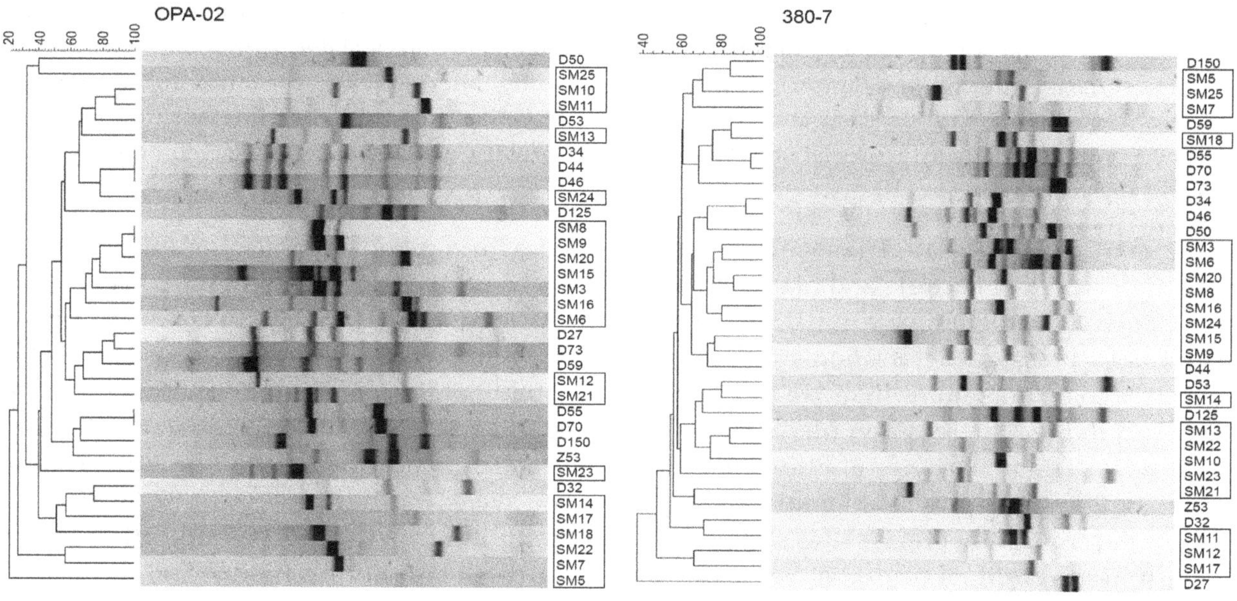

(a) (b)

Figure 5. The genotyping of *Stenotrophomonas* spp. isolates. Random amplified polymorphic DNA (RAPD) assay was performed as described in the Materials and Methods section using OPA-02 (**a**) and 380-7 (**b**) primers (Table S2). The isolates bracketed in black rectangles were isolated from clinical sources.

The genotyping of *Stenotrophomonas* spp. from clinical and environmental origins showed that isolates could not be grouped into clusters based on their origin of isolation. However, uneven distribution of TAs (i.e., the types of TAs and total TA number per isolate) among the isolates indicated that there were certain traits that distinguished the isolates of different origins. Therefore, we decided to compare other virulence-related traits of the isolates and see if the presence of TAs could be considered as a marker for virulence of the strain.

One of the characteristics that makes *S. maltophilia* a dangerous opportunistic pathogen is its innate resistance to antibiotics [31]. We tested the resistance of the isolates against a set of antibiotics (trimethoprime, chloramfenicol, gentamicine, ceftazidime, meropenem, and tetracyclin), including trimethoprime-sulfametoxazole and ciprofloxacine, which are commonly used against *S. maltophilia* infections. Resistance and sensitivity values were determined according EUCAST clinical breakpoints for *S. maltophilia* or *Pseudomonas aeruginosa*, if such data for *S. maltophilia* was not available. We observed significantly higher resistance profiles among the isolates of the clinical origin (Figure 6). Tetracyclin, ceftazidime, meropenem, and gentamicin were effective against most environmental isolates, while the majority of the clinical ones showed resistance.

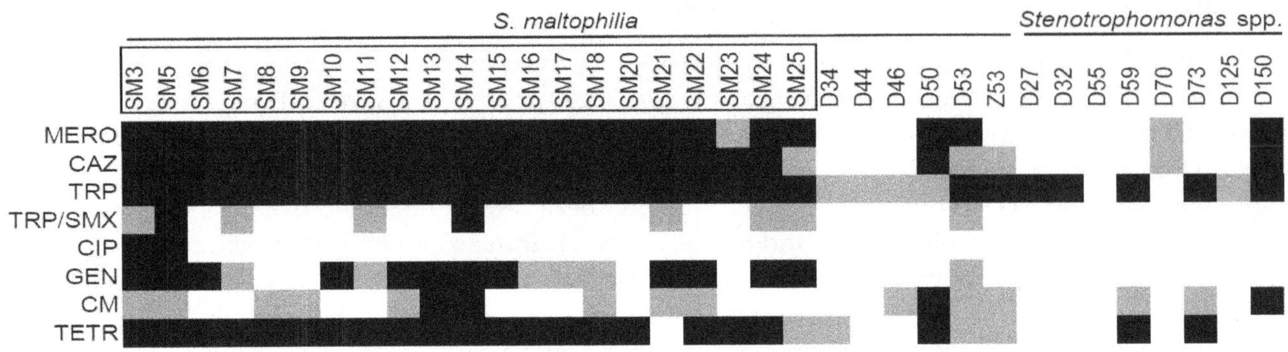

Figure 6. The resistance to antibiotics of *Stenotrophomonas* spp. of clinical and environmental origin. The isolates with names bracketed in black rectangles were isolated from clinical sources. Black squares indicate resistance; grey squares indicate intermediate resistance to antibiotic. The antibiotics are as indicated: meropenem (MERO), ceftazidime (CAZ), trimethoprime (TRP), trimethoprime-sulfametoxazole (TRP/SMX), ciprofloxacine (CIP), gentamicine (GEN), chloramphenicol (CM), and tetracyclin (TETR).

The next virulence-related trait we assessed was biofilm formation, a complex community structure composed of adhesive bacteria and matrix components. Various bacteria can form biofilms on medical devices, such as catheters or intubation tubes, since it increases pathogen survival during stress (e.g., antibiotic pressure, desiccation, or laminar flow) [32]. Previous studies have shown the ability of *S. maltophilia* to form biofilms [33]. We first checked the ability of the isolates to form biofilms at 37 °C, as described in the Materials and Methods section, since this was an expected temperature inside the infected host. The majority of *S. maltophilia* from clinical origins formed strong biofilms (Figure 7, Figure S1). Only two environmental isolates displayed weak biofilm growth at 37 °C (Figure 7). However, the environmental isolates showed severely reduced growth at 37 °C, indicating that their inability to form biofilms could be due to suboptimal growth conditions. Therefore, we tested the ability of all the isolates to form biofilms at 28 °C, a temperature that resembled environmental (outside the host) conditions. No change in biofilm formation was observed for the clinical strains; however, several of the environmental strains increased their biofilm formation under permissive temperature. Thus, the results demonstrated a clear differentiation of the strains of different in ability to form biofilm at both 37 °C and 28 °C.

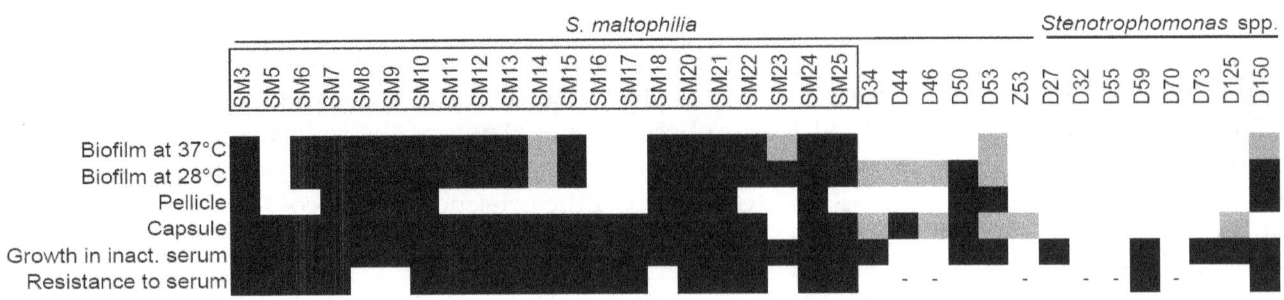

Figure 7. Biofilm formation and capsule-related characteristics of *Stenotrophomonas* spp. isolates. The biofilm, pellicle, capsule-forming phenotype, growth in inactivated serum and resistance to serum were determined as described in the Material and Methods section. The isolates with names bracketed in black rectangles were isolated from clinical sources. Black squares indicate the ability to form biofilm, pellicle, the presence of capsular polysaccharides (CPS), or the ability to grow in serum; grey squares indicate the ability to form weak biofilms or weak capsule-forming phenotype. "-" indicates that the isolates were not exposed to non-inactivated serum due to their inability to survive in inactivated serum.

We then analyzed the ability of the isolates to form pellicle, a type of biofilm, which formed at the air–liquid interface. This specific structure is associated with the increased virulence of bacteria [34]. As can be seen in Figure 7, only the isolates with biofilm forming capacity were able to form pellicle, though this structure was not common to all such isolates and was more frequent in the clinical strains.

The ability to form capsule is considered as an important virulence trait in bacteria, together with resistance to antibiotics, biofilm formation, adhesiveness, desiccation, and other features [35–37]. The role of capsular exopolysaccharides in *S. maltophilia* virulence is still unknown. Therefore, we were interested to discover if bacterial isolates could produce polysaccharides that were indicative of capsule formation. The capsular polysaccharides were extracted and analyzed as described in the Materials and Methods section. The vast majority of clinical *S. maltophilia* isolates (95%) produced high molecular mass polysaccharides, which were associated with capsular polysaccharides (CPS). This phenotype was demonstrated by only a third (36%) of environmental *Stenotrophomonas* spp. isolates (Figure 7, Figure S2). However, *S. maltophilia* environmental isolates tended to produce capsule-associated polysaccharides more often than *Stenotrophomonas* of other species.

The ability to produce exopolysaccharide capsule is indicative of a strain's ability to avoid killing by a host's immune system [36,38]. We aimed to test if isolates differed in their ability to survive serum-mediated killing. First, we tested the ability of isolates to grow in a heat-treated serum with an inactivated complement system. All of the clinical isolates were able to grow, yet only half of the environmental isolates grew in the presence of inactivated serum (Figure 7). Next, we assessed the growth of *Stenotrophomonas* isolates in the presence of the active serum. The majority of clinical *S. maltophilia* strains were resistant to complement components when grown in an active serum, whereas only two of the environmental isolates showed resistance, thereby demonstrating an inability of environmental strains to resist a host immune system and cause an infection.

3. Discussion

In this study, we aimed to investigate the prevalence and profiles of TA systems in a newly emerging opportunistic pathogen, *S. maltophilia,* as well as to compare the TA presence with the manifestation of the virulence features of clinical and environmental isolates.

While the role of plasmid-borne TA systems is known as plasmid maintenance [39], the biological function of chromosome encoded TA systems is far less clear. It has been shown that chromosome encoded TA systems help stabilize genome integrity by preventing large scale genome reductions, which keeps genes needed for survival in unfavorable conditions [40,41] and causes growth arrest that helps bacteria survive environmental stress [39]. The most widespread type II TA system families affect the translation apparatus of a cell by cleaving or otherwise disrupting the functionality of mRNA (*relBE* (including *higBA*) [42,43], *hicAB* [44], and *mazEF* [45]) or tRNA (*vapBC* [46] and *hipBA* [29]). It has also been reported that TA systems function as virulence factors. Thus, a *higBA* TA system in *Pseudomonas aeruginosa* influences different virulence factor expression and biofilm formation [20]. *Mycobacterium tuberculosis MazEF* type TA systems play a role in pathogen intracellular survival inside the host [47]. A *Staphylococcus aureus SavRS* TA system functions as a regulator of virulence traits [48]. Moreover, previous studies have shown that the number of TA systems is enlarged in pathogenic strains compared with closely related nonpathogenic bacteria (*M. tuberculosis* encodes 88 putative TA systems [49], while non-harmful *M. smegmatis* encodes only three TA modules [50]). In this study, we investigated whether the same trend was characteristic to TA systems in the opportunistic pathogen *S. maltophilia*. The data observed on *vapBC* and *relBE* families, which are two major type II TA families [51,52], were in line with the observations on the prevalence of TA systems.

The amplicon detection-based screen of the TA systems in bacterial isolates revealed unexpected results, since *relBE1*, *relBE2,* and COG3832-ArsR TA systems were found solely in clinical *S. maltophilia* isolates and were not detected in environmental samples. Moreover, some preferences were observed for *hicAB* and *hipBA* TAs. The dissimilarity between observations of bioinformatic TA analysis in genomes available in the databases and our detection might be explained by the different geographic locations of bacterial sources. The bacteria analyzed in our study were isolated from soil in Lithuania, while the origin of *S. maltophilia* genomes in the databases varied greatly. However, the genotyping assay proved that the analyzed isolates did not belong to the same clonal group and showed high genetic diversity. Similar results were previously reported for *S. maltophilia* [53,54], supporting our findings about genotypic diversity among clinical and environmental *S. maltophilia* isolates from Lithuania.

In the present study, the putative COG3832-ArsR TA system was found in nearly all analyzed *S. maltophilia* genomes. COG3832-ArsR operon was previously proposed to be a TA system on the basis of comparative prokaryotic genomic analysis [30]. This module was found among 10 of the most induced TA systems in drug tolerant persister cells for *M. tuberculosis* [23], though functional analysis of this operon was not done in any TA system research. However, *S. maltophilia* COG3832-ArsR did not show toxicity in the kill-rescue assay and therefore its attribution to TA systems remains questionable. Altogether, functional analysis revealed that four of the seven analyzed *S. maltophilia* TA systems were functionally active TA operons in *E. coli*. The confirmed TAs belonged to previously described classes. To exclude the possibility of host incompatibility, we transferred the inactive toxins into *S. maltophilia*, yet no toxicity was observed. Interestingly, we identified two types of the same *S. maltophilia hicAB* TA systems via bioinformatic analysis in bacterial isolates. Previous studies have reported that TA system genes may contain mutations, which can result in nonfunctional pseudogene formation [55]. Loss of function mutations in virulence-associated genes can indicate genomic plasticity and adaptation to different environments [56], suggesting that *hicAB* TA systems may be favorable or, otherwise, unfavorable when living under particular conditions.

In our study, only the chromosome sequences of *S. maltophilia* were used for TA system prediction, however, the detected TA systems might be located on both chromosomes and mobile elements. TA systems are known to be transmitted via horizontal gene transfer [21,57]. It was shown that in *S. maltophilia*, trimethoprim/sulfamethoxazole resistance genes were found in class 1 integron and plasmid [58], indicating the ability *S. maltophilia* to acquire genetic information through horizontal gene transfer. Therefore, it is possible that some of the analyzed TA systems can spread via mobile elements, yet further studies are needed to confirm this possibility.

In this study, we demonstrated that clinical and environmental *Stenotrophomonas* isolates possessed not only distinct TA profiles but other important virulence-related traits. *S. maltophilia* is notorious for its multidrug resistance associated with intrinsic resistance factors, such as low membrane permeability, multidrug resistance efflux pumps, antibiotic-modifying enzymes, and ability to acquire resistance via mutations or acquisition of resistance genes through horizontal gene transfer [31]. Our study revealed a high antibiotic resistance of clinical *S. maltophilia* isolates, all of which displayed resistance to at least five out of eight antimicrobials. Environmental *Stenotrophomonas* isolates were more sensitive to antibiotics compared to bacteria of clinical origins, including meropenem (though *S. maltophilia* is known to have metallo-β-lactamases, providing resistance to various β-lactam antibiotics including carbapenems) [59,60]. The data on the antibiotic resistance level of clinical and environmental *S. maltophilia* isolates are contradictory [60–62], although it has been proposed that some mechanisms of antibiotic resistance occur in the environment and could remain in clinical isolates [63].

Biofilm formation is one of the best characterized *S. maltophilia* virulence traits. *S. maltophilia* forms biofilms on glass and different types of plastic [33], including medical devices (indwelling venous catheters [64], prosthetic valves [65], and lens implants [66]). Biofilm formation on biotic surfaces (e.g., human epithelial respiratory cells) has also been reported [67,68]. In this study, we showed that clinical and environmental *S. maltophilia* isolates differed in biofilm formation abilities, which was in line with previous studies [9,69]. Pellicle is a type of biofilm, formed at the air–liquid interface [70] and protecting bacterial population from environmental stresses. Further, it can increase survival rates by increasing accessibility to oxygen [71]. Less than a half of both clinical and environmental isolates were able to form pellicle, indicating that this phenotype may not be directly linked to virulence traits in *S. maltophilia*.

Although the ability of *S. maltophilia* to form capsule has not been described, studies of capsule functions in other bacterial pathogens showed its crucial role in the protection from unfavorable environmental conditions, including antibiotic treatment, host immune response, and resistance to disinfectants and heat [72–75]. Based on the separation of exopolysaccharides on SDS-PAGE gel, in some of the strains we observed the fraction of high molecular mass polysaccharides, as well as fractions resembling lipopolysaccharides [76]. Typically, such fractions represent capsular exopolysaccharides [37]. We demonstrated that CPS associated high molecular mass exopolysaccharides were detected in nearly all clinical isolates, while only 36% of environmental *Stenotrophomonas* spp. isolates displayed this phenotype. However, the majority of environmental isolates belonging to the *S. maltophilia* species did synthesize extracellular polysaccharides. Therefore, further studies are needed to investigate the importance of capsules in environmental isolates. Nevertheless, the presence of the capsule phenotypes among clinical *S. maltophilia* isolates supports the idea that capsule is beneficial to helping unfavorable conditions and host immune responses survive [38]. The hospital-related strains were also distinguished by increased resistance to serum-induced stress, especially when compared to strains of environmental origin. The ability of isolates to grow at 37 °C (temperature of the host) and resist the complement system of the active serum indicates the increased infectivity potential of these strains. Some of the environmental isolates did not grow even in the heat-inactivated serum, which indicates the inability of these strains to multiply inside the host even before the encounter with the immune system. This could hint at the inability of environmental isolates to persist and grow due to improper nutrients and blood serum composition.

Based on the phenotypes of clinical and environmental *S. maltophilia* and genotyping results, we can assume that the environment could be the primary source of opportunistic *S. maltophilia* infections. However, further development of virulence potential might include an exchange of genetic information and switching on potential virulence genes.

4. Conclusions

Both clinical and environmental *S. maltophilia* genomes contain a variety of type II TA systems, the most common of which belong to *vapBC*, *relBE*, *hipBA*, *hicAB*, and putative COG3832-ArsR families. While bioinformatics assay did not show differences in TA distribution between clinical and environmental TAs, the screening of our collection of isolates did. *relBE* and COG3832-ArsR operons were present solely in clinical *S. maltophilia* isolates, while *hipBA* was more frequent in the environmental isolates. Together with different TA composition, the clinical *S. maltophilia* isolates exhibited clearly stronger biofilm formation, as well as increased antibiotic and serum resistance compared to environmental isolates.

5. Materials and Methods

5.1. Bioinformatic Assays

In total, 21 complete *S. maltophilia* genomes from NCBI database (date of accession 2018-10-13) (Table S3) were analyzed using the TADB 2.0 database tool TA-finder [24]. To identify TA systems in *S. maltophilia* genomes, a basic local alignment search tool (BLAST) [77] was used. The Clustal Omega tool [78] was used for sequence alignments.

5.2. The Bacteria Used in the Study and Growth Conditions

Plasmids and strains used in this study are listed in Table 2. Clinical and environmental *Stenotrophomonas* spp. isolates are listed in Table 3. *Stenotrophomonas* spp. and *E. coli* were grown in an LB medium, unless indicated otherwise. *E. coli* and clinical *S. maltophilia* isolates were grown at 37 °C. Environmental *S. maltophilia* and *Stenotrophomonas* spp. isolates were grown at 28 °C, unless indicated otherwise. DNA was transformed to *E. coli* by heat shock [79] and electroporation was used for

S. maltophilia [80]. Antibiotics (ampicillin 100 mg/L, chloramphenicol 30 mg/L, and kanamycin 60 mg/L) were used to maintain plasmids when needed.

Table 2. Strains and plasmids used in this study.

Strain/Plasmid	Description	Reference
E. coli JM107	*endA1 glnV44 thi-1 relA1 gyrA96* Δ*(lac-proAB)* *[F′ traD36 proAB*$^+$ *lacI*^{q} *lacZ*Δ*M15]* *hsdR17*$(R_K^-\ m_K^+)\ \lambda^-$	[81]
E. coli BW25113 F′	*proA*$^+$*B*$^+$*lacIq*Δ*lacZ)M15 zzf::mini-Tn10* (KanR)	[82]
pBAD30	Expression plasmid	[83]
pUHE 25-2(cat)	Expression plasmid	[82]
pBAD1_gfp	*gfp* gene cloned into pBAD-MCS-1 vector	[84]

Table 3. Clinical and environmental *Stenotrophomonas* spp. isolates used in this study.

Isolate	Genus/Species	Origin	Source [1]	Isolation Year
D27	*S. maltophilia/rhizophila*	Environmental	Soil (Conventional wheat)	2016
D32	*Stenotrophomonas* sp.	Environmental	Soil (Conventional wheat)	2016
D34	*S. maltophilia*	Environmental	Soil (Conventional wheat)	2016
D44	*S. maltophilia*	Environmental	Soil (Conventional wheat)	2016
D46	*S. maltophilia*	Environmental	Soil (Conventional wheat)	2016
D50	*S. maltophilia*	Environmental	Soil (Conventional wheat)	2016
D53	*S. maltophilia*	Environmental	Soil (Conventional wheat)	2016
D55	*Stenotrophomonas* sp.	Environmental	Soil (Organic rapeseed)	2016
D59	*Stenotrophomonas* sp.	Environmental	Soil (Organic rapeseed)	2016
D70	*S. rhizophila*	Environmental	Soil (Organic rapeseed)	2016
D73	*Stenotrophomonas* sp.	Environmental	Soil (Organic rapeseed)	2016
D125	*Stenotrophomonas* sp.	Environmental	Soil (Organic maize)	2016
D150	*Stenotrophomonas* sp.	Environmental	Soil (Conventional maize)	2016
Z53	*S. maltophilia*	Environmental	Fish (Carp)	2016
SM3	*S. maltophilia*	Clinical	NCI	2017
SM5	*S. maltophilia*	Clinical	VCCH LM	2018
SM6	*S. maltophilia*	Clinical	VCCH LM	2018
SM7	*S. maltophilia*	Clinical	VCCH LM	2018
SM8	*S. maltophilia*	Clinical	NCI	2019
SM9	*S. maltophilia*	Clinical	NCI	2019
SM10	*S. maltophilia*	Clinical	VCCH LM	2019
SM11	*S. maltophilia*	Clinical	VUH SK PD	2019
SM12	*S. maltophilia*	Clinical	VUH SK PD	2019
SM13	*S. maltophilia*	Clinical	VUH SK PD	2019
SM14	*S. maltophilia*	Clinical	VCCH LM	2019
SM15	*S.maltophilia*	Clinical	KCPHC	2019
SM16	*S. maltophilia*	Clinical	VCCH LM	2019
SM17	*S. maltophilia*	Clinical	NCI	2019
SM18	*S. maltophilia*	Clinical	NCI	2019
SM20	*S. maltophilia*	Clinical	NCI	2019
SM21	*S. maltophilia*	Clinical	NCI	2019
SM22	*S. maltophilia*	Clinical	NCI	2019
SM23	*S. maltophilia*	Clinical	NCI	2019
SM24	*S. maltophilia*	Clinical	NCI	2019
SM25	*S. maltophilia*	Clinical	NCI	2019

[1] NCI—National Cancer Institute (Lithuania); VCCH LM—Vilnius City Clinical Hospital, Laboratory of Microbiology; VUH SK PD—Vilnius University Hospital Santaros Klinikos, Pediatrics Department, Division of Infectious Diseases; KCPHC—Kaunas City Public Health Center. The isolates from soil were isolated and described previously [85].

5.3. Detection of TA Systems in Stenotrophomonas

For the detection of *S. maltophilia* TA systems, primers targeting conservative gene fragments were designed (Table S2) and PCR was performed using DreamTaq polymerase (Thermo Fisher Scientific).

5.4. Cloning of TA System Genes

All enzymes used for cloning were from Thermo Fisher Scientific and were used according to the manufacturer's recommendations. For gene cloning, a high-fidelity Phusion polymerase was used with the primers listed in Table S2. A *hicAB* TA system was cloned using SM6, SM9, SM11, SM12, SM13, SM16, SM20, and SM24 isolate DNA as the template. For *relBE2* and *higBA* the isolate SM11 was used, for *relBE1, hipBA, vapBC,* and COG3832-ArsR the isolate SM9 DNA was used as a template. For TA functional analysis in *E. coli*, all toxin genes were cloned into pBAD30 plasmid; antitoxin genes were cloned into pUHcat plasmid. Both vectors were hydrolyzed with SphI, then overhangs were blunted with T4 polymerase. Final restriction was performed with HindIII. To clone the toxins into a broad host range pBAD1 vector, pBAD1_GFP plasmid was used. First, the pBAD1_GFP vector was hydrolyzed with XbaI and other steps were performed as described above.

5.5. Kill-Rescue Assay

The kill-rescue assay was performed as described previously [86]. Briefly, *E. coli* BW25113 F′ strains harboring plasmids pBAD_tox and pUHEcat_antitox were grown overnight, diluted 1:500 into fresh medium supplemented with 0.2% glucose, and grown to its early exponential phase (optical density at 600 nm (OD_{600}) = 0.12). The expression of toxin and/or antitoxin was then induced with arabinose (0.002% and 0.2%) and/or IPTG (1 mM), respectively. Growth was measured as OD_{600} in a Tecan Infinite M200 Pro plate reader at 37 °C with shaking.

5.6. Random Amplified Polymorphic DNA (RAPD) Assay

For the RAPD assay, DNA amplification was performed at the same time on the same thermocycler in a final reaction volume of 25 μL, containing 2.5 μL of 10× polymerase reaction buffer with KCl: 0.5 μL 10 mM dNTP, 1.25 μL 50 mM $MgCl_2$, 2 μL 10 pmol/μL OPA-02 or 380-7 primer, 0.15 μL DreamTaq polymerase, and 5 μL of bacteria lysate.

Cycling conditions are as follows: initial denaturation at 95 °C for 2 min followed by 39 cycles of denaturation at 95 °C for 30 s, annealing at 36 °C/34 °C for 1 min (OPA-02 and 380-7 primers, respectively), extension at 72 °C for 2 min and final extension at 72 °C for 5 min.

PCR products were loaded on a 1% (*w/v*) agarose gel with 0.5 mg/mL of ethidium bromide (1 × TAE buffer, 7 V/cm voltage, 1.5 h). Electrophoresis gels were visualized with Bio Rad Molecular Imager and analyzed using GelCompar II software (Applied Maths) with the Dice coefficient set at 1% and band tolerance set at 1.5% using the UPGMA method.

5.7. The Evaluation of Antibiotic Resistance

Antimicrobial resistance analysis was performed by growing *Stenotrophomonas* isolates on agarose LB plates with selected concentrations of antimicrobial agents for 16 h at 28 °C. Control growth was considered as growth of isolates on the LB plates without antimicrobial agents. The isolate was considered as resistant to tested antibiotics only if growth visually looked the same as the control. In the case of significantly weaker growth, the isolate was referred to as an intermediate resistant and in the absence of growth, the isolate was considered as sensitive.

5.8. Biofilm Formation Assay

Biofilm formation experiments were performed as described elsewhere [38]. Briefly, overnight *Stenotrophomonas* spp. cultures (grown in TSB medium (Oxoid) at 37 °C for 16 h) were inoculated into the 150 μL TSB medium containing wells of 96 U-form polystyrene plate (30-fold dilution), then incubated stationary at 28 °C or 37 °C for 24 h. The OD_{600} of planktonic culture was measured and wells were washed 3 times with PBS buffer. Adherent cells were stained with 0.1% crystal violet dye for 15 min, then washed 5 times with the PBS buffer. Dye was dissolved in 96% ethanol for OD_{580}

measurements. The OD_{580}/OD_{600} ratio was calculated to normalize the number of biofilm forming cells to the total cell number.

5.9. Pellicle Formation Assay

The pellicle formation was performed as previously described [87]. Briefly, overnight cultures *Stenotrophomonas* spp. (grown 16 h in TSB medium at 37 °C) were inoculated into the TSB medium containing wells of a flat-bottom 12 well polystyrene microplate in a total volume of 3 mL (1000-fold dilution). The cultures were incubated stationary at 28 °C for 30 h. Results were evaluated qualitatively by monitoring the presence/absence of bacterial biomass formed in the air–liquid contact.

5.10. Extraction of Capsular Polysaccharides

The bacteria were grown on LB agar plates for 16 h and suspended in the PBS buffer. Capsular polysaccharide extraction and analysis were performed as described earlier [38]. Briefly, polysaccharides were released by vortexing for 30 s. After centrifugation at 9000× *g* for 10 min, polysaccharides were precipitated in 75% ice-cold ethanol. The pellet was resuspended in SDS sample buffer and boiled for 5 min. Samples were loaded on the 12% SDS-PAGE gels. After electrophoresis, gels were stained overnight with 0.1% (*w*/*v*) Alcian Blue.

5.11. Serum Resistance

Serum resistance was evaluated by measuring bacterial growth in LB medium, heat inactivated FBS (fetal bovine serum (Gibco, 12657029)), and active FBS. FBS was inactivated by incubation at 56 °C for 30 min with constant shaking. Overnight cultures were inoculated at ×1000 dilution to LB medium and 80% FBS (20% of LB medium) or inactive 80% FBS. Growth curves were measured at 37 °C with periodic shaking every 20 min using a Tecan Infinite M200 Pro microplate reader.

Author Contributions: J.A. and E.S. conceived and designed the experiments; L.K. and J.S. performed the experiments; L.K., J.S. and J.A. analyzed the data; J.A. and L.K. wrote the paper. All authors have read and agreed to the published version of the manuscript.

Acknowledgments: We would like to thank Modestas Ružauskas for *Stenotrophomonas* spp., Jolanta Miciulevičienė, R. Plančiūnienė and K. Sužiedėlis for the clinical *S. maltophilia* isolates.

References

1. Brooke, J.S. Stenotrophomonas maltophilia: An Emerging Global Opportunistic Pathogen. *Clin. Microbiol. Rev.* **2012**, *25*, 2–41. [CrossRef]
2. Ryan, R.P.; Monchy, S.; Cardinale, M.; Taghavi, S.; Crossman, L.; Avison, M.B.; Berg, G.; van der Lelie, D.; Dow, J.M. The versatility and adaptation of bacteria from the genus Stenotrophomonas. *Nat. Rev. Microbiol.* **2009**, *7*, 514–525. [CrossRef]
3. Falagas, M.E.; Kastoris, A.C.; Vouloumanou, E.K.; Rafailidis, P.I.; Kapaskelis, A.M.; Dimopoulos, G. Attributable mortality of Stenotrophomonas maltophilia infections: A systematic review of the literature. *Future Microbiol.* **2009**, *4*, 1103–1109. [CrossRef] [PubMed]
4. WHO | World Health Organization. Available online: https://www.who.int/drugresistance/AMR_Importance/en/ (accessed on 28 July 2020).

5. Pompilio, A.; Savini, V.; Fiscarelli, E.; Gherardi, G.; Di Bonaventura, G. Clonal Diversity, Biofilm Formation, and Antimicrobial Resistance among Stenotrophomonas maltophilia Strains from Cystic Fibrosis and Non-Cystic Fibrosis Patients. *Antibiotics* **2020**, *9*, 15. [CrossRef] [PubMed]
6. Trifonova, A.; Strateva, T. *Stenotrophomonas maltophilia*—A low-grade pathogen with numerous virulence factors. *Infect. Dis.* **2019**, *51*, 168–178. [CrossRef] [PubMed]
7. Adamek, M.; Overhage, J.; Bathe, S.; Winter, J.; Fischer, R.; Schwartz, T. Genotyping of Environmental and Clinical Stenotrophomonas maltophilia Isolates and their Pathogenic Potential. *PLoS ONE* **2011**, *6*, e27615. [CrossRef]
8. Bostanghadiri, N.; Ghalavand, Z.; Fallah, F.; Yadegar, A.; Ardebili, A.; Tarashi, S.; Pournajaf, A.; Mardaneh, J.; Shams, S.; Hashemi, A. Characterization of Phenotypic and Genotypic Diversity of Stenotrophomonas maltophilia Strains Isolated From Selected Hospitals in Iran. *Front. Microbiol.* **2019**, *10*. [CrossRef]
9. Pompilio, A.; Pomponio, S.; Crocetta, V.; Gherardi, G.; Verginelli, F.; Fiscarelli, E.; Dicuonzo, G.; Savini, V.; D'Antonio, D.; Di Bonaventura, G. Phenotypic and genotypic characterization of Stenotrophomonas maltophiliaisolates from patients with cystic fibrosis: Genome diversity, biofilm formation, and virulence. *BMC Microbiol.* **2011**, *11*, 159. [CrossRef]
10. Madi, H.; Lukić, J.; Vasiljević, Z.; Biočanin, M.; Kojić, M.; Jovčić, B.; Lozo, J. Genotypic and Phenotypic Characterization of Stenotrophomonas maltophilia Strains from a Pediatric Tertiary Care Hospital in Serbia. *PLoS ONE* **2016**, *11*, e0165660. [CrossRef]
11. Valdezate, S.; Vindel, A.; Martin-Davila, P.; Del Saz, B.S.; Baquero, F.; Canton, R. High Genetic Diversity among Stenotrophomonas maltophilia Strains Despite Their Originating at a Single Hospital. *J. Clin. Microbiol.* **2004**, *42*, 693–699. [CrossRef]
12. Lira, F.; Berg, G.; Martínez, J.L. Double-Face Meets the Bacterial World: The Opportunistic Pathogen Stenotrophomonas maltophilia. *Front. Microbiol.* **2017**, *8*. [CrossRef] [PubMed]
13. Dzidic, S.; Bedeković, V. Horizontal gene transfer-emerging multidrug resistance in hospital bacteria. *Acta Pharmacol. Sin.* **2003**, *24*, 519–526. [PubMed]
14. Brown, S.P.; Cornforth, D.M.; Mideo, N. Evolution of virulence in opportunistic pathogens: Generalism, plasticity, and control. *Trends Microbiol.* **2012**, *20*, 336–342. [CrossRef] [PubMed]
15. Harms, A.; Brodersen, D.E.; Mitarai, N.; Gerdes, K. Toxins, Targets, and Triggers: An Overview of Toxin-Antitoxin Biology. *Mol. Cell* **2018**, *70*, 768–784. [CrossRef]
16. Lee, K.-Y.; Lee, B.-J. Structure, Biology, and Therapeutic Application of Toxin-Antitoxin Systems in Pathogenic Bacteria. *Toxins* **2016**, *8*, 305. [CrossRef]
17. Yamaguchi, Y.; Park, J.-H.; Inouye, M. Toxin-Antitoxin Systems in Bacteria and Archaea. *Annu. Rev. Genet.* **2011**, *45*, 61–79. [CrossRef]
18. Lobato-Márquez, D.; Díaz-Orejas, R.; García-del Portillo, F. Toxin-antitoxins and bacterial virulence. *FEMS Microbiol. Rev.* **2016**, *40*, 592–609. [CrossRef]
19. Wang, Y.; Wang, H.; Hay, A.J.; Zhong, Z.; Zhu, J.; Kan, B. Functional RelBE-Family Toxin-Antitoxin Pairs Affect Biofilm Maturation and Intestine Colonization in Vibrio cholerae. *PLoS ONE* **2015**, *10*, e0135696. [CrossRef]
20. Wood, T.L.; Wood, T.K. The HigB/HigA toxin/antitoxin system of Pseudomonas aeruginosa influences the virulence factors pyochelin, pyocyanin, and biofilm formation. *Microbiol. Open* **2016**, *5*, 499–511. [CrossRef]
21. Leplae, R.; Geeraerts, D.; Hallez, R.; Guglielmini, J.; Drèze, P.; Van Melderen, L. Diversity of bacterial type II toxin–antitoxin systems: A comprehensive search and functional analysis of novel families. *Nucleic Acids Res.* **2011**, *39*, 5513–5525. [CrossRef]
22. Georgiades, K.; Raoult, D. Genomes of the Most Dangerous Epidemic Bacteria Have a Virulence Repertoire Characterized by Fewer Genes but More Toxin-Antitoxin Modules. *PLoS ONE* **2011**, *6*, e17962. [CrossRef] [PubMed]
23. Sala, A.; Bordes, P.; Genevaux, P. Multiple Toxin-Antitoxin Systems in Mycobacterium tuberculosis. *Toxins* **2014**, *6*, 1002–1020. [CrossRef]
24. Xie, Y.; Wei, Y.; Shen, Y.; Li, X.; Zhou, H.; Tai, C.; Deng, Z.; Ou, H.-Y. TADB 2.0: An updated database of bacterial type II toxin–antitoxin loci. *Nucleic Acids Res.* **2018**, *46*, D749–D753. [CrossRef] [PubMed]
25. Sevin, E.W.; Barloy-Hubler, F. RASTA-Bacteria: A web-based tool for identifying toxin-antitoxin loci in prokaryotes. *Genome Biol.* **2007**, *8*, R155. [CrossRef] [PubMed]

26. Akarsu, H.; Bordes, P.; Mansour, M.; Bigot, D.-J.; Genevaux, P.; Falquet, L. TASmania: A bacterial Toxin-Antitoxin Systems database. *PLoS Comput. Biol.* **2019**, *15*, e1006946. [CrossRef] [PubMed]
27. Butt, A.; Müller, C.; Harmer, N.; Titball, R.W. Identification of type II toxin–antitoxin modules in Burkholderia pseudomallei. *FEMS Microbiol. Lett.* **2013**, *338*, 86–94. [CrossRef]
28. Xu, J.; Zhang, N.; Cao, M.; Ren, S.; Zeng, T.; Qin, M.; Zhao, X.; Yuan, F.; Chen, H.; Bei, W. Identification of Three Type II Toxin-Antitoxin Systems in Streptococcus suis Serotype 2. *Toxins* **2018**, *10*, 467. [CrossRef]
29. Coussens, N.P.; Daines, D.A. Wake me when it's over—Bacterial toxin-antitoxin proteins and induced dormancy. *Exp. Biol. Med.* **2016**, *241*, 1332–1342. [CrossRef]
30. Makarova, K.S.; Wolf, Y.I.; Koonin, E.V. Comprehensive comparative-genomic analysis of type 2 toxin-antitoxin systems and related mobile stress response systems in prokaryotes. *Biol. Direct* **2009**, *4*, 19. [CrossRef]
31. Sánchez, M.B. Antibiotic resistance in the opportunistic pathogen Stenotrophomonas maltophilia. *Front. Microbiol.* **2015**, *6*, 658. [CrossRef]
32. Bjarnsholt, T. The role of bacterial biofilms in chronic infections. *Apmis. Suppl.* **2013**, 1–51. [CrossRef] [PubMed]
33. Di Bonaventura, G.; Spedicato, I.; D'Antonio, D.; Robuffo, I.; Piccolomini, R. Biofilm formation by Stenotrophomonas maltophilia: Modulation by quinolones, trimethoprim-sulfamethoxazole, and ceftazidime. *Antimicrob. Agents Chemother.* **2004**, *48*, 151–160. [CrossRef] [PubMed]
34. Armitano, J.; Méjean, V.; Jourlin-Castelli, C. Gram-negative bacteria can also form pellicles. *Environ. Microbiol. Rep.* **2014**, *6*, 534–544. [CrossRef] [PubMed]
35. Cress, B.F.; Englaender, J.A.; He, W.; Kasper, D.; Linhardt, R.J.; Koffas, M.A.G. Masquerading microbial pathogens: Capsular polysaccharides mimic host-tissue molecules. *FEMS Microbiol. Rev.* **2014**, *38*, 660–697. [CrossRef]
36. Niu, T.; Guo, L.; Luo, Q.; Zhou, K.; Yu, W.; Chen, Y.; Huang, C.; Xiao, Y. Wza gene knockout decreases Acinetobacter baumannii virulence and affects Wzy-dependent capsular polysaccharide synthesis. *Virulence* **2020**, *11*, 1–13. [CrossRef]
37. Geisinger, E.; Isberg, R.R. Antibiotic modulation of capsular exopolysaccharide and virulence in Acinetobacter baumannii. *PLoS Pathog.* **2015**, *11*, e1004691. [CrossRef]
38. Skerniškytė, J.; Krasauskas, R.; Péchoux, C.; Kulakauskas, S.; Armalytė, J.; Sužiedėlienė, E. Surface-Related Features and Virulence Among Acinetobacter baumannii Clinical Isolates Belonging to International Clones I and II. *Front. Microbiol.* **2018**, *9*, 3116. [CrossRef]
39. Van Melderen, L.; De Bast, M.S. Bacterial Toxin–Antitoxin Systems: More Than Selfish Entities? *PLoS Genet.* **2009**, *5*, e1000437. [CrossRef]
40. Rowe-Magnus, D.A. Comparative Analysis of Superintegrons: Engineering Extensive Genetic Diversity in the Vibrionaceae. *Genome Res.* **2003**, *13*, 428–442. [CrossRef]
41. Szekeres, S.; Dauti, M.; Wilde, C.; Mazel, D.; Rowe-Magnus, D.A. Chromosomal toxin-antitoxin loci can diminish large-scale genome reductions in the absence of selection: Chromosomal addiction loci minimize gene loss. *Mol. Microbiol.* **2007**, *63*, 1588–1605. [CrossRef]
42. Gotfredsen, M.; Gerdes, K. The Escherichia coli relBE genes belong to a new toxin-antitoxin gene family. *Mol. Microbiol.* **1998**, *29*, 1065–1076. [CrossRef] [PubMed]
43. Schureck, M.A.; Repack, A.; Miles, S.J.; Marquez, J.; Dunham, C.M. Mechanism of endonuclease cleavage by the HigB toxin. *Nucleic Acids Res.* **2016**, *44*, 7944–7953. [CrossRef] [PubMed]
44. Makarova, K.S.; Grishin, N.V.; Koonin, E.V. The HicAB cassette, a putative novel, RNA-targeting toxin-antitoxin system in archaea and bacteria. *Bioinformatics* **2006**, *22*, 2581–2584. [CrossRef] [PubMed]
45. Zhang, Y.; Zhu, L.; Zhang, J.; Inouye, M. Characterization of ChpBK, an mRNA interferase from Escherichia coli. *J. Biol. Chem.* **2005**, *280*, 26080–26088. [CrossRef] [PubMed]
46. Winther, K.S.; Gerdes, K. Enteric virulence associated protein VapC inhibits translation by cleavage of initiator tRNA. *Proc. Natl. Acad. Sci. USA* **2011**, *108*, 7403–7407. [CrossRef]
47. Tiwari, P.; Arora, G.; Singh, M.; Kidwai, S.; Narayan, O.P.; Singh, R. MazF ribonucleases promote Mycobacterium tuberculosis drug tolerance and virulence in guinea pigs. *Nat. Commun.* **2015**, *6*, 1–13. [CrossRef]
48. Wen, W.; Liu, B.; Xue, L.; Zhu, Z.; Niu, L.; Sun, B. Autoregulation and Virulence Control by the Toxin-Antitoxin System SavRS in Staphylococcus aureus. *Infect. Immun.* **2018**, *86*. [CrossRef]

49. Ramage, H.R.; Connolly, L.E.; Cox, J.S. Comprehensive Functional Analysis of Mycobacterium tuberculosis Toxin-Antitoxin Systems: Implications for Pathogenesis, Stress Responses, and Evolution. *PLoS Genet.* **2009**, *5*. [CrossRef]
50. Frampton, R.; Aggio, R.B.M.; Villas-Bôas, S.G.; Arcus, V.L.; Cook, G.M. Toxin-Antitoxin Systems of Mycobacterium smegmatis Are Essential for Cell Survival. *J. Biol. Chem.* **2012**, *287*, 5340–5356. [CrossRef]
51. Lopes, A.P.Y.; Lopes, L.M.; Fraga, T.R.; Chura-Chambi, R.M.; Sanson, A.L.; Cheng, E.; Nakajima, E.; Morganti, L.; Martins, E.A.L. VapC from the Leptospiral VapBC Toxin-Antitoxin Module Displays Ribonuclease Activity on the Initiator tRNA. *PLoS ONE* **2014**, *9*, e101678. [CrossRef]
52. Chan, W.T.; Espinosa, M.; Yeo, C.C. Keeping the Wolves at Bay: Antitoxins of Prokaryotic Type II Toxin-Antitoxin Systems. *Front. Mol. Biosci.* **2016**, *3*. [CrossRef]
53. Strateva, T.; Trifonova, A.; Savov, E.; Dimov, S. An update on the antimicrobial susceptibility and molecular epidemiology of Stenotrophomonas maltophilia in Bulgaria: A 5-year study (2011–2016). *Infect. Dis.* **2019**, *51*, 387–391. [CrossRef] [PubMed]
54. Travassos, L.H.; Pinheiro, M.N.; Coelho, F.S.; Sampaio, J.L.M.; Merquior, V.L.C.; Marques, E.A. Phenotypic properties, drug susceptibility and genetic relatedness of Stenotrophomonas maltophilia clinical strains from seven hospitals in Rio de Janeiro, Brazil. *J. Appl. Microbiol.* **2004**, *96*, 1143–1150. [CrossRef] [PubMed]
55. Schneider, B.; Weigel, W.; Sztukowska, M.; Demuth, D.R. Identification and functional characterization of type II toxin/antitoxin systems in Aggregatibacter actinomycetemcomitans. *Mol. Oral Microbiol.* **2018**, *33*, 224–233. [CrossRef] [PubMed]
56. Bliven, K.A.; Maurelli, A.T. Antivirulence Genes: Insights into Pathogen Evolution through Gene Loss. *Infect. Immun.* **2012**, *80*, 4061–4070. [CrossRef] [PubMed]
57. Lima-Mendez, G.; Alvarenga, D.O.; Ross, K.; Hallet, B.; Melderen, L.V.; Varani, A.M.; Chandler, M. Toxin-Antitoxin Gene Pairs Found in Tn3 Family Transposons Appear To Be an Integral Part of the Transposition Module. *mBio* **2020**, *11*. [CrossRef]
58. Hu, L.-F.; Chang, X.; Ye, Y.; Wang, Z.-X.; Shao, Y.-B.; Shi, W.; Li, X.; Li, J.-B. Stenotrophomonas maltophilia resistance to trimethoprim/sulfamethoxazole mediated by acquisition of sul and dfrA genes in a plasmid-mediated class 1 integron. *Int. J. Antimicrob. Agents* **2011**, *37*, 230–234. [CrossRef]
59. Walsh, T.R.; Macgowan, A.P.; Bennett, P.M. Sequence Analysis and Enzyme Kinetics of the L2 Serine β-Lactamase from Stenotrophomonas maltophilia. *Antimicrob. Agents Chemother.* **1997**, *41*, 5. [CrossRef]
60. Yang, Z.; Liu, W.; Cui, Q.; Niu, W.; Li, H.; Zhao, X.; Wei, X.; Wang, X.; Huang, S.; Dong, D.; et al. Prevalence and detection of Stenotrophomonas maltophilia carrying metallo-Î²-lactamase blaL1 in Beijing, China. *Front. Microbiol.* **2014**, *5*. [CrossRef]
61. Youenou, B.; Favre-Bonté, S.; Bodilis, J.; Brothier, E.; Dubost, A.; Muller, D.; Nazaret, S. Comparative Genomics of Environmental and Clinical Stenotrophomonas maltophilia Strains with Different Antibiotic Resistance Profiles. *Genome Biol. Evol.* **2015**, *7*, 2484–2505. [CrossRef]
62. Deredjian, A.; Alliot, N.; Blanchard, L.; Brothier, E.; Anane, M.; Cambier, P.; Jolivet, C.; Khelil, M.N.; Nazaret, S.; Saby, N.; et al. Occurrence of Stenotrophomonas maltophilia in agricultural soils and antibiotic resistance properties. *Res. Microbiol.* **2016**, *167*, 313–324. [CrossRef] [PubMed]
63. Berg, G.; Eberl, L.; Hartmann, A. The rhizosphere as a reservoir for opportunistic human pathogenic bacteria. *Environ. Microbiol.* **2005**, *7*, 1673–1685. [CrossRef] [PubMed]
64. Muder, R.R.; Harris, A.P.; Muller, S.; Edmond, M.; Chow, J.W.; Papadakis, K.; Wagener, M.W.; Bodey, G.P.; Steckelberg, J.M. Bacteremia due to Stenotrophomonas (Xanthomonas) maltophilia: A prospective, multicenter study of 91 episodes. *Clin. Infect. Dis. Off. Publ. Infect. Dis. Soc. Am.* **1996**, *22*, 508–512. [CrossRef] [PubMed]
65. Khan, I.A.; Mehta, N.J. Stenotrophomonas maltophilia endocarditis: A systematic review. *Angiology* **2002**, *53*, 49–55. [CrossRef]
66. Horio, N.; Horiguchi, M.; Murakami, K.; Yamamoto, E.; Miyake, Y. Stenotrophomonas maltophilia endophthalmitis after intraocular lens implantation. *Graefes Arch. Clin. Exp. Ophthalmol. Albrecht Von Graefes Arch. Klin. Exp. Ophthalmol.* **2000**, *238*, 299–301. [CrossRef]
67. de Vidipó, L.A.; de Marques, E.A.; Puchelle, E.; Plotkowski, M.-C. Stenotrophomonas maltophilia Interaction with Human Epithelial Respiratory Cells In Vitro. *Microbiol. Immunol.* **2001**, *45*, 563–569. [CrossRef]

68. De Oliveira-Garcia, D.; Dall'Agnol, M.; Rosales, M.; Azzuz, A.C.G.S.; Alcántara, N.; Martinez, M.B.; Girón, J.A. Fimbriae and adherence of Stenotrophomonas maltophilia to epithelial cells and to abiotic surfaces. *Cell. Microbiol.* **2003**, *5*, 625–636. [CrossRef]
69. Zhuo, C.; Zhao, Q.; Xiao, S. The Impact of spgM, rpfF, rmlA Gene Distribution on Biofilm Formation in Stenotrophomonas maltophilia. *PLoS ONE* **2014**, *9*, e108409. [CrossRef]
70. Branda, S.S.; Vik, Å.; Friedman, L.; Kolter, R. Biofilms: The matrix revisited. *Trends Microbiol.* **2005**, *13*, 20–26. [CrossRef]
71. Yamamoto, K.; Arai, H.; Ishii, M.; Igarashi, Y. Trade-off between oxygen and iron acquisition in bacterial cells at the air–liquid interface. *FEMS Microbiol. Ecol.* **2011**, *77*, 83–94. [CrossRef]
72. Campa, M.; Bendinelli, M.; Friedman, H. *Pseudomonas Aeruginosa as an Opportunistic Pathogen*; Springer Science & Business Media: New York, NY, USA, 2012; ISBN 978-1-4615-3036-7.
73. Tipton, K.A.; Dimitrova, D.; Rather, P.N. Phase-Variable Control of Multiple Phenotypes in Acinetobacter baumannii Strain AB5075. *J. Bacteriol.* **2015**, *197*, 2593–2599. [CrossRef]
74. Chin, C.Y.; Tipton, K.A.; Farokhyfar, M.; Burd, E.M.; Weiss, D.S.; Rather, P.N. A high-frequency phenotypic switch links bacterial virulence and environmental survival in Acinetobacter baumannii. *Nat. Microbiol.* **2018**, *3*, 563–569. [CrossRef] [PubMed]
75. Boulnois, G.J.; Roberts, I.S. Genetics of Capsular Polysaccharide Production in Bacteria. In *Proceedings of the Bacterial Capsules*; Jann, K., Jann, B., Eds.; Springer: Berlin/Heidelberg, Germany, 1990; pp. 1–18.
76. McKay, G.A.; Woods, D.E.; MacDonald, K.L.; Poole, K. Role of phosphoglucomutase of Stenotrophomonas maltophilia in lipopolysaccharide biosynthesis, virulence, and antibiotic resistance. *Infect. Immun.* **2003**, *71*, 3068–3075. [CrossRef] [PubMed]
77. Altschul, S.F.; Gish, W.; Miller, W.; Myers, E.W.; Lipman, D.J. Basic local alignment search tool. *J. Mol. Biol.* **1990**, *215*, 403–410. [CrossRef]
78. Sievers, F.; Higgins, D.G. Clustal Omega, accurate alignment of very large numbers of sequences. *Methods Mol. Biol.* **2014**, *1079*, 105–116. [CrossRef] [PubMed]
79. Nishimura, A.; Morita, M.; Nishimura, Y.; Sugino, Y. A rapid and highly efficient method for preparation of competent Escherichia coli cells. *Nucleic Acids Res.* **1990**, *18*, 6169. [CrossRef]
80. Ye, X.; Dong, H.; Huang, Y.-P. Highly efficient transformation of Stenotrophomonas maltophilia S21, an environmental isolate from soil, by electroporation. *J. Microbiol. Methods* **2014**, *107*, 92–97. [CrossRef]
81. Yanisch-Perron, C.; Vieira, J.; Messing, J. Improved M13 phage cloning vectors and host strains: Nucleotide sequences of the M13mpl8 and pUC19 vectors. *Gene* **1985**, *33*, 103–119. [CrossRef]
82. Motiejūnaitė, R.; Armalytė, J.; Markuckas, A.; Sužiedėlienė, E. *Escherichia coli dinJ-yafQ* genes act as a toxin-antitoxin module. *FEMS Microbiol. Lett.* **2007**, *268*, 112–119. [CrossRef]
83. Guzman, L.M.; Belin, D.; Carson, M.J.; Beckwith, J. Tight regulation, modulation, and high-level expression by vectors containing the arabinose PBAD promoter. *J. Bacteriol.* **1995**, *177*, 4121–4130. [CrossRef]
84. Petkevičius, V.; Vaitekūnas, J.; Stankevičiūtė, J.; Gasparavičiūtė, R.; Meškys, R. Catabolism of 2-Hydroxypyridine by *Burkholderia* sp. Strain MAK1: A 2-Hydroxypyridine 5-Monooxygenase Encoded by *hpdABCDE* Catalyzes the First Step of Biodegradation. *Appl. Environ. Microbiol.* **2018**, *84*. [CrossRef] [PubMed]
85. Armalytė, J.; Skerniškytė, J.; Bakienė, E.; Krasauskas, R.; Šiugždinienė, R.; Kareivienė, V.; Kerzienė, S.; Klimienė, I.; Sužiedėlienė, E.; Ružauskas, M. Microbial Diversity and Antimicrobial Resistance Profile in Microbiota From Soils of Conventional and Organic Farming Systems. *Front. Microbiol.* **2019**, *10*, 892. [CrossRef] [PubMed]
86. Jurėnaitė, M.; Markuckas, A.; Sužiedėlienė, E. Identification and Characterization of Type II Toxin-Antitoxin Systems in the Opportunistic Pathogen Acinetobacter baumannii. *J. Bacteriol.* **2013**, *195*, 3165–3172. [CrossRef] [PubMed]
87. Krasauskas, R.; Skerniškytė, J.; Armalytė, J.; Sužiedėlienė, E. The role of Acinetobacter baumannii response regulator BfmR in pellicle formation and competitiveness via contact-dependent inhibition system. *BMC Microbiol.* **2019**, *19*, 241. [CrossRef]

6

Role of $p38_{alpha/beta}$ MAP Kinase in Cell Susceptibility to *Clostridium sordellii* Lethal Toxin and *Clostridium difficile* Toxin B

Ilona Schelle, Janina Bruening, Mareike Buetepage and Harald Genth *

Institute for Toxicology, Hannover Medical School, Carl-Neuberg-Str. 1, D-30625 Hannover, Germany; ilona.schelle@t-online.de (I.S.); janina.bruening@twincore.de (J.B.); mbuetepage@ukaachen.de (M.B.)

* Correspondence: genth.harald@mh-hannover.de.

Abstract: Lethal Toxin from *Clostridium sordellii* (TcsL), which is casually involved in the toxic shock syndrome and in gas gangrene, enters its target cells by receptor-mediated endocytosis. Inside the cell, TcsL mono-O-glucosylates and thereby inactivates Rac/Cdc42 and Ras subtype GTPases, resulting in actin reorganization and an activation of p38 MAP kinase. While a role of p38 MAP kinase in TcsL-induced cell death is well established, data on a role of p38 MAP kinase in TcsL-induced actin reorganization are not available. In this study, TcsL-induced Rac/Cdc42 glucosylation and actin reorganization are differentially analyzed in $p38_{alpha}{}^{-/-}$ MSCV empty vector MEFs and the corresponding cell line with reconstituted $p38_{alpha}$ expression ($p38_{alpha}{}^{-/-}$ MSCV $p38_{alpha}$ MEFs). Genetic deletion of $p38_{alpha}$ results in reduced susceptibility of cells to TcsL-induced Rac/Cdc42 glucosylation and actin reorganization. Furthermore, SB203580, a pyridinyl imidazole inhibitor of $p38_{alpha/beta}$ MAP kinase, also protects cells from TcsL-induced effects in both $p38^{-/-}$ MSCV empty vector MEFs and in $p38_{alpha}{}^{-/-}$ MSCV $p38_{alpha}$ MEFs, suggesting that inhibition of $p38_{beta}$ contributes to the protective effect of SB203580. In contrast, the effects of the related *C. difficile* Toxin B are responsive neither to SB203580 treatment nor to $p38_{alpha}$ deletion. In conclusion, the protective effects of SB203580 and of $p38_{alpha}$ deletion are likely not based on inhibition of the toxins' glucosyltransferase activity rather than on inhibited endocytic uptake of specifically TcsL into target cells.

Keywords: endocytosis; Ras; *C. difficile* Toxin B; mono-O-glucosylation; p21-activated kinase; actin

---

1. Introduction

Toxin-producing strains of *C. difficile* and *C. sordellii* cause intestinal infections, including *C. difficile*-associated diarrhea (CDAD) in humans and horses, and *C. sordellii*-induced hemorrhagic enteritis and enterotoxemia in cattle, sheep, and other ruminants [1–4]. The major virulence factors involved in these infections are toxin A (TcdA) and toxin B (TcdB) of *C. difficile*, and lethal toxin (TcsL) and hemorrhagic toxin (TcsH) from *C. sordellii*. These single chained toxins exhibit an AB-like toxin structure with the C-terminal delivery domain mediating cell entry of the N-terminal glucosyltransferase domain by receptor-mediated endocytosis [5,6]. The endocytosed glucosyltransferase domain associates with membrane phosphatidylserine facilitating mono-O-glucosylation of small GTPases of the Rho and Ras subfamilies in a monovalent and divalent metal ion-dependent manner [7–9]. The acceptor amino acid of toxin-catalyzed mono-O-glucosylation is Thr-35 in Rac1, Cdc42, and (H/K/N)Ras and Thr-37 in Rho(A/B/C). Mono-O-glucosylated Rho/Ras GTPases are incapable of coupling to their regulatory and effector protein and thus are functionally inactive [10–13]. Treatment of cultured cells with the glucosylating toxins results in a breakdown of the actin-based cytoskeleton ("cytopathic effect") and (at higher toxin concentrations) in cell death ("cytotoxic effect") based on inhibition of Rho/Ras-dependent signaling pathways regulating actin dynamics [14], cell-matrix binding [15,16], cell cycle progression, and cell survival [17–19].

The family of p38 MAPKs encompasses the four isoforms $p38_{alpha}$, $p38_{beta}$, $p38_{gamma}$, and $p38_{delta}$. $p38_{alpha}$ and $p38_{beta}$ are ubiquitously expressed, while $p38_{gamma}$ and $p38_{delta}$ exhibit a more restricted expression patterns. The best characterized isoform is $p38_{alpha}$, which is involved in the regulation of the main cellular functions, including actin dynamics, differentiation, and cell death and survival [20,21]. TcsL (as well as TcdA and TcdB) has been shown to activate the family of mitogen-activated protein kinases (MAPKs) involving ERKs (extracellular signal-regulated kinases), JNKs (c-Jun-*N*-terminal kinases) and p38 MAPKs [22]. MAP kinase signaling cascades have been shown to be involved in the cytopathic as well as the cytotoxic effects of the glucosylating toxins. In particular, TcsL-induced activation of JNK has been suggested to facilitate TcsL-catalyzed GTPase substrate glucosylation and subsequently TcsL-induced actin reorganization [22]. $p38_{alpha}$ MAP kinase signaling has been implicated in TcsL-induced expression of the cell death-regulating GTPase RhoB [23]. Finally, TcsL-induced cell death has been shown to be responsive to inhibition by SB203580, a pyridinyl imidazole inhibitor of $p38_{alpha/beta}$ MAP kinase, suggesting a role of $p38_{alpha/beta}$ in TcsL-induced cell death [24].

In this study, TcsL-induced Rac/Cdc42 glucosylation and subsequent actin reorganization were analyzed for an involvement of $p38_{alpha}$ in murine embryonic fibroblasts (MEFs). In this study, TcsL-induced Rac/Cdc42 glucosylation and actin reorganization are differentially analyzed in $p38_{alpha}{}^{-/-}$ murine stem cell virus empty vector MEFs ($p38_{alpha}{}^{-/-}$ MSCV EV MEFs) and the corresponding cell line with reconstituted $p38_{alpha}$ expression ($p38_{alpha}{}^{-/-}$ MSCV $p38_{alpha}$ MEFs) [25,26]. We show that genetic deletion of $p38_{alpha}$ as well as SB203580 protects cells from TcsL-catalyzed glucosylation of Rac/Cdc42 subtype GTPases and TcsL-induced actin reorganization.

2. Results

2.1. Prevention of TcsL-Induced Actin Re-Organization upon Inhibition of $p38_{alpha/beta}$

TcsL time-dependently induced actin reorganization in $p38_{alpha}$-proficient $p38_{alpha}{}^{-/-}$ MSCV $p38_{alpha}$ MEFs (Figure 1), with TcsL treatment for 4 h being sufficient for almost complete cell rounding (Figure 1). In contrast, TcsL-induced rounding of $p38_{alpha}$-deficient $p38_{alpha}{}^{-/-}$ MSCV EV MEFs was clearly delayed, suggesting a reduced susceptibility of $p38_{alpha}{}^{-/-}$ MSCV EV MEFs to TcsL. Next, TcsL-induced cell rounding was analyzed in $p38_{alpha}{}^{-/-}$ MSCV $p38_{alpha}$ MEFs treated with SB203580, a pyridinyl imidazole inhibitor of $p38_{alpha/beta}$ MAP kinase [27]. SB203080 concentration-dependently reduced TcsL-induced cell rounding, with a SB203580 concentration of 10 μM being sufficient for almost complete prevention of TcsL-induced cell rounding (Figure 2A,B). A pronounced protective effect of SB203580 was also observed in a time-dependent experiment (Figure 2C). SB203580 (10 μM) alone did not change fibroblast morphology (Figure 2A). Finally, the protective effect of $p38_{alpha}$ inhibition was analyzed in TcsL concentration-dependent experiments (Figure 3). Either genetic deletion of $p38_{alpha}$ or SB203580 treatment delayed TcsL-induced cell rounding. Interestingly, SB203580 treatment of $p38_{alpha}$-deficient $p38_{alpha}{}^{-/-}$ MSCV EV MEFs further delayed TcsL-induced cell rounding, suggesting a role of $p38_{beta}$ inhibition in the reduced susceptibility to TcsL.

TcdB is highly related to TcsL (identity of 75% at amino acid level), as both TcdB and TcsL enter their target cells by receptor-mediated endocytosis and both cause changed actin dynamics by mono-O-glucosylation of small GTPases. Interestingly, neither genetic deletion (Figure 4A,B) nor treatment with the $p38_{alpha/beta}$ inhibitor SB203580 (Figure 4C,D) changed the kinetics of TcdB-induced actin reorganization. $p38_{alpha/beta}$ inhibition thus mediates protection of fibroblasts from TcsL-(not TcdB-) induced actin reorganization.

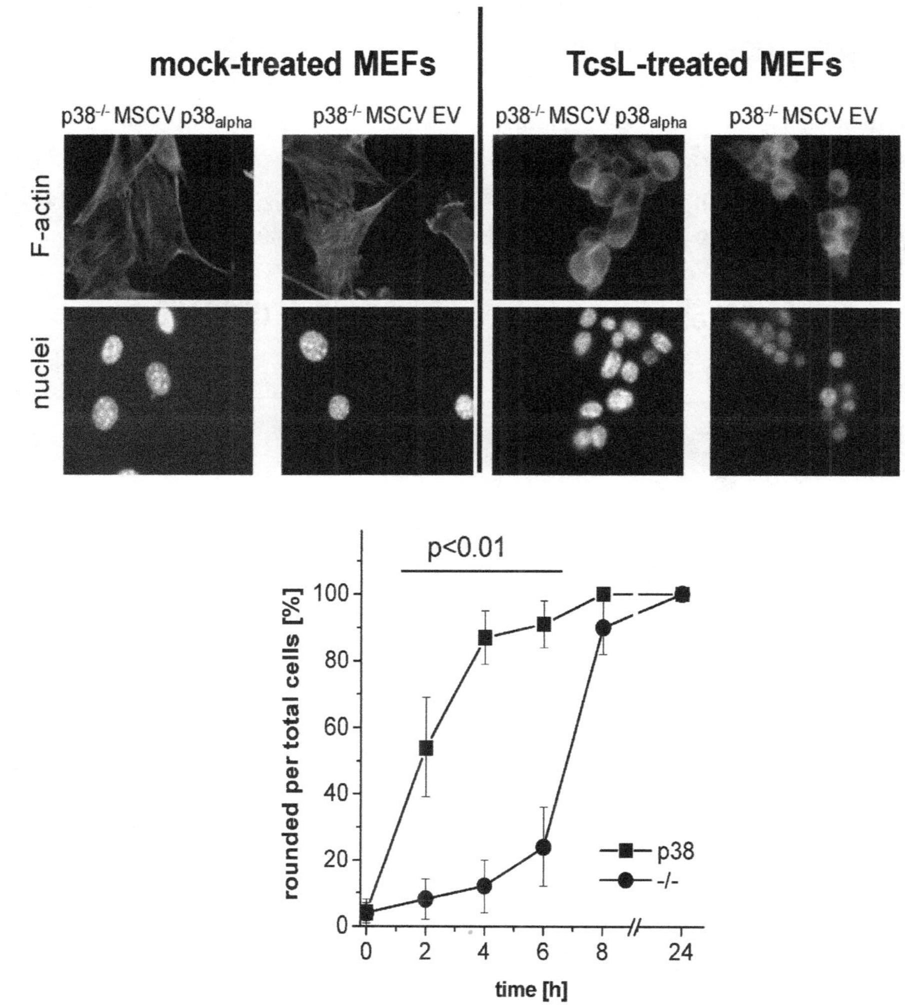

Figure 1. Effects of genetic deletion of $p38_{alpha}$ on TcsL-induced changes of cell morphology (time-dependency). $p38_{alpha}{}^{-/-}$ MSCV empty vector (EV) MEFs and the corresponding cell line with reconstituted $p38_{alpha}$ expression ($p38_{alpha}{}^{-/-}$ MSCV $p38_{alpha}$ MEFs) were treated with TcsL (1 μg/mL) for the indicated times. Cells were then washed, fixed, permeabilized, and stained with rhodamine-phalloidin and DAPI. Cell morphology was visualized using fluorescence microscopy (20× amplification). TcsL-induced changes of the morphology were time-dependently quantified in terms of the number of rounded per total cells. Six representative microscopic fields were chosen and 300 cells total were counted for characteristic cell rounding. Values are the mean ± SD from three independent experiments performed in triplicates. $p < 0.01$ indicates significant differences comparing $p38_{alpha}$-proficient with $p38_{alpha}$-deficient cells using Student's *t*-test.

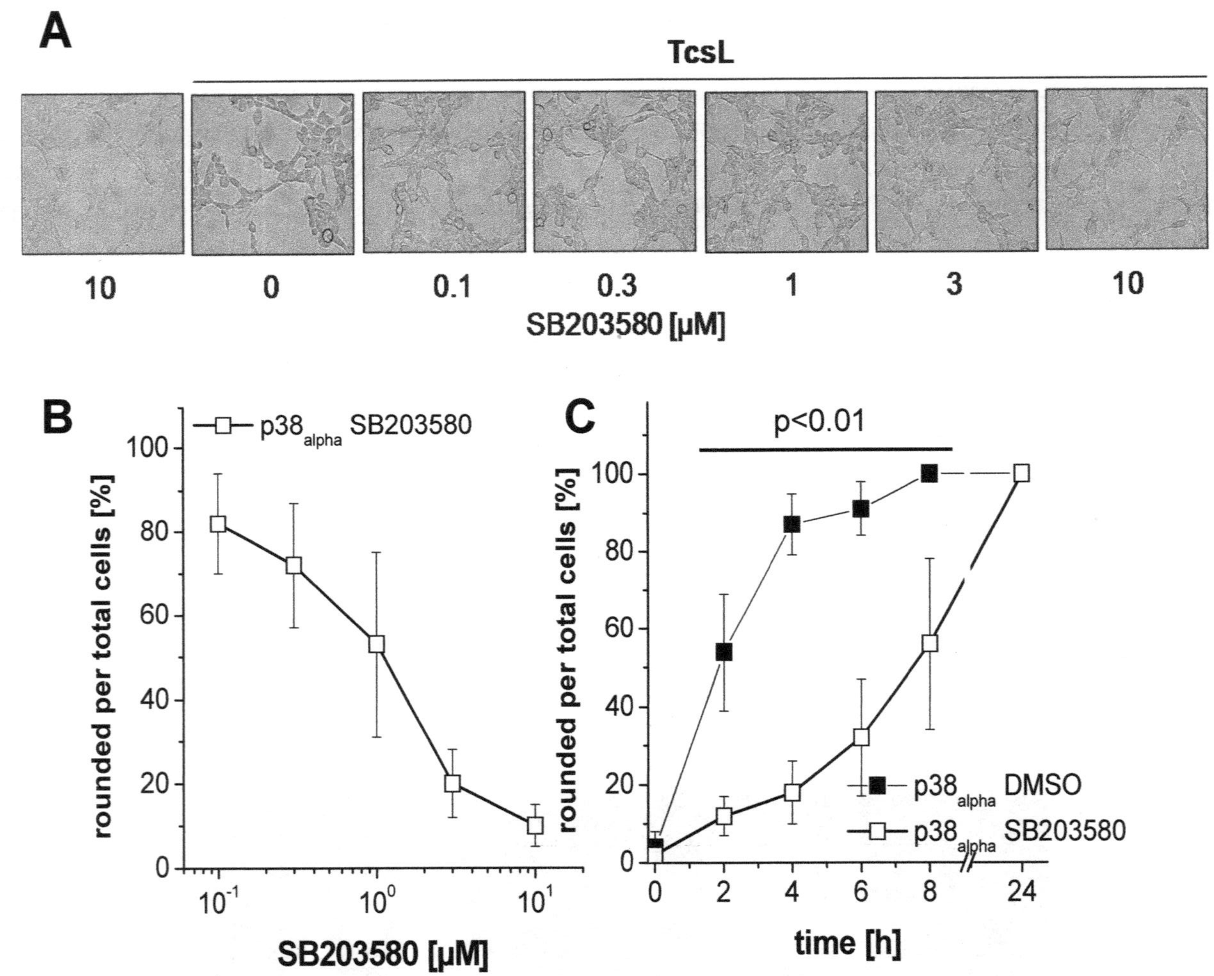

Figure 2. Effects of SB203580 treatment on TcsL-induced changes of cell morphology: (**A**) $p38_{alpha}{}^{-/-}$ MSCV $p38_{alpha}$ MEFs were treated with TcsL (1 μg/mL) or buffer in the presence of the indicated concentrations of SB203580 for 4 h. Cell morphology was visualized using phase contrast microscopy (10× amplification). (**B**) TcsL-induced changes of the morphology were quantified in terms of the number of rounded per total cells. Six representative microscopic fields were chosen and 300 cells total were counted for characteristic cell rounding. Values are the mean ± SD from three independent experiments performed in triplicates. (**C**) $p38_{alpha}{}^{-/-}$ MSCV $p38_{alpha}$ MEFs were treated with TcsL (1 μg/mL) in the presence of SB203580 (10 μM) or DMSO alone for the indicated times. TcsL-induced changes of the morphology were quantified in terms of the number of rounded per total cells. Values are the mean ± SD from three independent experiments performed in triplicates. $p < 0.01$ indicates significant differences comparing SB203580-treated with DMSO-treated cells using Student's *t*-test.

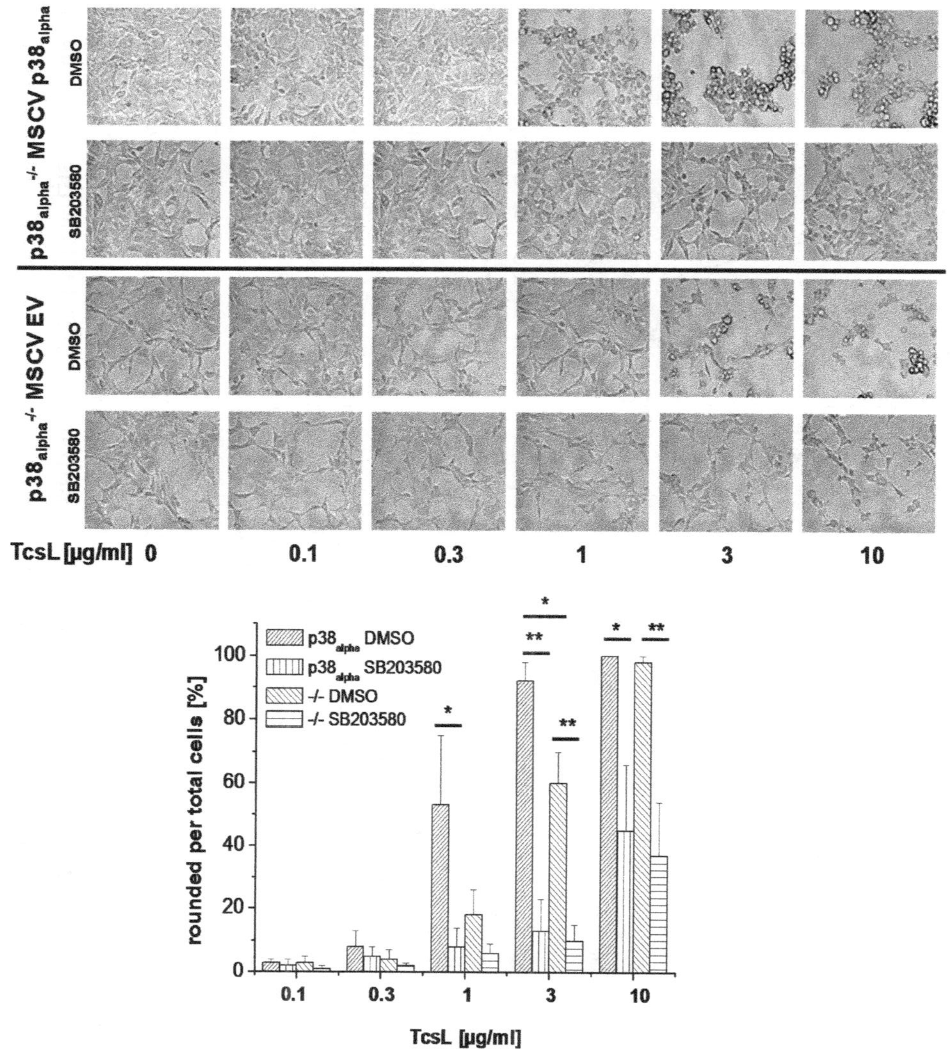

Figure 3. Effects of genetic deletion of $p38_{alpha}$ and of SB203580 treatment on TcsL-induced changes of cell morphology (concentration-dependency). $p38_{alpha}{}^{-/-}$ MSCV $p38_{alpha}$ MEFs and $p38_{alpha}{}^{-/-}$ MSCV empty vector (EV) MEFs were treated with the indicated concentrations of TcsL in the presence of SB203580 (10 µM) or DMSO alone for 4 h. Cell morphology was visualized using phase contrast microscopy (10× amplification). TcsL-induced changes of the morphology of $p38^{-/-}$ MSCV $p38_{alpha}$ MEFs and of $p38^{-/-}$ MSCV empty vector MEFs were quantified in terms of the number of rounded per total cells. Six representative microscopic fields were chosen and 300 cells total were counted for characteristic cell rounding. Values are the mean ± SD from three independent experiments performed in triplicates. */** indicates significant differences, $p < 0.05/p < 0.005$ as analyzed using Student's *t*-test.

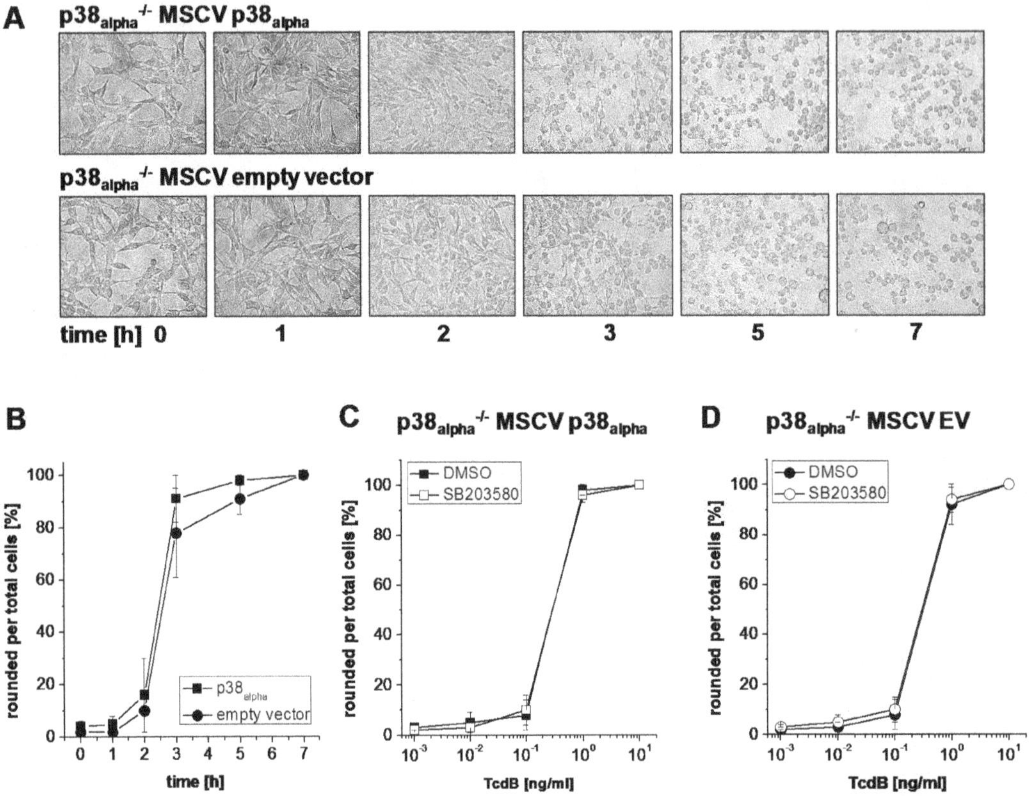

Figure 4. Effects of genetic deletion of $p38_{alpha}$ and of SB203580 treatment on TcdB-induced changes of cell morphology: (**A**,**B**) $p38_{alpha}{}^{-/-}$ MSCV $p38_{alpha}$ MEFs and $p38_{alpha}{}^{-/-}$ MSCV empty vector (EV) MEFs were treated with TcdB (1 ng/mL) for the indicated times; and (**C**,**D**) $p38_{alpha}{}^{-/-}$ MSCV $p38_{alpha}$ MEFs and $p38_{alpha}{}^{-/-}$ MSCV empty vector MEFs were treated with the indicated concentrations of TcdB in the presence of SB203580 (10 μM) or DMSO alone for 4 h. Cell morphology was visualized using phase contrast microscopy (10× amplification). TcdB-induced changes of the morphology were quantified in terms of the number of rounded per total cells. Six representative microscopic fields were chosen and 300 cells total were counted for characteristic cell rounding. Values are the mean ± SD from three independent experiments performed in triplicates.

2.2. Prevention of TcsL-Induced Glucosylation of Rac/Cdc42 Subtype GTPases upon Inhibition of $p38_{alpha/beta}$

Actin reorganization has been attributed to TcsL-/TcdB-catalyzed glucosylation of Rac/Cdc42 subtype GTPases and the subsequent loss of cell-matrix binding [15,28,29]. For the analysis of TcsL-catalyzed glucosylation of Rac/Cdc42, lysates from $p38_{alpha}{}^{-/-}$ MSCV $p38_{alpha}$ MEFs were analyzed by immunoblot analysis using the Rac1(clone 102) antibody. This antibody is specific for non-glucosylated Rac/Cdc42 subtype GTPases [30,31]. Once Rac/Cdc42 subtype GTPases are glucosylated by TcsL, the antibody does not detect its epitope, resulting in a lost signal. TcsL-treated $p38_{alpha}{}^{-/-}$ MSCV $p38_{alpha}$ MEFs exhibited time-dependent glucosylation of Rac/Cdc42 subtype GTPases (Figure 5A). The cellular level of Rac1 was not changed upon TcsL treatment (as analyzed using the Rac1(clone 23A8) antibody) (Figure 5A), confirming that decreasing detection of Rac/Cdc42 subtype GTPases by the Rac1(Mab 102) antibody was due to glucosylation but not due to degradation. TcsL-catalyzed Rac/Cdc42 glucosylation results in dephosphorylation of the FA component p21-associated kinase1/2 (PAK1/2), which is a Rac/Cdc42 effector protein [15,16]. TcsL treatment time-dependently resulted in decreasing levels of pS144/141-PAK1/2 (Figure 5A), indicating PAK1/2 deactivation [32]. The levels of total PAK2 also time-dependently decreased in TcsL-treated cells (Figure 5), showing that PAK1/2 deactivation was based on both dephosphorylation and degradation. TcsL-catalyzed Rac/Cdc42 glucosylation (i.e., inactivation) was thus reflected by deactivation of its effector kinase PAK1/2. Finally, p38 MAP kinase activity was tracked in terms of phosphorylation of its downstream target MAPKAPK2. TcsL treatment resulted in a transient increase of pT222-MAPKAPK2, indicating transient MAPKAPK2 activation. In $p38_{alpha}$-deficient $p38^{-/-}$ MSCV empty vector MEFs, in which MAPKAPK2 activation was hardly observed, Rac/Cdc42 glucosylation was clearly delayed as compared with $p38_{alpha}{}^{-/-}$ MSCV $p38_{alpha}$ MEFs (Figure 5). In a TcsL concentration-dependent experiment, TcsL turned out to be at least three fold more potent in inducing Rac/Cdc42 glucosylation (Figure 6) and PAK1/2 deactivation (Figure 6) in $p38^{-/-}$ MSCV $p38_{alpha}$ MEFs as compared with $p38^{-/-}$ MSCV empty vector MEFs. In contrast to TcsL, the kinetics of TcdB-induced Rac/Cdc42 glucosylation and of PAK1/2 deactivation were comparable in $p38_{alpha}{}^{-/-}$ MSCV $p38_{alpha}$ and $p38^{-/-}$ MSCV empty vector MEFs (Figure 7). TcdB-catalyzed Rac/Cdc42 glucosylation was thus not susceptible to genetic $p38_{alpha}$ inhibition (Figure 7), consistent with above observations on TcdB-induced actin reorganization (Figure 4). These observations show that genetic $p38_{alpha}$ inhibition protects fibroblasts specifically from TcsL-(not TcdB-) induced cytopathic effects.

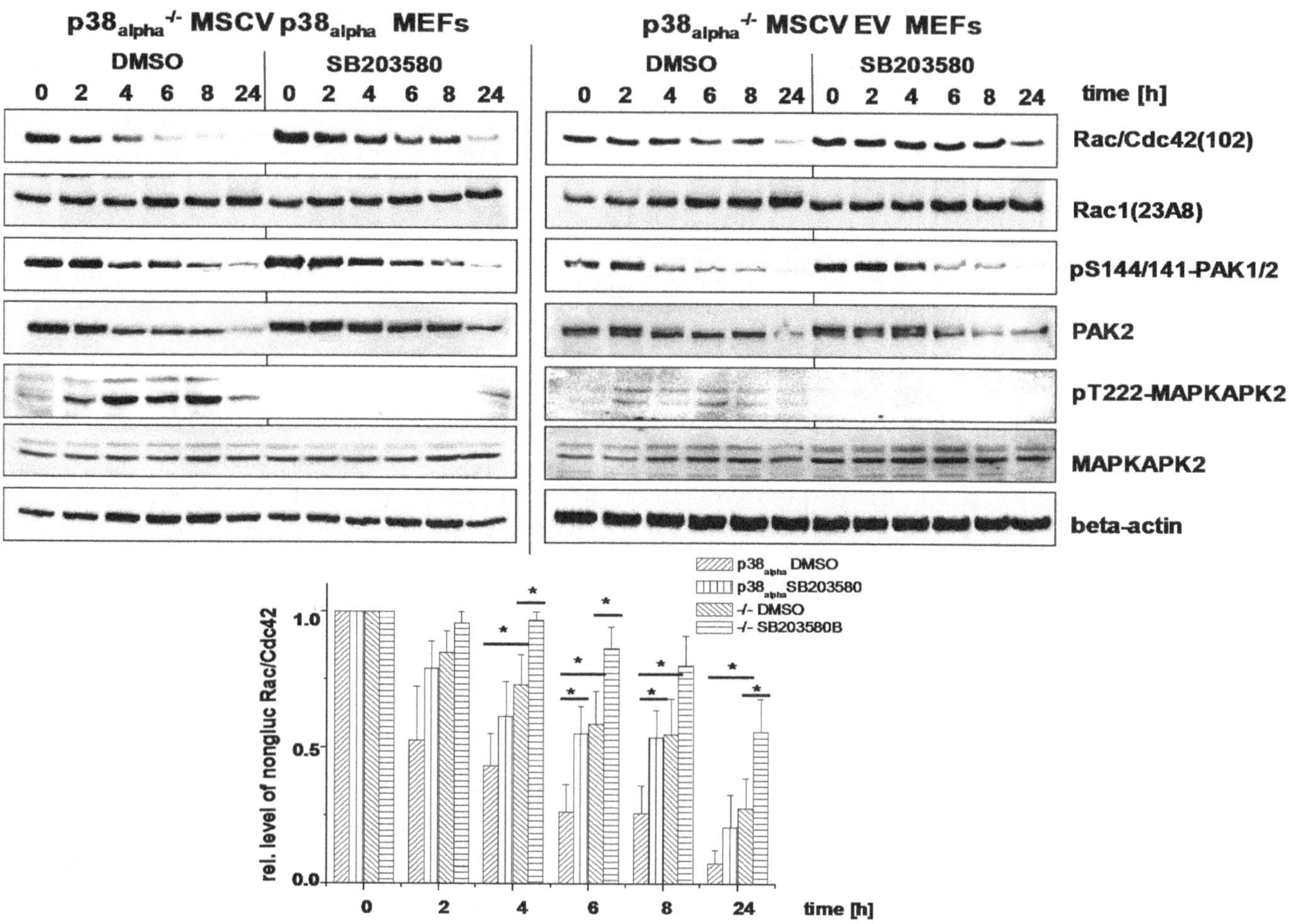

Figure 5. Effects of genetic deletion of $p38_{alpha}$ and of SB203580 treatment on TcsL-catalyzed Rac/Cdc42 glucosylation (time-dependency). $p38_{alpha}{}^{-/-}$ MSCV $p38_{alpha}$ MEFs and $p38_{alpha}{}^{-/-}$ MSCV empty vector (EV) MEFs were treated with TcsL (1 µg/mL) in the presence of SB203580 (10 µM) or DMSO alone for the indicated times. The cellular levels of non-glucosylated Rac/Cdc42, total Rac1, pS144/141-PAK1/2, PAK2, pT222-MAPKAPK2, MAPKAPK2, and beta-actin were analyzed by immunoblotting using the indicated antibodies. Quantifications of immunoblots were performed using Kodak software and relative amounts of non-glucosylated Rac/Cdc42 versus the total levels of Rac1, respectively, are expressed as mean ± SD of three independent experiments. * indicates significant differences, $p < 0.05$, as analyzed using Student's t-test.

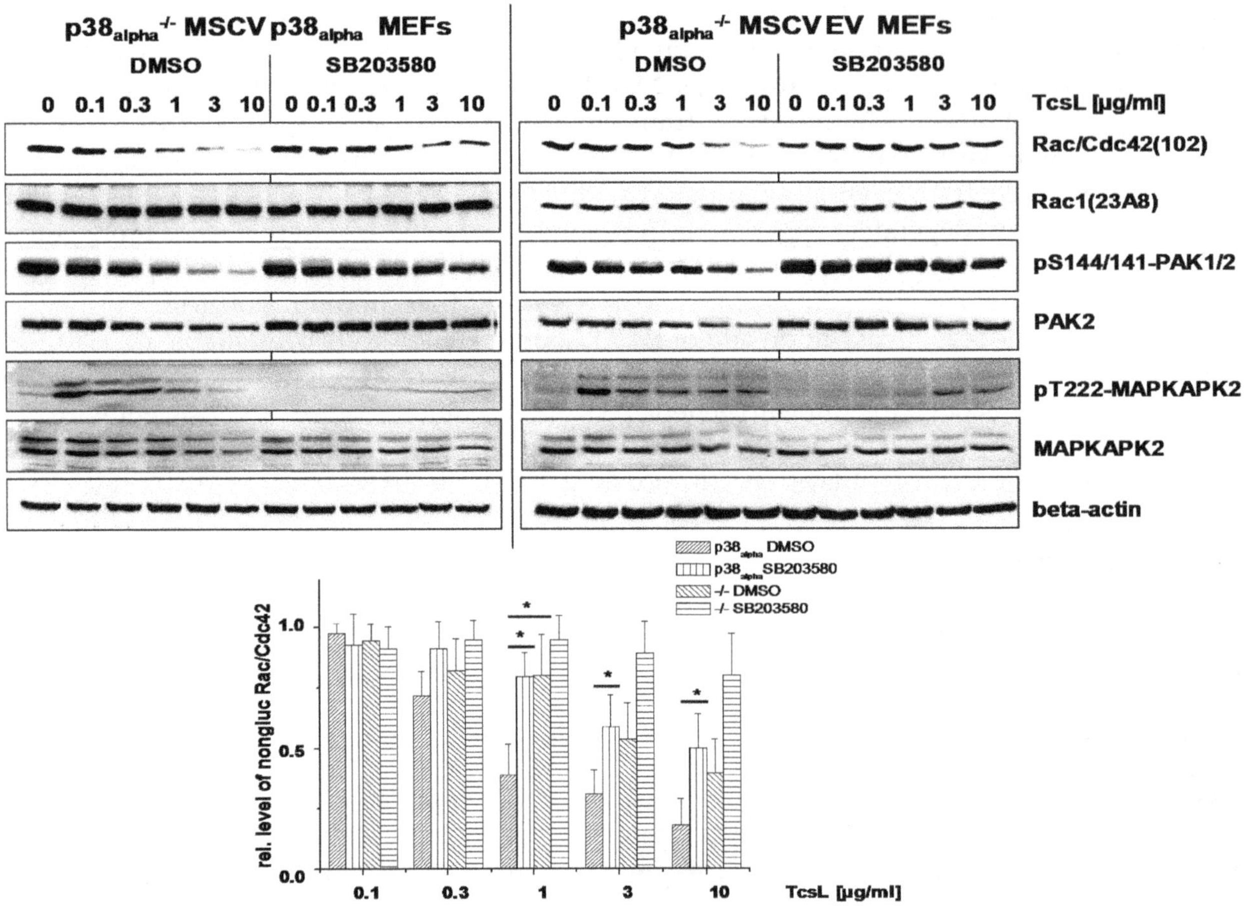

Figure 6. Effects of genetic deletion of $p38_{alpha}$ and of SB203580 treatment on TcsL-catalyzed Rac/Cdc42 glucosylation (TcsL concentration-dependency). $p38_{alpha}{}^{-/-}$ MSCV $p38_{alpha}$ MEFs and $p38_{alpha}{}^{-/-}$ MSCV empty vector (EV) MEFs were treated with the indicated concentrations of TcsL in the presence of SB203580 (10 μM) or DMSO alone for 4 h. The cellular levels of non-glucosylated Rac/Cdc42, total Rac1, pS144/141-PAK1/2, PAK2, and beta-actin were analyzed by immunoblotting using the indicated antibodies. Quantifications of immunoblots were performed using Kodak software and relative amounts of non-glucosylated Rac/Cdc42 versus the total levels of Rac1, respectively, are expressed as mean ± SD of three independent experiments. * indicates significant differences, $p < 0.05$, as analyzed using Student's *t*-test.

TcsL-induced Rac/Cdc42 glucosylation were next analyzed upon pharmacological inhibition of $p38_{alpha/beta}$. SB203580 treatment of either $p38_{alpha}{}^{-/-}$ MSCV $p38_{alpha}$ and $p38_{alpha}{}^{-/-}$ MSCV empty vector MEFs resulted in an almost complete loss of pT222-MAPKAPK2, confirming effective p38 inhibition (Figures 5 and 6). TcsL-catalyzed Rac/Cdc42 glucosylation and PAK1/2 deactivation were responsive to SB203580 treatment in both $p38_{alpha}{}^{-/-}$ MSCV $p38_{alpha}$ and $p38_{alpha}{}^{-/-}$ MSCV empty vector MEFs, both in time- (Figure 5) and concentration-dependent (Figure 6) experiments. The observation that cell rounding and Rac/Cdc42 glucosylation in $p38_{alpha}{}^{-/-}$ MSCV empty vector MEFs are responsive to SB203580 suggests that the protective effects of SB203580 involve inhibition of both $p38_{alpha}$ and $p38_{beta}$. Taken together, $p38_{alpha/beta}$ inhibition-mediated protection of fibroblasts from TcsL-induced actin reorganization coincides with protection from TcsL-catalyzed Rac/Cdc42 glucosylation.

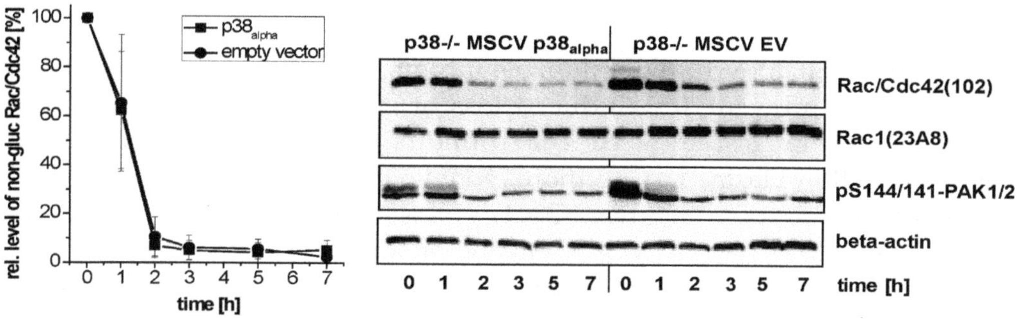

Figure 7. Effects of genetic deletion of $p38_{alpha}$ on TcdB-catalyzed Rac/Cdc42 glucosylation (time-dependency). $p38_{alpha}{}^{-/-}$ MSCV $p38_{alpha}$ MEFs and $p38_{alpha}{}^{-/-}$ MSCV empty vector (EV) MEFs were treated with TcdB (1 ng/mL) for the indicated times. The cellular levels of non-glucosylated Rac/Cdc42, total Rac1, pS144/141-PAK1/2, and beta-actin were analyzed by immunoblotting using the indicated antibodies. Quantifications of immunoblots were performed using Kodak software and relative amounts of non-glucosylated Rac/Cdc42 versus the total levels of Rac1, respectively, are expressed as mean ± SD of three experiments.

2.3. *SB203580 Preserves TcsL-Induced Loss of Epithelial Barrier Function*

C. sordellii-associated disease include necrotic and hemorrhagic enteritis, whereby TcsL has been shown to alter epithelial permeability [33,34]. To check if SB203580-mediated inhibition of TcsL might be useful with regard to disease treatment, TcsL-induced loss of epithelial barrier function was next analyzed in terms of the loss of the transepithelial resistance (TER) of a Madin-Darby canine kidney (MDCK-C7) monolayer [9,35]. TcsL treatment time-dependently decreased the TER of the MDCK-C7 monolayer (Figure 8A). In the presence of SB203580, TcsL-induced loss of TER was markedly attenuated (Figure 8A). SB203580 alone did not affect the TER (Figure 8A). In contrast, TcdB-induced loss of the TER of the MDCK-C7 monolayer was not responsive to SB203580 treatment (Figure 8B), consistent with above observations showing that TcdB-induced actin reorganization was not responsive to SB203580 treatment (Figure 4). SB203580 treatment thus might be useful in the light of treatment of the TcsL-induced loss of epithelial barrier function.

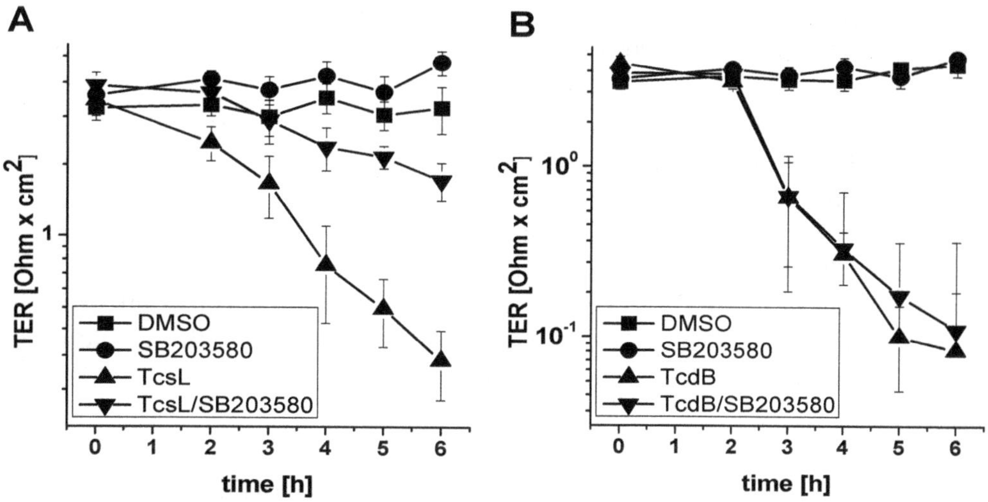

Figure 8. SB203580 preserves TcsL-induced loss of epithelial barrier function. Madin-Darby canine kidney (MDCK-C7) monolayers grown on Transwell filter inserts were treated with TcsL (30 µg/mL (**A**)) and TcdB (30 ng/mL (**B**)) in the presence of SB203580 (10 µM) or DMSO alone as indicated. Transepithelial electrical resistance (TER) was monitored for the indicated times. TER values are given as means ± SD of three independent experiments.

3. Discussion

In this study, SB203580, a pyridinyl imidazole inhibitor of $p38_{alpha/beta}$ MAP kinase, has been presented to efficaciously prevent TcsL-induced loss of epithelial barrier function of MDCK-C7 monolayers and to prevent TcsL-induced cell rounding, Rac/Cdc42 glucosylation, and PAK deactivation in murine fibroblasts. Furthermore, genetic deletion of $p38_{alpha}$ is also sufficient for preventing TcsL-induced cell rounding, Rac/Cdc42 glucosylation, and PAK deactivation in murine fibroblasts. Interestingly, $p38_{alpha}{}^{-/-}$ MSCV empty vector MEFs turn out to be still sensitive to SB203580, suggesting that the protective effects of SB203580 involves inhibition of both $p38_{alpha}$ and $p38_{beta}$. How does p38 inhibition mediate the protective effect against TcsL?

TcdB and TcsL both are mono-O-glucosyltransferases that modify small GTPases. Their N-terminal glucosyltransferase domains are structurally and functionally highly related, as they share Rac/Cdc42 as substrate GTPases (with threonine-35 being the acceptor amino acid) and UDP-glucose as a sugar donor [36]. From the observation that TcdB-induced Rac/Cdc42 glucosylation is insensitive to p38 inhibition, it can be concluded that SB203580 does not interfere with the intracellular glucosyltransferase activity of the toxins. This leads to the new hypothesis that $p38_{alpha/beta}$ inhibition affects endocytic uptake of TcsL into target cells. In fact, members of the p38 MAP kinase family are important regulators of endocytosis, as they control endocytic trafficking via the GDI-Rab5 complex [37]. In particular, p38 MAP kinase regulates stress-induced internalization of the epidermal growth factor receptor (EGFR) and μ opioid receptor endocytosis [38–40]. Against this background, inhibition of $p38_{alpha/beta}$ might prevent cell entry of specifically TcsL (not TcdB) by receptor-mediated endocytosis. This hypothetic model implies that TcsL and TcdB enter their target cells by exploiting distinct cell surface receptors, with the TcsL cell surface receptors being internalized in a $p38_{alpha/beta}$-dependent and the TcdB cell surface receptors being internalized in a $p38_{alpha/beta}$-independent manner.

In a former study, the responsiveness of TcsL-induced effects such as apoptotic cell death to inhibition by SB203580 has been interpreted in terms of an involvement of $p38_{alpha/beta}$ in TcsL-induced cell death [24]. The observations of this study have led to the conclusion that the protective effect of SB203580 is based on rather inhibition of TcsL uptake than on a role of p38 in the cytopathic effects of TcsL. Against this background, the responsiveness of TcsL-induced cell death to inhibition by SB203580 must be re-interpreted in terms of reduced TcsL uptake and subsequently reduced GTPase substrate glucosylation as the cause of cell death inhibition. In autophagy research, the pyridinyl imidazole inhibitors SB203580 and SB202190 have been shown to interfere with the autophagic flux independently of p38 MAP kinase [41]. These observations have led to the recommendation that pyridinyl imidazole class inhibitors should not be used as pharmacological tools in the analysis of MAPK11-MAPK14/p38-dependence [42]. The latter recommendation seems to be applicable also in the field of protein toxins.

The protective effect of the pyridinyl imidazole SB203580 is most interesting with regard to the development of non-antibiotic treatment for diseases caused by toxigenic *C. sordellii*. Further research will address the characterization of those pathways mediating the protective effect against TcsL upon inhibition of $p38_{alpha/beta}$. Furthermore, a screening of additional pyridinyl imidazole compounds capable of inhibiting the effects of TcsL is under way in our laboratory.

4. Conclusions

- Genetic deletion of $p38_{alpha}$ or treatment with SB203580 protects MEFs from Rac/Cdc42 glucosylation and actin reorganization induced by TcsL (not by the related TcdB).
- Treatment with SB203580 protects epithelial monolayer from loss of epithelial barrier function induced by TcsL (not by TcdB).
- The protective effects of SB203580 treatment and of $p38_{alpha}$ deletion are likely based on inhibition of endocytic uptake of TcsL rather than on inhibition of the toxins' glucosyltransferase activity.

5. Materials and Methods

5.1. Materials

The following reagents were obtained from commercial sources: SB203580 (4-(4-fluorophenyl)-2-(4-methylsulfinylphenyl)-5-(4-pyridyl)imidazole) (Calbiochem, Darmstadt, Germany); DAPI (40.6-diamidino-2-phenylindole) (Sigma-Aldrich, St. Louis, MO, USA); and rhodamine-conjugated phalloidin (Sigma-Aldrich, St. Louis, MO, USA).

Toxins: TcsL was prepared from *C. sordellii* IP82, which is the same strain as 6018, and TcdB from *C. difficile* VPI10463. Toxins were produced and purified yielding only one band on SDS-PAGE as previously described [43,44]. In brief, a dialysis bag containing 900 mL of 0.9% NaCl in a total volume of 4 liters of brain heart infusion (Difco, BD Life Sciences, Heidelberg, Germany) was inoculated with 100 mL of an overnight culture of *C. sordellii* or *C. difficile*. The culture was grown under microaerophilic conditions at 37 °C for 72 h. Bacteria were removed from the dialysis bag solution by centrifugation. Proteins from the culture supernatant from were precipitated by ammonium sulfate (Merck Millipore, Darmstadt, Germany) at 70% saturation. The precipitated proteins were dissolved in 50 mM Tris-HCl pH 7.5 buffer and extensively dialyzed against 50 mM Tris-HCl pH 7.5 buffer for 24 h. The protein solution was loaded onto an anion exchange column (MonoQ, GE Healthcare Europe, Freiburg, Germany). Either TcsL or TcdB were eluted with 50 mM Tris-HCl, pH 7.5, at 500–600 mM NaCl and were subsequently dialyzed against buffer (50 mM Tris-HCl pH 7.5, 15 mM NaCl). The absence of TcdA (which eluted at 150–200 mM NaCl) in TcdB preparations was checked by immunoblot analysis.

5.2. Cell Culture and Preparation of Lysates

$p38^{-/-}$ MSCV empty vector MEFs and the corresponding $p38^{-/-}$ MSCV $p38_{alpha}$ MEFs (kindly provided by Dr. Angel Nebreda, Institute for Research in Biomedicine, Barcelona, Spain) were cultivated in Dulbecco's modified essential medium supplemented with 10% FCS, 100 μg/mL penicillin, 100 U/mL streptomycin and 1 mM sodium pyruvate at 37 °C and 5% CO_2 according to standard protocols [45]. Cells sub-confluently seeded in 3.5-cm dishes were treated with TcsL, TcdB, and SB203580 for different times and concentrations as noted in the figures. Thereby, cells were pretreated with 10 μM SB202580 dissolved in DMSO (final DMSO concentration in the medium 2%) for 20 min and subsequently treated with the toxins or buffer. Upon incubation time, the cells were rinsed with 5 mL of ice-cold phosphate-buffered saline and scraped off in 200 μL of Laemmli lysis buffer per dish. The cells were disrupted mechanically by sonification (five times on ice). The lysate were submitted to immunoblot analysis.

5.3. Immunoblot Analysis

Cells lysates were separated on 15% polyacrylamide gels and transferred onto nitrocellulose for 2 h at 250 mA, followed by blocking with 5% (*w*/*v*) nonfat dried milk for 1 h. Blots were incubated with the appropriate primary antibody with dilution according to the manufacturers' instructions (beta-actin, Mab AC-40, Sigma-Aldrich, St. Louis, MO, USA; dilution 1:5000); MAPKAPK-2 (Cell signaling 3042, dilution 1:1000); pT222-MAPKAPK-2 (Cell signaling 3316, dilution 1:1000); PAK2 (Cell signaling 2608, dilution 1:1000); phospho-S144/141-PAK1/2 (Abcam ab40795; dilution 1:2000); Rac1 (BD Transduction Laboratories 610650, clone 102; dilution 1:1000); Rac1(Millipore 05-389, clone 23A8; dilution 1:1000) in buffer B (50 mM Tris-HCl, pH 7.2, 150 mM NaCl, 5 mM KCl, 0.05% (*w*/*v*) Tween 20) for 18 h and subsequently for 2 h with a horseradish peroxidase-conjugated secondary antibody (mouse: Rockland 610-1034-121; dilution 1:3000; rabbit Rockland 611-1302; dilution 1:3000). For the chemiluminescence reaction, ECL Femto (Fisher Scientific, Schwerte, Germany) was used. The signals were analyzed densitometrically using the KODAK 1D software (2004, Rochester, MN, USA).

5.4. Visualization of the Actin Cytoskeleton

$p38_{alpha}^{-/-}$ MSCV empty vector MEFs and the corresponding $p38_{alpha}^{-/-}$ MSCV $p38_{alpha}$ MEFs were grown on coverslips, fixed with formaldehyde (4%) in PBS and permeabilized with 0.1% Triton X-100 (Sigma-Aldrich, St. Louis, MO, USA). The actin cytoskeleton and the nuclei were labeled using rhodamine-phalloidin and DAPI, respectively. Fluorescence images were recorded using a Zeiss Axiovert M (Jena, Germany).

5.5. Transepithelial Resistance of Epithelial Monolayers

Madin-Darby canine kidney (MDCK-C7) cells were cultured under standard conditions (37 °C, 5% CO_2) as described [35]. Briefly, MDCK-C7 cells were cultured in minimum essential medium (MEM) enriched with Earle's salts, non-essential amino acids, glutamic acid and 10% fetal calf serum (Biochrom, Berlin, Germany) and split twice weekly using standard culture techniques. MDCK-C7 cells were seeded onto 12 well filter transwell inserts (pore size 0.4 µM, BD Life Sciences, Heidelberg, Germany). The transepithelial electrical resistance (TER) was determined by a Voltohmmeter equipped with Endom 24 chamber (EVOM, World Precision Instruments, Berlin, Germany). MDCK-C7 monolayers were cultivated up to an initial resistance of >2 $k\Omega \cdot cm^2$. The toxins and SB203580 (final DMSO concentration in the medium 2%) were applied on the basolateral site of the monolayer and toxin-induced loss of TER was analyzed in a time-dependent manner.

Acknowledgments: $p38^{-/-}$ MSCV empty vector MEFs and the corresponding $p38^{-/-}$ MSCV $p38_{alpha}$ MEFs were kindly provided by Angel Nebreda, Institute for Research in Biomedicine, Barcelona, Spain. This work was supported by Niedersachsen Vorab "Epidemiology and systems biology of the bacterial pathogen *Clostridium difficile*" grant number VWZN3215 to H.G.

Author Contributions: I.S., J.B., M.B., and H.G. conceived and designed the experiments and analyzed the data; I.S., M.B., and J.B. performed the experiments; and H.G. wrote the paper.

References

1. Hodges, K.; Hecht, G. Bacterial infections of the small intestine. *Curr. Opin. Gastroenterol.* **2013**, *29*, 159–163. [CrossRef] [PubMed]
2. Peniche, A.G.; Savidge, T.C.; Dann, S.M. Recent insights into *Clostridium difficile* pathogenesis. *Curr. Opin. Infect. Dis.* **2013**, *26*, 447–453. [CrossRef] [PubMed]
3. Diab, S.S.; Songer, G.; Uzal, F.A. *Clostridium difficile* infection in horses: A review. *Vet. Microbiol.* **2013**, *167*, 42–49. [CrossRef] [PubMed]
4. Al-Mashat, R.R.; Taylor, D.J. Production OD diarrhoea and enteric lesions in calves by the oral inoculation of pure cultures of *Clostridium sordellii. Vet. Rec.* **1983**, *112*, 141–146. [CrossRef] [PubMed]
5. Papatheodorou, P.; Zamboglou, C.; Genisyuerek, S.; Guttenberg, G.; Aktories, K. Clostridial glucosylating toxins enter cells via clathrin-mediated endocytosis. *PLoS ONE* **2010**, *5*, e10673. [CrossRef] [PubMed]
6. Jank, T.; Belyi, Y.; Aktories, K. Bacterial glycosyltransferase toxins. *Cell. Microbiol.* **2015**, *17*, 1752–1765. [CrossRef] [PubMed]
7. Varela, C.C.; Haustant, G.M.; Baron, B.; England, P.; Chenal, A.; Pauillac, S.; Blondel, A.; Popoff, M.R. The Tip of the four N-Terminal alpha-helices of *Clostridium sordellii* lethal toxin contains the interaction site with membrane phosphatidylserine facilitating small GTPases glucosylation. *Toxins (Basel)* **2016**, *8*, E90. [CrossRef] [PubMed]
8. LaFrance, M.E.; Farrow, M.A.; Chandrasekaran, R.; Sheng, J.; Rubin, D.H.; Lacy, D.B. Identification of an epithelial cell receptor responsible for *Clostridium difficile* TcdB-induced cytotoxicity. *Proc. Natl. Acad. Sci. USA* **2015**, *112*, 7073–7078. [CrossRef] [PubMed]
9. Genth, H.; Schelle, I.; Just, I. Metal Ion Activation of *Clostridium sordellii* Lethal Toxin and *Clostridium difficile* Toxin B. *Toxins (Basel)* **2016**, *8*, E109. [CrossRef] [PubMed]

10. Sehr, P.; Joseph, G.; Genth, H.; Just, I.; Pick, E.; Aktories, K. Glucosylation and ADP-ribosylation of Rho proteins—Effects on nucleotide binding, GTPase activity, and effector-coupling. *Biochemistry* **1998**, *37*, 5296–5304. [CrossRef] [PubMed]
11. Genth, H.; Aktories, K.; Just, I. Monoglucosylation of RhoA at Threonine-37 blocks cytosol-membrane cycling. *J. Biol. Chem.* **1999**, *274*, 29050–29056. [CrossRef] [PubMed]
12. Herrmann, C.; Ahmadian, M.R.; Hofmann, F.; Just, I. Functional consequences of monoglucosylation of H-Ras at effector domain amino acid threonine-35. *J. Biol. Chem.* **1998**, *273*, 16134–16139. [CrossRef] [PubMed]
13. Vetter, I.R.; Hofmann, F.; Wohlgemuth, S.; Herrmann, C.; Just, I. Structural consequences of mono-glucosylation of Ha-Ras by *Clostridium sordellii* lethal toxin. *J. Mol. Biol.* **2000**, *301*, 1091–1095. [CrossRef] [PubMed]
14. May, M.; Wang, T.; Muller, M.; Genth, H. Difference in F-actin depolymerization induced by toxin B from the *Clostridium difficile* strain VPI 10463 and toxin B from the variant *Clostridium difficile* serotype F strain 1470. *Toxins (Basel)* **2013**, *5*, 106–119. [CrossRef] [PubMed]
15. Geny, B.; Grassart, A.; Manich, M.; Chicanne, G.; Payrastre, B.; Sauvonnet, N.; Popoff, M.R. Rac1 inactivation by lethal toxin from *Clostridium sordellii* modifies focal adhesions upstream of actin depolymerization. *Cell. Microbiol.* **2010**, *12*, 217–232. [CrossRef] [PubMed]
16. Genth, H.; Pauillac, S.; Schelle, I.; Bouvet, P.; Bouchier, C.; Varela-Chavez, C.; Just, I.; Popoff, M.R. Haemorrhagic toxin and lethal toxin from *Clostridium sordellii* strain vpi9048: Molecular characterization and comparative analysis of substrate specificity of the large clostridial glucosylating toxins. *Cell. Microbiol.* **2014**, *16*, 1706–1721. [CrossRef] [PubMed]
17. Lica, M.; Schulz, F.; Schelle, I.; May, M.; Just, I.; Genth, H. Difference in the biological effects of *Clostridium difficile* toxin B in proliferating and non-proliferating cells. *Naunyn Schmiedebergs Arch. Pharmacol.* **2011**, *383*, 275–283. [CrossRef] [PubMed]
18. Wohlan, K.; Goy, S.; Olling, A.; Srivaratharajan, S.; Tatge, H.; Genth, H.; Gerhard, R. Pyknotic cell death induced by *Clostridium difficile* TcdB: Chromatin condensation and nuclear blister are induced independently of the glucosyltransferase activity. *Cell. Microbiol.* **2014**, *16*, 1678–1692. [CrossRef] [PubMed]
19. D'Auria, K.M.; Donato, G.M.; Gray, M.C.; Kolling, G.L.; Warren, C.A.; Cave, L.M.; Solga, M.D.; Lannigan, J.A.; Papin, J.A.; Hewlett, E.L. Systems analysis of the transcriptional response of human ileocecal epithelial cells to *Clostridium difficile* toxins and effects on cell cycle control. *BMC. Syst. Biol.* **2012**, *6*, 2.
20. Arechederra, M.; Priego, N.; Vazquez-Carballo, A.; Sequera, C.; Gutierrez-Uzquiza, A.; Cerezo-Guisado, M.I.; Ortiz-Rivero, S.; Roncero, C.; Cuenda, A.; Guerrero, C.; Porras, A. p38 MAPK down-regulates fibulin 3 expression through methylation of gene regulatory sequences: Role in migration and invasion. *J. Biol. Chem.* **2015**, *290*, 4383–4397. [CrossRef] [PubMed]
21. Cerezo-Guisado, M.I.; del Reino, P.; Remy, G.; Kuma, Y.; Arthur, J.S.; Gallego-Ortega, D.; Cuenda, A. Evidence of p38gamma and p38delta involvement in cell transformation processes. *Carcinogenesis* **2011**, *32*, 1093–1099. [CrossRef] [PubMed]
22. Geny, B.; Popoff, M.R. Activation of a c-Jun-NH2-terminal kinase pathway by the lethal toxin from *Clostridium sordellii*, TcsL-82, occurs independently of the toxin intrinsic enzymatic activity and facilitates small GTPase glucosylation. *Cell. Microbiol.* **2009**, *11*, 1102–1113. [CrossRef] [PubMed]
23. Huelsenbeck, J.; May, M.; Schulz, F.; Schelle, I.; Ronkina, N.; Hohenegger, M.; Fritz, G.; Just, I.; Gerhard, R.; Genth, H. Cytoprotective effect of the small GTPase RhoB expressed upon treatment of fibroblasts with the Ras-glucosylating *Clostridium sordellii* lethal toxin. *FEBS Lett.* **2012**, *586*, 3665–3673. [CrossRef] [PubMed]
24. Schulz, F.; Just, I.; Genth, H. Prevention of *Clostridium sordellii* lethal toxin-induced apoptotic cell death by tauroursodeoxycholic acid. *Biochemistry* **2009**, *48*, 9002–9010. [CrossRef] [PubMed]
25. Ronkina, N.; Menon, M.B.; Schwermann, J.; Arthur, J.S.; Legault, H.; Telliez, J.B.; Kayyali, U.S.; Nebreda, A.R.; Kotlyarov, A.; Gaestel, M. Stress induced gene expression: A direct role for MAPKAP kinases in transcriptional activation of immediate early genes. *Nucleic Acids Res.* **2011**, *39*, 2503–2518. [CrossRef] [PubMed]
26. Dolado, I.; Swat, A.; Ajenjo, N.; de, V.G.; Cuadrado, A.; Nebreda, A.R. p38alpha MAP kinase as a sensor of reactive oxygen species in tumorigenesis. *Cancer Cell.* **2007**, *11*, 191–205. [CrossRef] [PubMed]
27. Cuadrado, A.; Nebreda, A.R. Mechanisms and functions of p38 MAPK signalling. *Biochem. J.* **2010**, *429*, 403–417. [CrossRef] [PubMed]

28. Halabi-Cabezon, I.; Huelsenbeck, J.; May, M.; Ladwein, M.; Rottner, K.; Just, I.; Genth, H. Prevention of the cytopathic effect induced by *Clostridium difficile* Toxin B by active Rac1. *FEBS Lett.* **2008**, *582*, 3751–3756. [CrossRef] [PubMed]
29. Genth, H.; Just, I. Functional implications of lethal toxin-catalysed glucosylation of (H/K/N)Ras and Rac1 in *Clostridium sordellii*-associated disease. *Eur. J. Cell Biol.* **2011**, *90*, 959–965. [CrossRef] [PubMed]
30. Genth, H.; Huelsenbeck, J.; Hartmann, B.; Hofmann, F.; Just, I.; Gerhard, R. Cellular stability of Rho-GTPases glucosylated by *Clostridium difficile* toxin B. *FEBS Lett.* **2006**, *580*, 3565–3569. [CrossRef] [PubMed]
31. Brandes, V.; Schelle, I.; Brinkmann, S.; Schulz, F.; Schwarz, J.; Gerhard, R.; Genth, H. Protection from *Clostridium difficile* toxin B-catalysed Rac1/Cdc42 glucosylation by tauroursodeoxycholic acid-induced Rac1/Cdc42 phosphorylation. *Biol. Chem.* **2012**, *393*, 77–84. [CrossRef] [PubMed]
32. May, M.; Schelle, I.; Brakebusch, C.; Rottner, K.; Genth, H. Rac1-dependent recruitment of PAK2 to G phase centrosomes and their roles in the regulation of mitotic entry. *Cell Cycle* **2014**, *13*, 2210–2220. [CrossRef] [PubMed]
33. Boehm, C.; Gibert, M.; Geny, B.; Popoff, M.R.; Rodriguez, P. Modification of epithelial cell barrier permeability and intercellular junctions by *Clostridium sordellii* lethal toxins. *Cell. Microbiol.* **2006**, *8*, 1070–1085. [CrossRef] [PubMed]
34. Popoff, M.R. Bacterial factors exploit eukaryotic Rho GTPase signaling cascades to promote invasion and proliferation within their host. *Small GTPases* **2014**, *5*, e28209. [CrossRef] [PubMed]
35. Zak, J.; Schneider, S.W.; Eue, I.; Ludwig, T.; Oberleithner, H. High-resistance MDCK-C7 monolayers used for measuring invasive potency of tumour cells. *Pflugers Arch.* **2000**, *440*, 179–183. [CrossRef] [PubMed]
36. Jank, T.; Giesemann, T.; Aktories, K. *Clostridium difficile* glucosyltransferase toxin B-essential amino acids for substrate binding. *J. Biol. Chem.* **2007**, *282*, 35222–35231. [CrossRef] [PubMed]
37. Felberbaum-Corti, M.; Cavalli, V.; Gruenberg, J. Capture of the small GTPase Rab5 by GDI: Regulation by p38 MAP kinase. *Methods Enzymol.* **2005**, *403*, 367–381. [PubMed]
38. Mace, G.; Miaczynska, M.; Zerial, M.; Nebreda, A.R. Phosphorylation of EEA1 by p38 MAP kinase regulates mu opioid receptor endocytosis. *EMBO J.* **2005**, *24*, 3235–3246. [CrossRef] [PubMed]
39. Zwang, Y.; Yarden, Y. p38 MAP kinase mediates stress-induced internalization of EGFR: Implications for cancer chemotherapy. *EMBO J.* **2006**, *25*, 4195–4206. [CrossRef] [PubMed]
40. Zhou, Y.; Tanaka, T.; Sugiyama, N.; Yokoyama, S.; Kawasaki, Y.; Sakuma, T.; Ishihama, Y.; Saiki, I.; Sakurai, H. p38-Mediated phosphorylation of Eps15 endocytic adaptor protein. *FEBS Lett.* **2014**, *588*, 131–137. [CrossRef] [PubMed]
41. Menon, M.B.; Kotlyarov, A.; Gaestel, M. SB202190-induced cell type-specific vacuole formation and defective autophagy do not depend on p38 MAP kinase inhibition. *PLoS ONE* **2011**, *6*, e23054. [CrossRef] [PubMed]
42. Menon, M.B.; Dhamija, S.; Kotlyarov, A.; Gaestel, M. The problem of pyridinyl imidazole class inhibitors of MAPK14/p38alpha and MAPK11/p38beta in autophagy research. *Autophagy* **2015**, *11*, 1425–1427. [CrossRef] [PubMed]
43. Genth, H.; Selzer, J.; Busch, C.; Dumbach, J.; Hofmann, F.; Aktories, K.; Just, I. New method to generate enzymatically deficient *Clostridium difficile* toxin B as an antigen for immunization. *Infect. Immun.* **2000**, *68*, 1094–1101. [CrossRef] [PubMed]
44. Popoff, M.R. Purification and characterization of *Clostridium sordellii* lethal toxin and cross-reactivity with *Clostridium difficile* cytotoxin. *Infect. Immun.* **1987**, *55*, 35–43. [PubMed]
45. Ambrosino, C.; Mace, G.; Galban, S.; Fritsch, C.; Vintersten, K.; Black, E.; Gorospe, M.; Nebreda, A.R. Negative feedback regulation of MKK6 mRNA stability by p38alpha mitogen-activated protein kinase. *Mol. Cell. Biol.* **2003**, *23*, 370–381. [CrossRef] [PubMed]

7

Mechanisms of Tolerance and Resistance to Chlorhexidine in Clinical Strains of *Klebsiella pneumoniae* Producers of Carbapenemase: Role of New Type II Toxin-Antitoxin System, PemIK

Ines Bleriot [1,2], Lucia Blasco [1,2], Mercedes Delgado-Valverde [3], Ana Gual-de-Torrella [3], Anton Ambroa [1], Laura Fernandez-Garcia [1,2], Maria Lopez [1,4], Jesus Oteo-Iglesias [2,4,5], Thomas K. Wood [6], Alvaro Pascual [2,3,4], German Bou [1,2,4], Felipe Fernandez-Cuenca [2,3,4,†] and Maria Tomas [1,2,4,*,†]

1. Microbiology Department-Research Institute Biomedical A Coruña (INIBIC), Hospital A Coruña (CHUAC), University of A Coruña (UDC), 15006 A Coruña, Spain; bleriot.ines@gmail.com (I.B.); luciablasco@gmail.com (L.B.); anton17@mundo-r.com (A.A.); laugemis@gmail.com (L.F.-G.); maria.lopez.diaz@sergas.es (M.L.); German.Bou.Arevalo@sergas.es (G.B.)
2. Study Group on Mechanisms of Action and Resistance to Antimicrobials (GEMARA) the Behalf of the Spanish Society of Infectious Diseases and Clinical Microbiology (SEIMC), 28003 Madrid, Spain; jesus.oteo@isciii.es(J.-O.I.); apascual@us.es (A.P.); felipefc@us.es (F.F.-C.)
3. Clinical Unit for Infectious Diseases, Department of Microbiology and Medicine, Microbiology and Preventive Medicine, Hospital Universitario Virgen Macarena, University of Seville, Biomedicine Insititute of Seville (IBIS), 41009 Seville, Spain; mercdss@gmail.com (M.D.-V.); gualdetorrella.ana@gmail.com (A.G.-d.-T.)
4. Spanish Network for Research in Infectious Diseases (REIPI), 41071 Seville, Spain
5. Reference and Research Laboratory for Antibiotic Resistance and Health Care Infections, National Centre for Microbiology, Institute of Health Carlos III, 28222 Majadahonda, Spain
6. Department of Chemical Engineering, Pennsylvania State University, University Park, PA 16801, USA; tuw14@psu.edu

* Correspondence: MA.del.Mar.Tomas.Carmona@sergas.es.

† These authors equally contributed to this work.

Abstract: Although the failure of antibiotic treatment is normally attributed to resistance, tolerance and persistence display a significant role in the lack of response to antibiotics. Due to the fact that several nosocomial pathogens show a high level of tolerance and/or resistance to chlorhexidine, in this study we analyzed the molecular mechanisms associated with chlorhexidine adaptation in two clinical strains of *Klebsiella pneumoniae* by phenotypic and transcriptomic studies. These two strains belong to ST258-KPC3 (high-risk clone carrying β-lactamase KPC3) and ST846-OXA48 (low-risk clone carrying β-lactamase OXA48). Our results showed that the *K. pneumoniae* ST258-KPC3CA and ST846-OXA48CA strains exhibited a different behavior under chlorhexidine (CHLX) pressure, adapting to this biocide through resistance and tolerance mechanisms, respectively. Furthermore, the appearance of cross-resistance to colistin was observed in the ST846-OXA48CA strain (tolerant to CHLX), using the broth microdilution method. Interestingly, this ST846-OXA48CA isolate contained a plasmid that encodes a novel type II toxin/antitoxin (TA) system, PemI/PemK. We characterized this PemI/PemK TA system by cloning both genes into the IPTG-inducible pCA24N plasmid, and found their role in persistence and biofilm formation. Accordingly, the ST846-OXA48CA strain showed a persistence biphasic curve in the presence of a chlorhexidine-imipenem combination, and these results were confirmed by the enzymatic assay (WST-1).

Keywords: tolerance; persistence; cross-resistance; toxin-antitoxin system; PemI/PemK; *Klebsiella pneumoniae*

Key Contribution: Combination of resistance, tolerance and persistence mechanisms seen in clinical isolates under biocide and antimicrobial stress. PemI/PemK system TA system discovered in the plasmid carrying β-lactamase OXA48.

---

1. Introduction

The increase in antimicrobial resistance due to the emergence of multi-drug resistant (MDR) pathogens is one of the world's greatest public health challenges, as it can lead to an era without effective antibiotics [1]. Recently, the World Health Organization (WHO) published a list of "priority pathogens", which includes those microorganisms considered a serious threat to human health. Some members of this list are carbapenem-resistant pathogens and are known under the acronym of ESKAPE, including among other species, *Klebsiella pneumoniae* [1–3]. *K. pneumoniae* is a Gram-negative, opportunistic bacteria pathogen associated with a wide range of diseases such as urinary tract infections, pneumoniae, septicemia, wounds, and soft tissue infections [4]. Carbapenem resistance is increasing rapidly worldwide, particularly among *K. pneumoniae*. The main carbapenem-resistance mechanism is acquisition of plasmid-encoded carbapenemases, which may belong to the molecular class A (i.e., KPC-type), B (i.e., imipenem (IMP)-type, VIM-type, NDM-type) and D (i.e., OXA-48-type). The high-risk clones of *K. pneumoniae*, in contrast to low-risk clones, have an extraordinary ability to persist and spread in the nosocomial environment, disseminating these carbapenemases and therefore being involved in nosocomial outbreaks [4].

Nevertheless, much less attention has been paid to the presence or occurrence of resistance to antiseptics and biocides, such as chlorhexidine (CHLX) [5], widely used in hospital settings. CHLX is a symmetric bis-biguanide molecule comprising two chloroguanide chains that are connected by a central hexamethylene chain, and carry two positive charges at physiological pH. CHLX is sparingly soluble in water, and thereby normally formulated with either acetate or gluconate to form water-soluble salts [6]. The antimicrobial effect of this compound is based on damaging the bacterial membrane, leading to the subsequent leakage of cytoplasmatic material. Therefore, mechanisms conferring resistance toward CHLX include multidrug efflux pumps and cell membrane changes [5]. Moreover, CHLX adaptation has been associated with the emergence of stable resistance to the last-resort antibiotic colistin (polymyxin E) [7–9].

In general, the failure of antibiotic treatments has been associated with resistance mechanisms. However, it has recently been noted that other mechanisms such as tolerance and persistence were also involved [10]. The recovery of persistent cells is one of the main causes of prolonged and recurrent infections, that can lead to the complete failure of antibiotic treatments [11]. In this context, it is important to distinguish between resistant, tolerant, and persistent bacteria [12]. The term resistance is generally used to describe the inherited ability of a bacterial population to grow in the presence of high concentrations of antibiotics, regardless of the duration of treatment [12], due to active defense mechanisms associated with mutations [10]. Whereas, the term tolerance is used to describe the ability, inherited or not, of a bacterial population to survive the transient exposure of high concentrations of antibiotics without causing changes in minimum inhibitory concentrations (MICs), due to the deceleration of essential biological processes [10–12]. It is important to emphasize that despite the slow-growth rate, tolerant bacteria keep a metabolically active state. In contrast to resistance and tolerance, persistence is characterized by the ability, not inherited, of a bacterial subpopulation (around 0.001–1%) [10] to resist antibiotics by growth arrest due to the inactivation of their metabolism and their non-replicative state, thus it is due to their dormant state. Persistent bacteria exhibit transient levels of tolerance to antibiotics that do not affect their MICs, so once the drug pressure is removed and their metabolism is reactivated, they can rapidly re-grow. Nowadays, it is known that multiple molecular mechanisms are involved in the formation of persistent bacteria such as the stringent response molecule (p)ppGpp, stress response, SOS response, quorum sensing, toxin-antitoxin (TA) systems, efflux pumps, the ROS response and energy metabolism, among others [11].

The involvement of TA systems in cell physiology, specifically in: (i) biofilm formation by regulating fimbriae [13,14], (ii) bacterial persistence, by generating slowly-growing cells tolerant to antibiotics and environmental changes [15–19], (iii) plasmid maintenance [20,21], (iv) general stress response [22], and (v) phage inhibition [23–25] is becoming clearer [18]. A TA system is a module of two genes encoding a stable toxin and an unstable antitoxin. Under normal growth conditions the antitoxin inhibits the toxin, but under stress conditions the antitoxin is degraded, leaving the toxin free to inhibit the basic cellular processes like DNA replication or protein synthesis, and also promoting plasmid maintenance, slow growth and latency [18]. These systems are widely distributed and found in the bacterial chromosome, plasmids, and bacteriophages [2].

In this context, this study provides a better comprehension of the molecular mechanisms associated with chlorhexidine adaptation (CA) in two clinical strains of *K. pneumoniae*, both of which produce carbapenemase: ST258-KPC3 (high-risk clone carrying β-lactamase KPC3) and ST846-OXA48 (low-risk clone carrying β-lactamase OXA48), from a phenotypic and transcriptomic point of view. It should be noted that international high-risk clones of *K. pneumoniae* are among the most common nosocomial pathogens. The success of these clones is due to their facility to spread their plasmids, which carry a considerable variety of antimicrobial resistance genes [26,27]. Thus, the study of this type of clones is of great clinical relevance. Moreover, this study aims to characterize a new toxin-antitoxin system (PemIK) located in a plasmid inside the ST846-OXA48CA strain, and to examine the possible role of this system in persistence and in biofilm formation.

2. Results

2.1. Results Subsection

2.1.1. Time-Killing Curve in the Presence of CHLX (10 × MIC)

The time-killing curves of the strains ST258-KPC3CA and ST846-OXA48CA in the presence of CHLX (10 × MIC) showed two different growth patterns (Figure 1). The strain ST258-KPC3CA showed a slight reduction in its bacterial population, occurring in the first two hours of CHLX exposure, decreasing from 6 LogCFU/mL (1.47×10^6 CFU/mL) to 4 LogCFU/mL (7.75×10^4 CFU/mL) at 2 h (Figure 1A). This slight reduction in the bacterial population occurs during the activation of defense mechanisms, such as efflux pumps, which can reduce the effective concentration of the drug in the cell. In contrast, the ST846-OXA48CA strain, dramatically reduced its bacterial population during the first four hours of CHLX exposure, decreasing from of a bacterial population of 6 LogCFU/mL (2.43×10^6 CFU/mL) to 2 LogCFU/mL (1.05×10^2 CFU/mL) at 4 h (Figure 1B). After this period, both bacterial strains grew again reaching respectively a bacterial population of 6 LogCFU/mL (6.63×10^6 CFU/mL) and 5 LogCFU/mL (2.75×10^5 CFU/mL) at 48 h. The curves have the characteristics of a resistant strain in the first case (ST258-KPC3CA) and a tolerant strain in the second one (ST846-OXA48CA), according to the definition of each mechanisms. Resistance is the ability of bacterial population to grow at a similar rate in the presence of an environmental stress, while the tolerance is the ability of a bacterial population to withstand the stress.

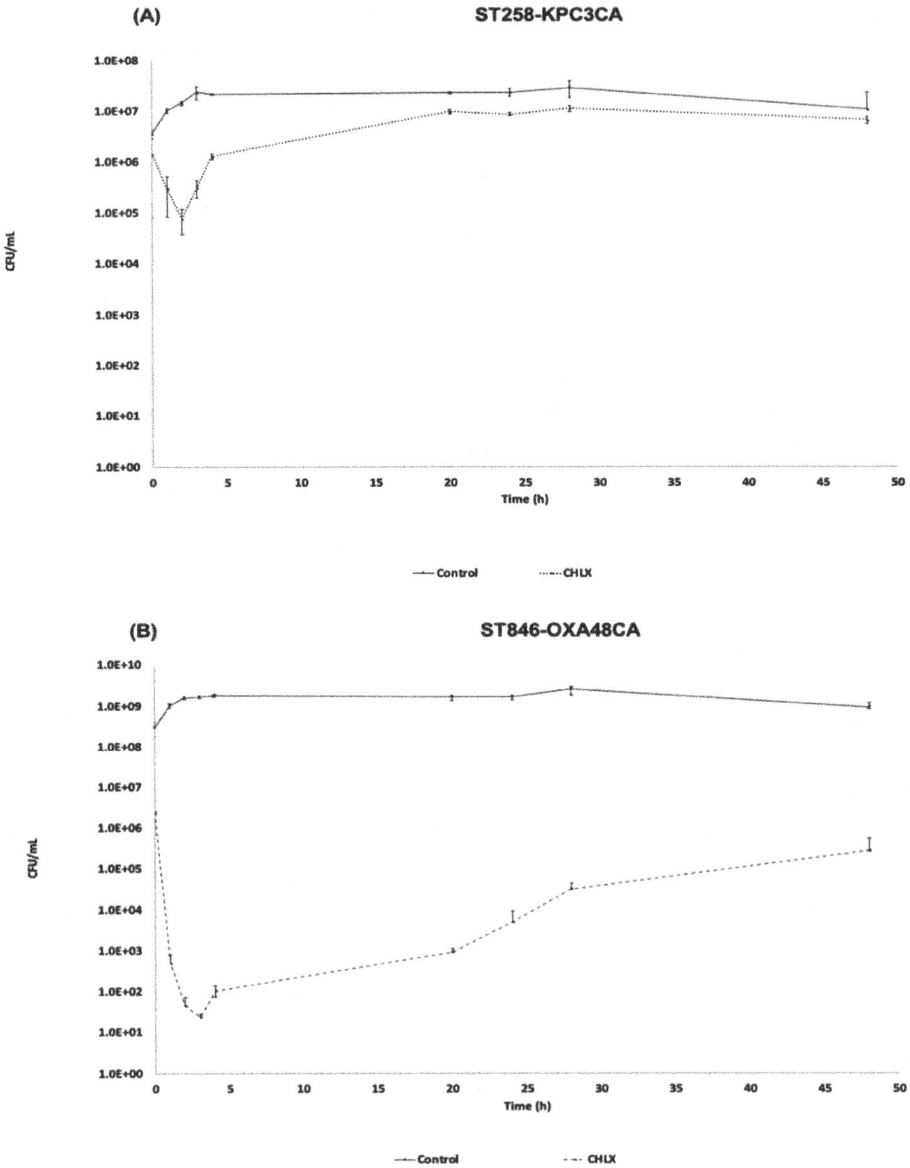

Figure 1. Time-killing curve in the presence of chlorhexidine (CHLX) (10 × minimum inhibitory concentration (MIC)) in *K. pneumoniae* chlorhexidine adaptation (CA) strains ST258-KPC3CA (**A**) and ST846-OXA48CA (**B**). The same strains without being exposed to biocide pressure are used as controls. The errors bar represents the standard deviation of the three replicates experiment.

2.1.2. Transcriptomic Study

All the transcriptomic results are deposited in the NCBI database as a GenBank BioProject (Code number: PRJNA609262) and GEO series (Code number: GSE147316). The transcriptomic profile from the ST258-KPC3CA isolate indicated a probable CHLX resistant profile. Indeed, this strain has a higher number of overexpressed genes (Log2fold change > 1.5), especially for transporters and efflux pumps such as the methyl viologen resistance gene *smvA* (Log2fold change: 3.635), which is involved in the cationic biocide resistance. However, strain ST846-OXA48CA showed what it was as a CHLX tolerant profile, with repressed genes (Log2FoldChange < 1.5) for efflux pumps, TA systems, SOS response, and ppGpp mechanisms (Table 1). This strain also showed high levels of expression of genes, *pmrD*, and *pmrK* (Log2fold change 2.360 and 1.570, respectively), characteristics of the colistin resistance. Therefore, these transcriptomic results corroborated the results obtained by the time-killing curves, showing activation of molecular mechanisms of resistance and tolerance molecular mechanisms in response to CHLX in ST258-KPC3CA and ST846-OXA48CA strains, respectively.

Table 1. Gene expression in response to CHLX in the strain ST258-KPC3CA and ST846-OXA48CA.

Mechanism	Gene [a]	Description	ST258-KPC3CA		ST846-OXA48CA	
			Log2 FoldChange	ID Gene	Log2 FoldChange	ID Gene
Transporter	smvA	Methyl viologen protein (cationic biocide resistance)	3.635	HGAILKPD_00917	1.209	EMNICGIE_00134
	actP	Acetate permease ActP (cation/acetate symporter)	2.724	HGAILKPD_00571	0.649	EMNICGIE_02128
	csbX	MFS superfamily	2.549	HGAILKPD_04325	0.618	EMNICGIE_04796
	lldP	L-lactate permease	2.486	HGAILKPD_04496	0.085	EMNICGIE_00243
	cysW	Ferric iron ABC transporter	2.181	HGAILKPD_02877	−0.266	EMNICGIE_02673
	potA	ABC transporter	1.749	HGAILKPD_02785	−0.351	EMNICGIE_03352
	pmrD	Signal transduction protein PmrD (colistin resistance)	-	-	2.360	EMNICGIE_04427
	pmrK	Polymyxin resistance protein PmrK (colistin resistance)	-	-	1.570	EMNICGIE_02839
ATP metabolism	atpD	ATP synthase beta chain	−0.209	HGAILKPD_02375	−0.232	EMNICGIE_00435
TA systems	ortT	Orphan toxin OrtT	0.731	HGAILKPD_02791	0.727	EMNICGIE_00095
	pemI	Programmed cell death antitoxin PemI	-	-	−0.100	EMNICGIE_05097
	pemK	Programmed cell death toxin PemK	-	-	−0.302	EMNICGIE_05098
(p)ppGpp	gppA	Guanosine-5′-triphosphate,3′-diphosphate pyrophosphatase	−0.765	HGAILKPD_02586	−0.399	EMNICGIE_03280
ROS response	cydA	Cytochrome d ubiquinol oxidase subunit I	1.318	HGAILKPD_03209	0.441	EMNICGIE_03423
	cybB	Cytochrome b561	0.456	HGAILKPD_02756	0.196	EMNICGIE_00060
SOS system	yedK	Putative SOS response-associated peptidase YedK	1.117	HGAILKPD_04848	0.003	EMNICGIE_04152
	yebG	DNA damage-inducible gene in SOS regulon	0.722	HGAILKPD_02193	0.455	EMNICGIE_04716

[a] All gene expression showed have a p-value < 0.05, and overexpression and repression were considered from as a Log2fold change of at least 1.5 and 0.5, respectively. (-) Not detected. The rows shaded differently reveal those genes of high relevance in this study, the cationic biocide resistance gene, the colistin resistance genes, and the novel TA system PemI/PemK. Background colour indicates the genes of most interest from this study.

2.1.3. Antimicrobial Susceptibly Testing

The antimicrobial susceptibility test was done for wild-type ST258-KPC3 and ST846-OXA48 wild-type and the two CA strains. According to adaptation to CHLX, an increase in MICs of CHLX was observed in both CA strains. However, no differences in the minimum inhibitory concentration (MIC) values were observed for the other antibiotics tested, except for the colistin, in that in the ST846-OXA48CA strain showed an increase in the MIC value of 32-fold, that corresponds to a resistance value (Table 2).

Table 2. MIC values (µg/mL) of different antibiotics for ST258-KPC3, ST258-KPC3CA, ST846-OXA48, and ST846-OXA48CA.

	MIC (µg/mL)														
Strain	**CHLX**	**CIP**	**TGC**	**TOB**	**IMP**	**MRP**	**GEN**	**CAZ**	**TZP**	**SAM**	**NET**	**DOX**	**AMK**	**MIN**	**CST**
ST258-KPC3	9.8	>32	2	64	4	8	4	>32	>32	1024	128	2	16	4	0.25
ST258-KPC3CA	39.1	>32	2	64	4	8	4	>32	>32	1024	128	2	16	4	0.25
ST846-OXA48	19.5	8	8	32	16	>32	32	>32	>32	128	16	8	2	8	0.5
ST846-OXA48CA	78.2	8	8	32	16	>32	32	>32	>32	128	16	8	2	8	16

CHLX, Chlorhexidine; CIP, Ciprofloxacin; TGC, Tigecycline; TOB, Tobramycin; IMP, Imipenem; MRP, Meropenem; GEN, Gentamycin; CAZ, Ceftazidime; TZP, Piperacillin-tazobactam; SAM, Sulbactam; NET, Netilmicin; AMK, Amikacin; MIN, Minociclin; CST, Colistin.

2.1.4. Characterization of the New TA System, PemI/PemK, Present in a Plasmid in the Strain ST846-OXA48CA

A PemI/PemK TA system, whose closest relative is the type II TA toxin-antitoxin system PemK/MazF family toxin belonging to Enterobacteriacea (Query: 78%; Identity: 99.35%; Code number: WP_077688581.1) and which have not been previously described in *K. pneumoniae*, it was identified by transcriptomic analysis (Table 1) as encoded by a plasmid of *K. pneumoniae* ST846-OXA48CA (Figure S1). This plasmid harbors several genes, such as *repA, dsbC, trbA, trbC, lusR*, CPBP metalloprotease, *umuD, umuC*, restriction endonuclease, *IS1, ssb, mobC, nikA, dotD*/TraH family lipoprotein, secretion systems type IV, *traO, traP, traQ, traW, traX, dotA/traY*, and *repC*, in addition to another TA system, the RelE/RelB TA system. This PemI/PemK TA system is composed of a 258 bp antitoxin gene (*pemI*) and a 333 bp toxin gene (*pemK*). To confirm that this system is a TA system, *pemI/pemK* and *pemK* genes alone were cloned into the overexpression vector pCA24N, widely used in the literature to overexpress TA systems [28,29], and transformed into the cured plasmid strain ST846-OXA48CA CP (i.e., lacking plasmids and therefore the plasmid that encodes the PemI/PemK TA system). The toxicity of this TA system was tested by growth curves overexpressing both *pemI/pemK* and *pemK* (Figure 2). Overexpression of *pemK* in ST846-OXA48CA CP/pCA24N (*pemK*) inhibited bacterial growth, while overexpression of the *pemI/pemK* system (ST846-OXA48CA CP/pCA24N (*pemIK*)) in the strain led to normal bacterial growth, which was slightly impaired compared to the empty plasmid. Therefore, the plasmid-based PemI/PemK TA system found in *K. pneumoniae* ST846-OXA48CA is functional.

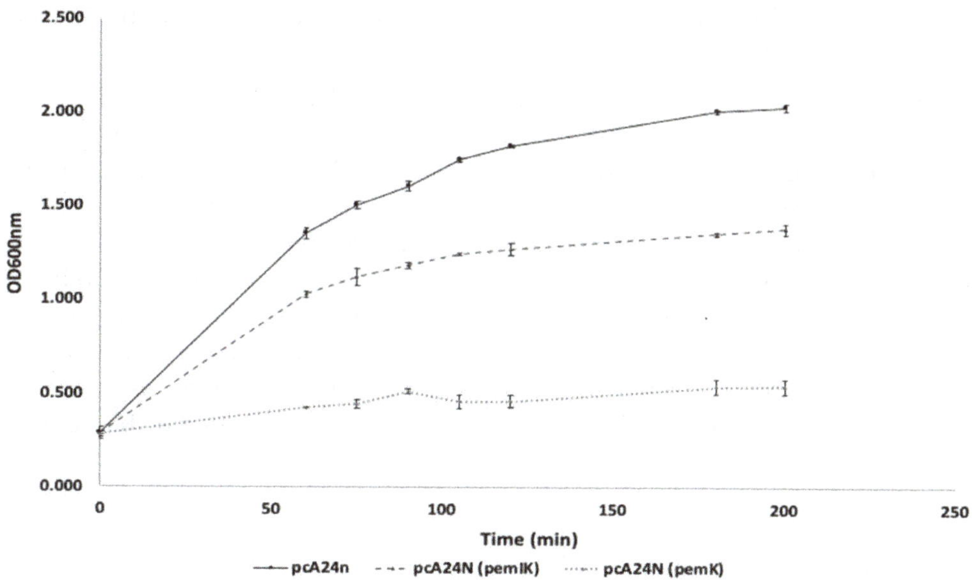

Figure 2. Growth curves of ST846-OXA48CA CP containing pCA24N plasmids with *pemIK* (dark grey, dashed line) and *pemK* (light grey, dotted line) in the presence of 1 mM IPTG. The strain ST846-OXA48CA CP is used as a control as it carries the empty plasmid pCA24N (black). The errors bar represents the standard deviation of the three experimental replicates.

2.1.5. Biofilm Formation Assay

Since TA systems have been associated with the arrest of bacterial growth and the formation of biofilms, we studied the effect of the PemI/PemK TA system and the PemK toxin on biofilm formation (Figure 3). Production of PemK toxin resulted in a significant decrease in biofilm formation compared to the control (ST846-OXA48CA CP/pCA24N) (p-value < 0.001). Moreover, production of PemI/PemK restored a similar phenotype as the control, lacking a significant difference in biofilm formation (p-value < 0.05). Therefore, the PemI/PemK TA system influences *K. pneumoniae* biofilm formation.

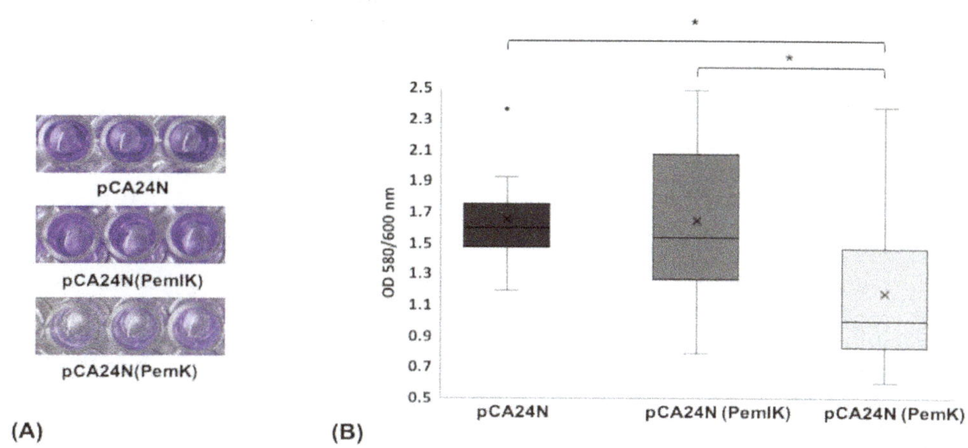

Figure 3. Biofilm formation assay. (**A**) Biofilm strained with 10% crystal violet was dissolved in 30% acetic acid. (**B**) Box and whisker plot of the optical density of biofilm produced by the strains ST846-OXA48CA CP/pCA24N, ST846-OXA48CA CP/pCA24N (PemIK), and ST846-OXA48CA CP/pCA24N (PemK). The biofilm formation was expressed as the ratio between OD580/600 nm, in order to normalize the data. Boxes indicate the lower and upper quartile. Horizontal lines in each box represents the median value of biofilm formation. The mean biofilm formation for each strain is indicated by a +. Vertical lines extending from each box represent the minimum and maximum biofilm formation. *, p-value < 0.05. All experiments were performed in triplicates.

2.1.6. Time-Killing Curve in the Presence of Imipenem or in Combination with Chlorhexidine for ST846-OXA48CA and ST846-OXA48CA CP

The time-killing curves of *K. pneumoniae* ST846-OXA48CA tolerant to CHLX (Figure 4A) were performed in the presence of imipenem (IMP) (50 × MIC) alone or in combination with CHLX (10 × MIC). A drastic reduction in the number of CFU was observed during the first four hours in the presence of IMP (50 × MIC), decreasing from 7 LogCFU/mL (9.35 × 10^7 CFU/mL) to 3 LogCFU/mL (1.50 × 10^3 CFU/mL). However, bacterial regrowth occurred after four hours, reaching similar levels of CFU/mL as the control (8 LogCFU/mL (6.83 × 10^8 CFU/mL) vs. 9 LogCFU/mL (1.73 × 10^9 CFU/mL)) at 28 h and exceeding it at 48 h (9 LogCFU/mL (1.25 × 10^9 CFU/mL) vs. 9 LogCFU/mL (1.07 × 10^9 CFU/mL)). In the case of the combination of IMP and CHLX, a greater CFU reduction than IMP alone was observed, with no CFU detected after 4 h. Nevertheless, a regrowth of the bacterial population was observed after 28 h. Thus, ST846-OXA48CA in presence of the combination of IMP and CHLX showed a characteristic behavior of the persistent subpopulation.

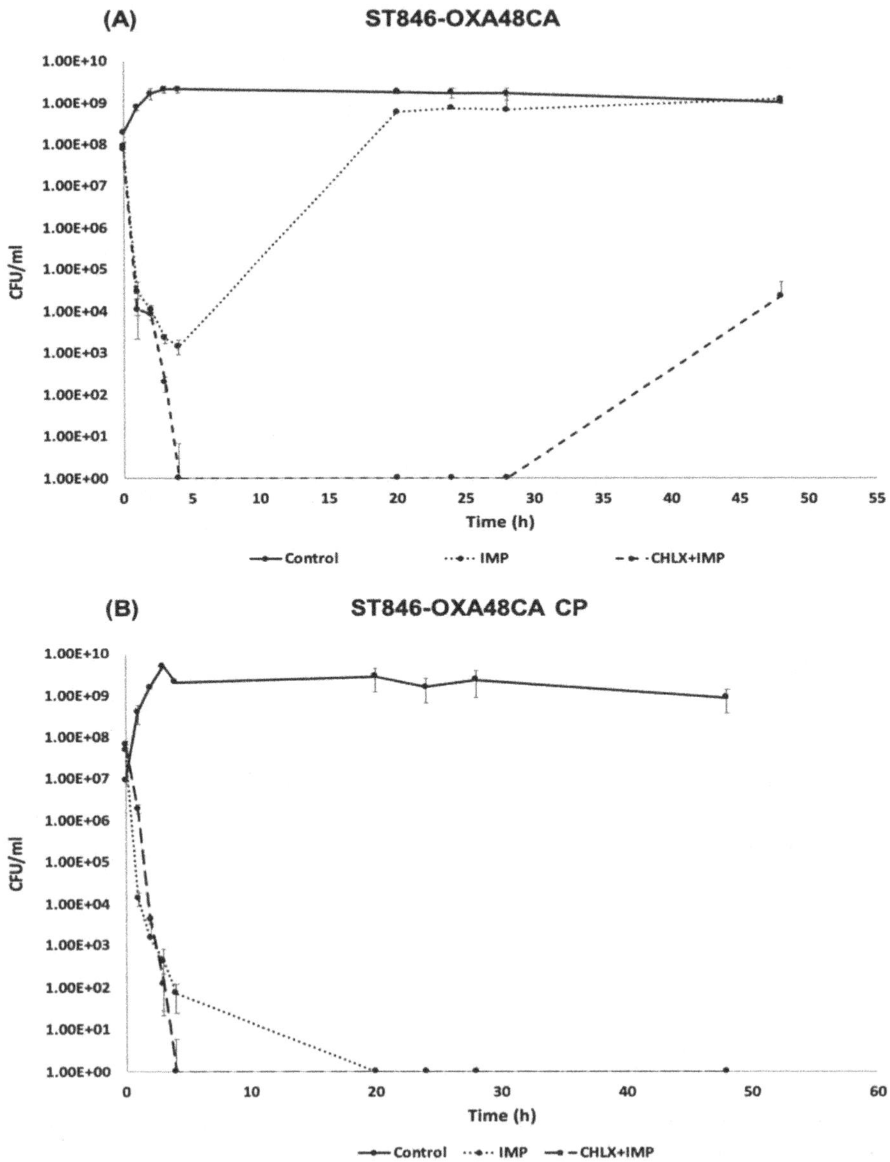

Figure 4. Time-killing curve in the presence of IMP (50 × MIC) and in presence of the combination of IMP (50 × MIC) and CHLX (10 × MIC) for the strains of *K. pneumoniae* ST846-OXA48CA (**A**) and ST846-OXA48CA CP (**B**). The controls are the strains without exposure to any stress (IMP, IMP + CHLX). The error bars represent the standard deviation of the three replicates of the experiment.

In the case of the ST846-OXA48CA CP strain (Figure 4B), in which the plasmid was removed by means of a curing agent, 3% sodium dodecyl sulfate (SDS) (10% *w/v* pH = 7.4), a drastic reduction in the number of CFU, was observed both with IMP alone as for in the combination of IMP and CHLX. In fact, in the case of IMP alone no CFUs were recovered at 20 h while, in the case of the combination, no CFUs were recovered at 4 h (Figure 4). Finally, the culture was considered dead as no regrowth was observed throughout the rest of the assay. Thus, the ST846-OXA48CA CP strain, unlike the ST846-OXA48CA strain, showed more sensitive behavior curve pattern in the presence of IMP alone and in the presence of the combination of drug. These results may suggest that the absence of the plasmid containing both the β-lactamase OXA48 and the TA system PemI/PemK could be a factor responsible for the absence of regrowth. Moreover, the lack of the TA system PemI/PemK could be implicated in the non-appearance of a persistent subpopulation in presence of the combination of IMP and CHLX, contrary to what happens in the strain ST846-OXA48 CA.

2.1.7. Enzymatic Analysis Using the Cell Proliferation Reagent WST-1

The results of the time-killing curves in the presence of the IMP and CHLX combination was confirmed by enzymatic analysis using the cell proliferation reagent WST-1 (Figure 5), which measures the omnipresent reducing agents NADH and NADPH as biochemical markers to evaluate the metabolic activity of the cell [30]. Indeed, ST846-OXA48CA lacks metabolic activity/cell proliferation at 24 h ($OD_{480\ nm} < 0.01$), whereas it presents a significant increase at 48 h ($OD_{480\ nm} > 0.4$; p-value < 0.0001), confirming regrowth in the bacterial culture. In contrast, the ST846-OXA48CA CP strain, despite showing significant differences (p-value < 0.002) between 24 and 48 h in terms of metabolic activity/cell proliferation, is considered as a dead culture since its $OD_{480\ nm}$ is less than 0.1.

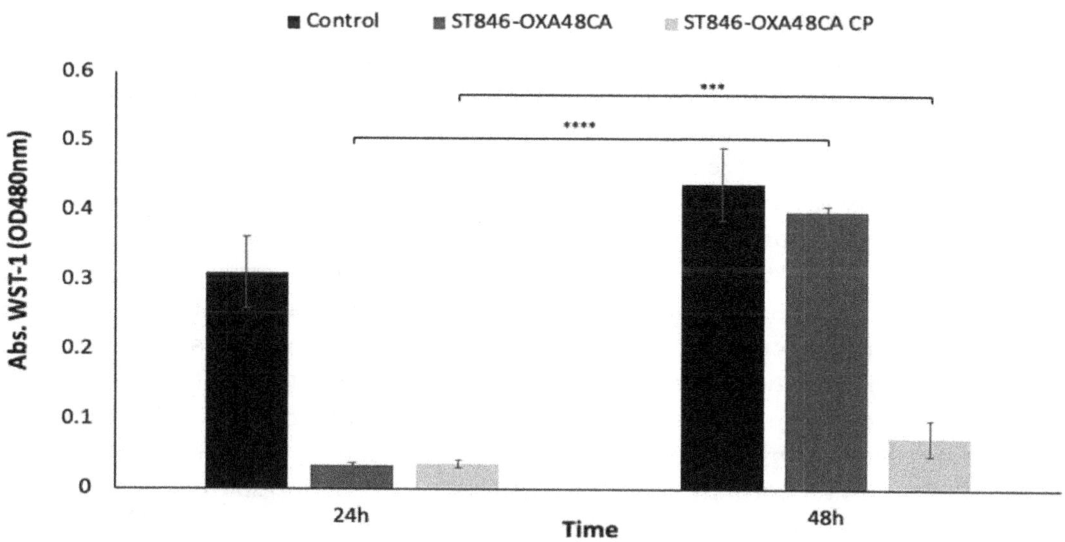

Figure 5. Enzymatic activity by the colorimetric assay (WST-1 based) of the strain *K. pneumoniae* ST846-OXA48CA and ST846-OXA48CA CP in the presence of the combination of IMP (50 × MIC) and CHLX (10 × MIC). The growth control is ST846-OXA48CA strain without antibiotic pressure. ***, p-value < 0.001 and ****, p-value < 0.0001. The errors bars represent the standard deviation of the three experiment replicates.

3. Discussion

Due to the emergence of MDR pathogens over the past few decades, public health officials faces new challenges, such as the alarming increase in antimicrobial resistance, as well as the emerging link between resistance strategies used by bacteria against antibiotics and biocides [18]. This last problem is even more worrisome due to the routinely and uncontrolled use of antiseptics and biocides in clinical practice [9]. One example of this is CHLX, a bis-biguanide antiseptic of cationic nature that has bactericidal activity through membrane disruption [31]. For these reasons it of great interest to decipher the molecular mechanisms involved in the adaptation to CHLX in clinical strains of *K. pneumoniae*, producers of carbapenemases.

In order to determine at the molecular level the effect of the adaptation to CHLX in a strains of *K. pneumoniae*, we performed a phenotypic study in the presence of CHLX (10 × MIC), which showed that the adaptation to CHLX led to the activation of two different molecular mechanisms in the clinical strains of *K. pneumoniae* ST258-KPC3CA and ST846-OXA48CA. In effect, the ST258-KPC3CA strain presented a growth curve typical of resistant bacteria, where a slight reduction in the bacterial population occurs during the time of activation of defense mechanisms [32] (e.g., efflux pumps, TA systems, quorum network), followed by a regrowth period similar to the control. In contrast, ST846-OXA48CA had a characteristic growth curve of tolerant bacteria, where the strain undergoes a drastic reduction or arrest of growth during the first four hours of exposure to the bactericide [31]. These results were corroborated by the transcriptomic study where the transcriptomic profile of the ST258-KPC3CA strain revealed the overexpression of a larger number of genes compared to ST846-OXA48CA strain, especially those related to the overexpression of efflux pumps, which are generally considered as a basic molecular mechanism associated with resistance [33]. In our study, the efflux pump gene that was overexpressed in ST258-KPC3CA was the methyl viologen resistance gene *smvA*. SmvA is responsible for resistance to CHLX in *K. pneumoniae* strains previously adapted to CHLX, due to the interruption of the *smvR* gene (*smvA* repressor) [9].

The antimicrobial susceptibility test revealed the appearance of cross-resistance to colistin (polymyxin E), in the ST846-OXA48CA strain tolerant to CHLX, increasing its MIC value 32-fold. These results were also corroborated by the transcriptomic studies, in which a high level of expression of the colistin resistance genes, *pmrD* and *pmrK* could be observed. This phenomenon of cross-resistance was also described previously, where five out of six strains of *K. pneumoniae* adapted to CHLX also presented resistance to colistin, increasing their MIC values from 2–4 mg/L to 64 mg/L [9]. This phenotype is due to the common biochemical characteristics of CHLX and colistin: in fact, both are cationic compounds with hydrophobic functions [6]. As for the other antibiotics tested, no change in the MIC value was observed in the strain adapted to CHLX compared to the wild-types.

In the previous study of Fernández-García et al. (2018), a combination of β-lactam antibiotic IMP with the CHLX biocide caused in some strains of *Acinetobacter baumannii* the formation of a subpopulation of persistent bacteria [12]. This phenomenon could be observed in the ST846-OXA48CA strain in the presence of the same combination. However, we have seen that the combination of this β-lactam antibiotic and the biocide did not lead to the appearance of a persistent subpopulation in ST846-OXA48 CA CP, cured of the plasmid, but rather led to the death of the bacterial culture after 4 h. This suggests that the presence of the TA systems carried by the plasmid, PemI/PemK and RelE/RelB, could be responsible of the emergence of a persistent subpopulation. As it has long been shown that TA systems are involved in the formation of persistent bacterial subpopulations [19,34,35]; indeed, TA systems are genetic elements composed of a toxin, which inhibits bacterial growth by interfering with essential cellular processes, and an antitoxin, which is able to neutralize the effect of the toxin in normal growth conditions [11,18,36]. The *pemI/pemK* genetic module present in the plasmid of *K. pneumoniae* ST846-OXA48CA, has never been described in *K. pneumoniae*. This TA system was characterized and overexpression assays confirmed that this module corresponds to a TA system. The overexpression of the *pemK* gene led to the inhibition of bacterial growth; however, overexpression of the *pemIK* module led to normal bacterial growth. The same phenomenon was observed in *Bacillus anthracis* where the overexpression of *pemK* in the pHCMC05 vector was severely toxic to the growth of *B. anthracis* cells [37].

In recent decades, many studies have shown that TA systems are associated with the formation of biofilms [22,38]. Biofilms are characterized by a dense multicellular community of microorganisms, constituted after the attachment of bacteria to a biotic or an abiotic surface [39]. In fact, the first TA system linked to biofilm formation was the MqsR/MqsA system of *Escherichia coli* [13]. The toxins have been described as modulators of biofilm formation [13,40]. In this study, we have seen that the overexpression of the PemK toxin contributed to a significant decrease in biofilm formation. The effect that we observed for the toxin is corroborated by the study of García-Contreras et al. (2008), [41], where the *hha*-deletion mutant significantly increased biofilm formation. In addition, the complemented *hha* mutant showed a consistently and dramatically inhibition of biofilm formation. Besides, Ma et al. (2019), described that the disruption of MazF toxin, in *Staphylococcus aureus*, led to an increase in biofilm formation in an *ica*-cluster dependent way, as the disruption of *mazF* produced an increase in the level of expression of *icaA*, *icaB*, and *icaC* genes [42]. Furthermore, in the study of Kim et al. (2009), the authors described that the overexpression of five toxins conduced to a decrease in biofilm formation at 8 h, although at 24 h they observed an increase in biofilm [14].

4. Conclusions

This is the first study that describes the different effects of the adaptation to CHLX in two clinical strains of *K. pneumoniae*, producers of carbapenemase, that become resistant (ST258-KPC3CA) and tolerant (ST846-OXA48CA) to CHLX. This adaptation has lead, in the case of ST846-OXA48CA strain, to the development of cross-resistance to colistin, an antibiotic of last resort in hospital infections.

Furthermore, this study is the first one to describe the relationship between the mechanisms of bacterial persistence and the combination of a β-lactam antibiotic (IMP) and a biocide (CHLX) in the clinical isolate of *K. pneumoniae* ST846-OXA48CA. Finally, a new PemI/PemK TA system was identified in a plasmid of the ST846-OXA48CA strain. Its subsequent characterization demonstrated it participates in the development of persister cells as well as the establishment of biofilms.

5. Materials and Methods

5.1. Bacterial Strains and CHLX Adaptation

Two clinical strains of *K. pneumoniae*, producers of carbapenemases harbored in plasmids, ST258-KPC3 (high-risk clone) and ST846-OXA48 (low-risk clone) were used in this study. These clinical strains belonging to different sequence types (ST) and are from urine and sputum samples, respectively. The attribution of the ST was carried out in the study of Esteban-Cantos et al. (2017) according to the scheme of the Pasteur Institute (http://bigsdb.web.pasteur.fr/Klebsiella/) [43].

In order to obtain the CHLX adapted strains, the clinical strains of *K. pneumoniae* clinical ST258-KPC3 and ST846-OXA48, were exposed for two weeks to $\frac{1}{4}$ of their MIC of CHLX (9.8 and 19.5 μg/mL), respectively in liquid media with aeration. The antibiotic was replaced every 24 h. After these two weeks exposure, the two CA strains were obtained with MIC values of 39.1 and 78.2 μg/mL, respectively.

Furthermore, to study the role of the plasmid (Figure S1) encoding the PemI/PemK TA system in the ST846-OXA48CA strain plasmids were removed from the strain. The cured plasmid ST846-OXA48CA CP strain of *K. pneumoniae* was generated following the protocol of El-Mansi et al. (2000) [44], where 3% sodium dodecyl sulfate (SDS) (10% *w*/*v* pH = 7.4) was used as a curing agent. To check the effective loss of the plasmid, PCR was carried out using the verification primers and, to corroborate the results, plasmid extractions were also performed and subsequently loaded on a 1% agarose gel.

All the bacterial strains and plasmids used in this study shown in the Table 3.

Table 3. Description of the bacterial strains and plasmids used in this study.

Strain or Plasmid	Main Characteristics	Source or Reference
ST258-KPC3	*K. pneumoniae* high-risk clone carrying β-lactamase KPC3	This study
ST258-KPC3 CA	*K. pneumoniae* high-risk clone carrying β-lactamase KPC3 adapted to CHLX	This study

Table 3. *Cont.*

ST846-OXA48	*K. pneumoniae* low-risk clone carrying β-lactamase OXA48	This study
ST846-OXA48 CA	*K. pneumoniae* low-risk clone carrying β-lactamase OXA48 adapted to CHLX	This study
ST846-OXA48 CA CP	*K. pneumoniae* low-risk clone carrying β-lactamase OXA48 adapted to CHLX and cured plasmid strain	This study
pCA24N	Expression plasmid Cm^R, LacIq	[45]
pCA24N (*pemIK*)	Expression plasmid pCA24N with the TA systems *pemIK*	This study
pCA24N (*pemK*)	Expression plasmid pCA24N with the TA systems *pemIK*	This study

5.2. Time-Killing Curve

The different time-killing curves were performed according to Hofsteenge et al. (2016) [8] in low-nutrient Luria-Bertani (LN-LB) broth (2 g/L tryptone, 1 g/L yeast extract and 5 g/L NaCl). The culture was incubated at 37 °C with shaking (180 rpm) until it reached the optical density at 600 nm ($OD_{600\ nm}$) of 0.6. At that moment, CHLX digluconate (10 × MIC) (Sigma-Aldrich, Darmstadt, Germany), IMP (50 × MIC) (Sigma-Aldrich, Darmstadt, Germany), or the combination of IMP (50 × MIC) and CHLX (10 × MIC), were added. Bacterial concentrations (CFU/mL) were determined at 0, 1, 2, 3, 4, 20, 24, 28, and 48 h by serial dilutions and plating on LB agar (10 g/L tryptone, 5 g/L yeast extract, 5 g/L NaCl, and 20 g/L agar). All experiments were performed in triplicates.

The MIC value for CHLX and IMP of the strains ST258-KPC3CA and ST846-OXA48CA are shown in Table 2 (MIC value of CHLX: 39.1 μg/mL and 78.2 μg/mL, respectively; and value of MIC of IMP of strain ST846-OXA48CA is 16 μg/mL). Therefore, 391 μg/mL and 782 μg/mL of CHLX was added in the time-killing curves in presence of CHLX digluconate (10 × MIC) in the strains ST258-KPC3CA and ST846-OXA48CA, respectively. In the case of the strains ST846-OXA48CA and ST846-OXA48CA CP, where the time-killing curves were performed in presence of (50 × MIC) IMP alone or in combination with (10 × MIC) CHLX, 800 μg/mL of IMP was added.

5.3. Transcriptomic Study

The strain ST258-KPC3, ST258-KPC3CA, ST846-OXA48, and ST846-OXA48CA were cultured on solid LB plates, with or without CHLX according to the requirement of the strains and incubated at 37 °C for 24 h. One colony was removed and inoculated in liquid LB medium, with (1/16 × MIC) CHLX for the CA strain and incubated overnight at 37 °C with shaking (180 rpm). The inoculum was diluted (1:100) in LB media with (1/4 × MIC) CHLX for CA strains and without CHLX for wild-type strains. Then, when the culture reached a logarithmic growth phase ($OD_{600\ nm}$: 0.6), the RNA was extracted using the High Pure RNA Isolation kit (Roche, Mannheim, Germany), following the kit instructions, and the extract was treated with DNase (Roche, Mannheim, Germany). The RNA was subsequently quantified in a Nanodrop ND-100 spectrophometer (NanoDrop Technologies, Waltham, MA, USA) and was analyzed with an Agilent 2100 Bioanalyzer with RNA 6000 Nano reagents and RNA Nano Chips (Agilent Technologies, Santa Clara, CA, USA) to determine the quality and the integrity of the samples. All extractions were carried out for quadruple. After RNA extraction of the four replicates of each strain, rRNA and tRNA were removed using the Ribo-Zero rRNA removal kits (bacteria) (Illumina, San diego, CA, USA). The rRNA depletion was checked using Agilent RNA ScreenTape Assay (TapeStation 4200 Agilent, Santa Clara, CA, USA) and the RNA was quantified by Qubit TM RNA HS Assay Kit (Thermo Fisher Scientific, Waltham, MA, USA). Then, the transcriptomic libraries were performed using the Ion RNAseq Kit v2 in combination with the Ion Xpress TM RNA-Seq Barcode 01-16 Kit. Sequencing was carried out by emulsion PCR with Ion Sphere Particles (IPs). The enrichment of the library carrying IPs and the subsequent loading of the chip was developed in the Ion Chef automated system using the Ion PI Hi-Q Chef kit (Thermo Fisher Scientific, Waltham, MA, USA) and Ion PI chips v3 (Thermo Fisher Scientific, Waltham, MA, USA). Chip analysis was carried out on the Ion Proton sequencer (Thermo Fisher Scientific, Waltham, MA, USA). The generated data was analyzed using the specific software of the Torrent Suite 5.6.2 platform (Thermo Fisher Scientific, Waltham, MA, USA). The resulting readings (approx. 60 million) were exported in a FASTQ file using the FileExporter 4.6.0.0 (Thermo Fisher Scientific, Waltham, MA, USA) plugin. The alignment of the sequences against their respective controls was performed with the STAR software. Subsequently, the aligned readings were counted using the htseq-count software (HTSeq 0.6.1.p2). All data was normalized using DESeq2.

5.4. Antimicrobial Susceptibility Test

MIC values of chlorhexidine, ciprofloxacin, tigecycline, tobramycin, imipenem, meropenem, gentamycin, piperacillin-tazobactam, ceftazidime, sulbactam, netilmicin, doxycycline, amikacin, minocycline, and colistin were determined by broth microdilution according to Clinical and Laboratory Standards Institute (CLSI) 2018 [46].

5.5. Construction of pCA24N (pemIK) and pCA24N (pemK)

The plasmids pCA24N (*pemIK*) and pCA24N (*pemK*) were constructed by amplifying the *pemI/pemK* module and the *pemK* gene in the expression vector pCA24N (cmR; LacIq) [45] inducible by IPTG (Fisher Scientific). The insertion of these genes was performed at the BseRI and NotI restriction sites under the control of T5-lac promotor (Table 4). Final constructions were verified by DNA sequencing.

Table 4. Oligonucleotide used for cloning and sequencing.

Primer Name	Sequences	Sense	Reference
	Oligonucleotide for clonation		
PemI_Fow(BseRI)	GAGGAGAAATTAACTATCATGCATACCACTCGACTG	5′-3′	This study
PemK_Fow(BseRI)	GAGGAGAAATTAACTATCATGGAAAGAGGGGAAATC	5′-3′	This study
PemK_Rev(NotI)	ATAAGAATGCGGCCGCCGCTCAGGTCAGGATGGTGGC	5′-3′	This study
	Oligonucleotide for sequencing		
pCA24N Up	GCCCTTTCGTCTTCAC	5′-3′	This study
pCA24N Down	GAACTCCATCTGGATTTGTT	5′-3′	This study

5.6. Toxicity Assay

The toxicity assay was performed as previously described Wood T.L. and Wood T.K. (2016) [47] with certain modifications. Overnight cultures of *K. pneumoniae* strains ST846-OXA48CA CP/pCA24N, ST846-OXA48CA CP/pCA24N (*pemIK*) and ST846-OXA48CA CP/pCA24N (*pemK*) were inoculated into 25 mL of LB broth medium with chloramphenicol (60 μg/mL; Sigma Aldrich, Darmstadt, Germany) to maintain the plasmid. IPTG (1 mM) was added when $OD_{600\ nm}$ was 0.3. For 200 min, the $OD_{600\ nm}$ was measured to determine growth evolution. All experiments were performed in triplicates.

5.7. Biofilm Formation Assay

The biofilm formation assay was performed in a 96-well polystyrene plate for 30 h. Briefly, cells of ST846-OXA48CA CP/pcA24N, ST846-OXA48CA CP/pCA24N (*pemIK*) and ST846-OXA48CA CP/pCA24N (*pemK*) were inoculated at $OD_{600\ nm}$ was 0.05 and incubated at 37 °C without shaking. After 6 h, IPTG at 1 mM was added and cells were incubated at 37 °C without shaking. 0.4 mM of IPTG was added every 8 h to avoid its degradation and to be able to observe the action of either the TA system or the toxin, as they are under an IPTG inducible promoter. Then, the cell density ($OD_{600\ nm}$)

and total biofilm ($OD_{580\ nm}$) were measured by using 10% crystal violet staining, and quantified in a NanoQuant plate reader. Normalized biofilm was calculated by dividing the total biofilm by the bacterial growth for each strain. All experiments were performed in triplicates.

5.8. Enzymatic Assay Using the Cell Proliferation Reagent WST-1

The cell proliferation/metabolic activity of the ST846-OXA48CA and ST846-OXA48CA CP strains in presence of the combination of IMP (50 × MIC) and CHLX (10 × MIC) was analyzed using a colorimetric enzymatic assay based on the water soluble tetrazolium salt (WST-1) reagent and electron mediators (Roche, Mannheim, Germany). Tetrazolium salts have become some of the most widely used tools in cell biology for measuring the metabolic activity of cells ranging from mammalian to microbial origin [48,49]. Briefly, the cultures of ST846-OXA48 and ST846-OXA48CA CP were incubated at 37 °C with shaking (180 rpm) until $OD_{600\ nm}$ was 0.6. At that moment the combination of IMP and CHLX, was added. After 24 and 48 h of antibiotic exposition and two washing, the culture cell ($OD_{600\ nm}$ = 0.1) was put in 96-well polystyrene plate (Corning Incorporated, NY, USA) and 10 µL of the reagent was added. After 1 h of incubation at 37 °C without shaking and 10 min with shaking (180 rpm), the optical density was measured at $OD_{480\ nm}$. The ST846-OXA48CA strain without antibiotic addition was taken as a control. The $OD_{480\ nm}$ of the medium culture (LN-LB) in the presence of WST-1 reagent was used to normalize all data. All experiments were performed in triplicates.

5.9. Statistical Analysis

Statistical analysis was based on the number of populations and comparisons. A Student's *t*-test was used to compare two populations. All statistical analysis were performed using the GraphPad (Prism 8) software.

Author Contributions: Conceptualization of the study was carried out by, I.B., L.B., M.D.-V., A.G.-d.-T., A.A., L.F.-G., M.L., Methodology, validation, formal analysis and writing, T.K.W., J.O.-I., A.P., G.B., F.F.-C., M.T., Supervision, analysis, writing and funding. All authors have read and agreed to the published version of the manuscript.

Acknowledgments: Spanish Network for the Research in Infectious Diseases (REIPI, RD16/0016/0001, RD16/CIII/0004/0002 and RD16/0016/0006) and by the Study Group on Mechanisms of Action and Resistance to Antimicrobials, GEMARA (SEIMC, http://www.seimc.org/).

References

1. Mulani, M.S.; Kamble, E.E.; Kumkar, S.N.; Tawre, M.S.; Pardesi, K.R. Emerging Strategies to Combat ESKAPE Pathogens in the Era of Antimicrobial Resistance: A Review. *Front. Microbiol.* **2019**, *10*, 539. [CrossRef] [PubMed]
2. Yang, Q.E.; Walsh, T.R. Toxin-antitoxin systems and their role in disseminating and maintaining antimicrobial resistance. *FEMS Microbiol. Rev.* **2017**, *41*, 343–353. [CrossRef] [PubMed]
3. Santajit, S.; Indrawattana, N. Mechanisms of Antimicrobial Resistance in ESKAPE Pathogens. *BioMed Res. Int.* **2016**, *2016*, 2475067. [CrossRef]
4. Cadavid, E.; Robledo, S.M.; Quiñones, W.; Echeverri, F. Induction of Biofilm Formation in. *Antibiotics* **2018**, *7*, 103. [CrossRef]
5. Cieplik, F.; Jakubovics, N.S.; Buchalla, W.; Maisch, T.; Hellwig, E.; Al-Ahmad, A. Resistance Toward Chlorhexidine in Oral Bacteria—Is There Cause for Concern? *Front. Microbiol.* **2019**, *10*, 587. [CrossRef]
6. Hashemi, M.M.; Holden, B.S.; Coburn, J.; Taylor, M.F.; Weber, S.; Hilton, B.; Zaugg, A.L.; McEwan, C.; Carson, R.; Andersen, J.L.; et al. Proteomic Analysis of Resistance of Gram-Negative Bacteria to Chlorhexidine

and Impacts on Susceptibility to Colistin, Antimicrobial Peptides, and Ceragenins. *Front. Microbiol.* **2019**, *10*, 210. [CrossRef]

7. Kampf, G. Acquired resistance to chlorhexidine—Is it time to establish an 'antiseptic stewardship' initiative? *J. Hosp. Infect.* **2016**, *94*, 213–227. [CrossRef]
8. Venter, H.; Henningsen, M.L.; Begg, S.L. Antimicrobial resistance in healthcare, agriculture and the environment: The biochemistry behind the headlines. *Essays Biochem.* **2017**, *61*, 1–10. [CrossRef]
9. Wand, M.E.; Bock, L.J.; Bonney, L.C.; Sutton, J.M. Mechanisms of Increased Resistance to Chlorhexidine and Cross-Resistance to Colistin following Exposure of Klebsiella pneumoniae Clinical Isolates to Chlorhexidine. *Antimicrob. Agents Chemother.* **2017**, *61*. [CrossRef]
10. Trastoy, R.; Manso, T.; Fernández-García, L.; Blasco, L.; Ambroa, A.; Pérez Del Molino, M.L.; Bou, G.; García-Contreras, R.; Wood, T.K.; Tomás, M. Mechanisms of Bacterial Tolerance and Persistence in the Gastrointestinal and Respiratory Environments. *Clin. Microbiol. Rev.* **2018**, *31*. [CrossRef]
11. Harms, A.; Maisonneuve, E.; Gerdes, K. Mechanisms of bacterial persistence during stress and antibiotic exposure. *Science* **2016**, *354*. [CrossRef] [PubMed]
12. Fernández-García, L.; Fernandez-Cuenca, F.; Blasco, L.; López-Rojas, R.; Ambroa, A.; Lopez, M.; Pascual, Á.; Bou, G.; Tomás, M. Relationship between Tolerance and Persistence Mechanisms in Acinetobacter baumannii Strains with AbkAB Toxin-Antitoxin System. *Antimicrob. Agents Chemother.* **2018**, *62*. [CrossRef] [PubMed]
13. Ren, D.; Bedzyk, L.A.; Thomas, S.M.; Ye, R.W.; Wood, T.K. Gene expression in Escherichia coli biofilms. *Appl. Microbiol. Biotechnol.* **2004**, *64*, 515–524. [CrossRef] [PubMed]
14. Kim, Y.; Wang, X.; Ma, Q.; Zhang, X.S.; Wood, T.K. Toxin-antitoxin systems in Escherichia coli influence biofilm formation through YjgK (TabA) and fimbriae. *J. Bacteriol.* **2009**, *191*, 1258–1267. [CrossRef]
15. Moyed, H.S.; Bertrand, K.P. hipA, a newly recognized gene of Escherichia coli K-12 that affects frequency of persistence after inhibition of murein synthesis. *J. Bacteriol.* **1983**, *155*, 768–775. [CrossRef]
16. Korch, S.B.; Henderson, T.A.; Hill, T.M. Characterization of the hipA7 allele of Escherichia coli and evidence that high persistence is governed by (p)ppGpp synthesis. *Mol. Microbiol.* **2003**, *50*, 1199–1213. [CrossRef]
17. Bokinsky, G.; Baidoo, E.E.; Akella, S.; Burd, H.; Weaver, D.; Alonso-Gutierrez, J.; García-Martín, H.; Lee, T.S.; Keasling, J.D. HipA-triggered growth arrest and β-lactam tolerance in Escherichia coli are mediated by RelA-dependent ppGpp synthesis. *J. Bacteriol.* **2013**, *195*, 3173–3182. [CrossRef]
18. Page, R.; Peti, W. Toxin-antitoxin systems in bacterial growth arrest and persistence. *Nat. Chem. Biol.* **2016**, *12*, 208–214. [CrossRef]
19. Wood, T.K.; Knabel, S.J.; Kwan, B.W. Bacterial persister cell formation and dormancy. *Appl. Environ. Microbiol.* **2013**, *79*, 7116–7121. [CrossRef]
20. Ogura, T.; Hiraga, S. Mini-F plasmid genes that couple host cell division to plasmid proliferation. *Proc. Natl. Acad. Sci. USA* **1983**, *80*, 4784–4788. [CrossRef]
21. Kamruzzaman, M.; Iredell, J. A ParDE-family toxin antitoxin system in major resistance plasmids of Enterobacteriaceae confers antibiotic and heat tolerance. *Sci. Rep.* **2019**, *9*, 9872. [CrossRef] [PubMed]
22. Wang, X.; Wood, T.K. Toxin-antitoxin systems influence biofilm and persister cell formation and the general stress response. *Appl. Environ. Microbiol.* **2011**, *77*, 5577–5583. [CrossRef] [PubMed]
23. Pecota, D.C.; Wood, T.K. Exclusion of T4 phage by the hok/sok killer locus from plasmid R1. *J. Bacteriol.* **1996**, *178*, 2044–2050. [CrossRef] [PubMed]
24. Hazan, R.; Engelberg-Kulka, H. Escherichia coli mazEF-mediated cell death as a defense mechanism that inhibits the spread of phage P1. *Mol. Genet. Genom.* **2004**, *272*, 227–234. [CrossRef]
25. Song, S.; Wood, T.K. Post-segregational Killing and Phage Inhibition Are Not Mediated by Cell Death Through Toxin/Antitoxin Systems. *Front. Microbiol.* **2018**, *9*, 814. [CrossRef]
26. Mathers, A.J.; Peirano, G.; Pitout, J.D. The role of epidemic resistance plasmids and international high-risk clones in the spread of multidrug-resistant Enterobacteriaceae. *Clin. Microbiol. Rev.* **2015**, *28*, 565–591. [CrossRef]
27. Domokos, J.; Damjanova, I.; Kristof, K.; Ligeti, B.; Kocsis, B.; Szabo, D. Multiple Benefits of Plasmid-Mediated Quinolone Resistance Determinants in. *Front. Microbiol.* **2019**, *10*, 157. [CrossRef]
28. Wang, X.; Lord, D.M.; Cheng, H.Y.; Osbourne, D.O.; Hong, S.H.; Sanchez-Torres, V.; Quiroga, C.; Zheng, K.; Herrmann, T.; Peti, W.; et al. A new type V toxin-antitoxin system where mRNA for toxin GhoT is cleaved by antitoxin GhoS. *Nat. Chem. Biol.* **2012**, *8*, 855–861. [CrossRef]

29. Fernandez-Garcia, L.; Kim, J.S.; Tomas, M.; Wood, T.K. Toxins of toxin/antitoxin systems are inactivated primarily through promoter mutations. *J. Appl. Microbiol.* **2019**, *127*, 1859–1868. [CrossRef]
30. Präbst, K.; Engelhardt, H.; Ringgeler, S.; Hübner, H. Basic Colorimetric Proliferation Assays: MTT, WST, and Resazurin. *Methods Mol. Biol.* **2017**, *1601*, 1–17. [CrossRef]
31. Gilbert, P.; Moore, L.E. Cationic antiseptics: Diversity of action under a common epithet. *J. Appl. Microbiol.* **2005**, *99*, 703–715. [CrossRef] [PubMed]
32. Munita, J.M.; Arias, C.A. Mechanisms of Antibiotic Resistance. *Microbiol. Spectr.* **2016**, *4*. [CrossRef] [PubMed]
33. Brauner, A.; Fridman, O.; Gefen, O.; Balaban, N.Q. Distinguishing between resistance, tolerance and persistence to antibiotic treatment. *Nat. Rev. Microbiol.* **2016**, *14*, 320–330. [CrossRef] [PubMed]
34. Keren, I.; Shah, D.; Spoering, A.; Kaldalu, N.; Lewis, K. Specialized persister cells and the mechanism of multidrug tolerance in Escherichia coli. *J. Bacteriol.* **2004**, *186*, 8172–8180. [CrossRef] [PubMed]
35. Tashiro, Y.; Kawata, K.; Taniuchi, A.; Kakinuma, K.; May, T.; Okabe, S. RelE-mediated dormancy is enhanced at high cell density in Escherichia coli. *J. Bacteriol.* **2012**, *194*, 1169–1176. [CrossRef]
36. Lewis, K. Persister cells. *Annu. Rev. Microbiol.* **2010**, *64*, 357–372. [CrossRef]
37. Agarwal, S.; Bhatnagar, R. Identification and characterization of a novel toxin-antitoxin module from Bacillus anthracis. *FEBS Lett.* **2007**, *581*, 1727–1734. [CrossRef]
38. Kędzierska, B.; Hayes, F. Emerging Roles of Toxin-Antitoxin Modules in Bacterial Pathogenesis. *Molecules* **2016**, *21*, 790. [CrossRef]
39. Martínez, L.C.; Vadyvaloo, V. Mechanisms of post-transcriptional gene regulation in bacterial biofilms. *Front. Cell. Infect. Microbiol.* **2014**, *4*, 38. [CrossRef]
40. González Barrios, A.F.; Zuo, R.; Hashimoto, Y.; Yang, L.; Bentley, W.E.; Wood, T.K. Autoinducer 2 controls biofilm formation in Escherichia coli through a novel motility quorum-sensing regulator (MqsR, B3022). *J. Bacteriol.* **2006**, *188*, 305–316. [CrossRef]
41. García-Contreras, R.; Zhang, X.S.; Kim, Y.; Wood, T.K. Protein translation and cell death: The role of rare tRNAs in biofilm formation and in activating dormant phage killer genes. *PLoS ONE* **2008**, *3*, e2394. [CrossRef] [PubMed]
42. Ma, D.; Mandell, J.B.; Donegan, N.P.; Cheung, A.L.; Ma, W.; Rothenberger, S.; Shanks, R.M.Q.; Richardson, A.R.; Urish, K.L. The Toxin-Antitoxin MazEF Drives Staphylococcus aureus Biofilm Formation, Antibiotic Tolerance, and Chronic Infection. *mBio* **2019**, *10*. [CrossRef] [PubMed]
43. Esteban-Cantos, A.; Aracil, B.; Bautista, V.; Ortega, A.; Lara, N.; Saez, D.; Fernández-Romero, S.; Pérez-Vázquez, M.; Navarro, F.; Grundmann, H.; et al. The Carbapenemase-Producing Klebsiella pneumoniae Population Is Distinct and More Clonal than the Carbapenem-Susceptible Population. *Antimicrob. Agents Chemother.* **2017**, *61*. [CrossRef]
44. El-Mansi, M.; Anderson, K.J.; Inche, C.A.; Knowles, L.K.; Platt, D.J. Isolation and curing of the *Klebsiella pneumonia* large indigenous plasmid using sodium dodecyl sulphate. *Res. Microbiol.* **2000**, *151*, 201–208. [CrossRef]
45. Kitagawa, M.; Ara, T.; Arifuzzaman, M.; Ioka-Nakamichi, T.; Inamoto, E.; Toyonaga, H.; Mori, H. Complete set of ORF clones of Escherichia coli ASKA library (a complete set of *E. coli* K-12 ORF archive): Unique resources for biological research. *DNA Res.* **2005**, *12*, 291–299. [CrossRef] [PubMed]
46. Clinical and Laboratory Standard Institute. *Performance Standards for Antimicrobial Sucesceptibilty Testing*, 28th ed.; CLSI suplement M100; Clinical and Laboratory Standard Institute: Wayne, PA, USA, 2018.
47. Wood, T.L.; Wood, T.K. The HigB/HigA toxin/antitoxin system of Pseudomonas aeruginosa influences the virulence factors pyochelin, pyocyanin, and biofilm formation. *Microbiologyopen* **2016**, *5*, 499–511. [CrossRef]
48. Berridge, M.V.; Herst, P.M.; Tan, A.S. Tetrazolium dyes as tools in cell biology: New insights into their cellular reduction. *Biotechnol. Annu. Rev.* **2005**, *11*, 127–152. [CrossRef]
49. Tsukatani, T.; Suenaga, H.; Higuchi, T.; Akao, T.; Ishiyama, M.; Ezoe, K.; Matsumoto, K. Colorimetric cell proliferation assay for microorganisms in microtiter plate using water-soluble tetrazolium salts. *J. Microbiol. Methods* **2008**, *75*, 109–116. [CrossRef]

8

The Fst/Ldr Family of Type I TA System Toxins: Potential Roles in Stress Response, Metabolism and Pathogenesis

Keith Weaver

Division of Basic Biomedical Sciences, Sanford School of Medicine, University of South Dakota, Vermillion, SD 57069, USA; kweaver@usd.edu

Abstract: The par_{pAD1} locus was the first type I toxin–antitoxin (TA) system described in Gram-positive bacteria and was later determined to be the founding member of a widely distributed family of plasmid- and chromosomally encoded TA systems. Indeed, homology searches revealed that the toxin component, Fst_{pAD1}, is a member of the Fst/Ldr superfamily of peptide toxins found in both Gram-positive and Gram-negative bacteria. Regulation of the Fst and Ldr toxins is distinct in their respective Gram-positive and Gram-negative hosts, but the effects of ectopic over-expression are similar. While, the plasmid versions of these systems appear to play the canonical role of post-segregational killing stability mechanisms, the function of the chromosomal systems remains largely obscure. At least one member of the family has been suggested to play a role in pathogenesis in *Staphylococcus aureus*, while the regulation of several others appear to be tightly integrated with genes involved in sugar metabolism. After a brief discussion of the regulation and function of the foundational par_{pAD1} locus, this review will focus on the current information available on potential roles of the chromosomal homologs.

Keywords: type I toxin–antitoxin system; small protein toxin structure; Fst/Ldr family

Key Contribution: This is an in-depth comparison of the sequence, structure, regulation and function of a widespread family of type I toxin-antitoxin system small-membrane protein toxins and their potential role in stress response, metabolism and pathogenesis.

---

1. Introduction

The par_{pAD1} locus of the *Enterococcus faecalis* pheromone-responsive conjugative plasmid pAD1 was the first type I toxin-antitoxin (TA-1) system identified and characterized in Gram-positive bacteria [1–3]. It consists of a ~230 nucleotide mRNA, designated RNA I_{pAD1} and encoding the 33 amino acid toxin of the system, Fst_{pAD1}, and the ~40 nucleotide sRNA antitoxin, RNA II_{pAD1}. The toxin and antitoxin RNAs are transcribed from convergent promoters and share a bidirectional factor-independent transcriptional terminator [3] (Figure 1A). The resultant complementary terminator stem-loops provide the site of initiation of complex formation between toxin and antitoxin RNAs with a U-turn motif in the toxin RNA playing a key role in forming the initial reversible kissing complex [4] (Figure 1B). In addition, the toxin and antitoxin RNAs are transcribed in opposite directions across a pair of DNA direct repeats, DRa and DRb, which provide a second region of complementarity between the RNAs. Interaction in this region suppresses translation of the toxin by sequestering the Fst_{pAD1} translation initiation site [4–6]. An intramolecular stem-loop within RNA I_{pAD1}, the 5′-SL in Figure 1B, sequesters the Fst_{pAD1} Shine–Dalgarno (SD) sequence until interaction with RNA II_{pAD1} can be completed [7]. Once irreversible interaction is established at all of the complementary regions, RNA I_{pAD1} and RNA II_{pAD1} form a stable complex that facilitates the accumulation of a translationally inactive pool of the toxin message [8]. RNA II_{pAD1} is preferentially degraded from the pool by an unknown mechanism, although an intramolecular "upstream helix" (5′-UH in Figure 1B) sequestering the extreme 5′ end of RNA I_{pAD1} contributes to its differential stability [5,9]. At steady state, the two RNAs are maintained at an approximately equimolar ratio and as long as the plasmid is maintained in the cell, the antitoxin is continually replenished by ongoing transcription [8]. If the plasmid is lost, RNA II_{pAD1} is degraded from the complex pool, Fst_{pAD1} is translated from RNA I_{pAD1}, and host cell growth is inhibited [2,10]. Thus, the par_{pAD1} locus functions as a typical post-segregational killing (PSK) or addiction module [11]. The salient features of the par_{pAD1} locus and the important structural elements of RNA I_{pAD1} and RNA II_{pAD1} are shown in Figure 1.

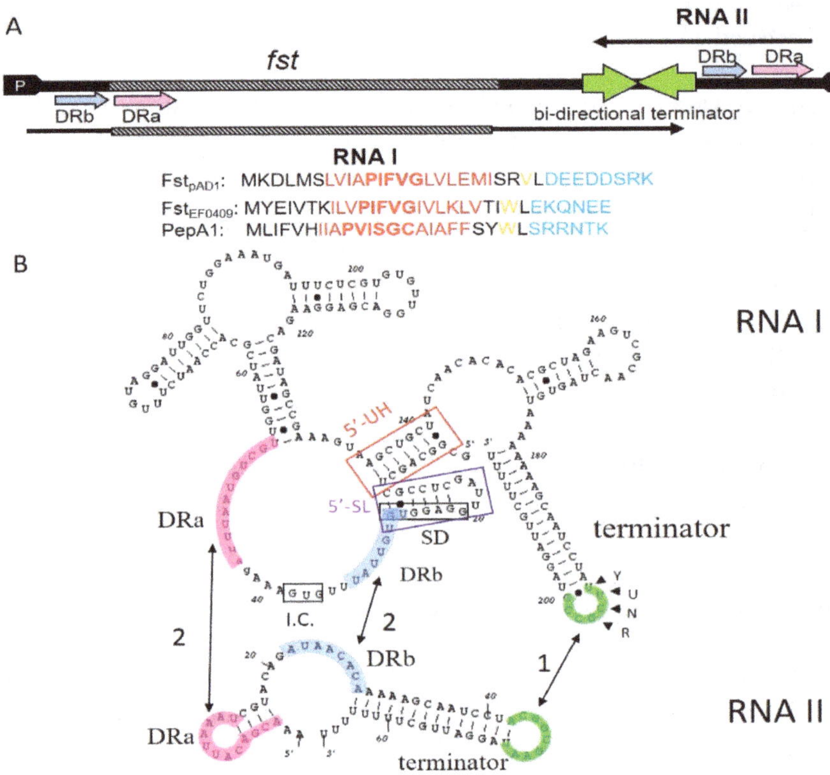

Figure 1. General organization of *par*-like operons and RNA–RNA regulatory interactions as defined in the prototypical par_{pAD1} system. (**A**) Genetic organization of *par* and conserved domains within Fst-like proteins. Convergent promoters (arrows labeled P on each end of the DNA map) produce two transcripts (arrows above and below the DNA map): the toxin message, RNA I, and the regulatory RNA, RNA II. RNAs are transcribed in opposite directions across a pair of direct repeats, DRa and DRb, and a bidirectional terminator providing regions of complementarity for interaction. RNA–RNA interactions at DRa and DRb prevent ribosome binding at the toxin coding sequence, *fst*. Shown below the map are the sequences of the two *E. faecalis* Fst paralogs, the plasmid-encoded Fst_{pAD1} and the chromosomally encoded Fst_{EF0409}, and the *S. aureus* PepA1 sequence. The conserved hydrophobic domain is shown in red with the PXXXG(C) motif bolded, the charged C-terminal tail is shown in blue and the location of the conserved tryptophan residue is show in orange. Fst_{pAD1} has a valine substitution at this position. (**B**) Map of RNA I-RNA II interaction sites. Interaction is initiated at the loop of the intrinsic terminators (green) stimulated by the YUNR U-turn motif in RNA I. Interaction then spreads to the DRa and DRb sequences that sequester the GUG initiation codon (I.C.) and the Shine–Dalgarno sequence (SD) for Fst. The interaction sites are color coordinated with those shown in the genetic map in Figure 1A. Intramolecular structures within RNA I sequester the ribosome binding site to further delay translation of *fst* (5′-SL) and stabilize the RNA I transcript (5′-UH). These figures were modified from [12].

Exhaustive searches later revealed that *par*-like loci are ubiquitous in the genomes of the Firmicutes (Gram-positive bacteria with a low G + C content) with hundreds of homologs of Fst_{pAD1} identified [12–15]. While many of these homologs were present on apparent mobile genetic elements (MGE), including phage and pathogenicity islands as well as plasmids, several loci were also identified on core genomes. Expanded examination of the surrounding sequences revealed that all of the structural features present in par_{pAD1} were highly conserved, including the convergent promoters, the bidirectional intrinsic terminator, the direct repeats and the regions of intramolecular complementarity in RNA I sequestering the 5′ end and the SD sequence, suggesting that the mechanisms of regulation of toxin expression were broadly conserved. However, not all homologs contained a U-turn motif in any of the predicted loop structures and examination of some of the regulatory details of the chromosomal loci suggest that there is variability in the mechanism of regulation (see below). Since the functions of the non-MGE-located loci must be distinct from par_{pAD1}, variability in regulatory details are to be expected. In spite of efforts made to characterize other par_{pAD1}-related loci, the only experimentally verified function of any member of the par_{pAD1} family remains the plasmid stabilization function of its founding member.

The purpose of this review is to provide a detailed comparison of the sequence, structure, regulation and effects of the toxins related to Fst_{pAD1}. This information will be used to speculate on the possible functions of the toxins in the stress response and pathogenesis of the host organisms.

2. Sequence and Structural Comparisons of the Fst/Ldr Family of Small Protein Toxins

Alignment of the Fst_{pAD1} homologs revealed a superfamily signature consisting of a highly conserved tryptophan located between a hydrophobic predicted transmembrane (TM) helix and a highly charged C-terminus [13] (Figure 1A). Interestingly, the conserved tryptophan is replaced by a valine in the Fst_{pAD1} prototype, indicating that it is not essential for function. Surprisingly, this signature was also present in the Ldr family of peptide toxins present in the Gram-negative enterobacterial TA-1 *ldr/rdl* loci [16]. In addition to the defined signature, the majority of the toxins contain a P/D/S/TXXXG(C) motif within the putative transmembrane domain (unpublished observation). The N-terminal residue of the motif varies by clade with the most common PXXXG motif present in *E. faecalis*, *Lactobacillus*, one *Streptococcus pneumoniae* and several *Staphylococcus* clades and the Ldr toxins. The DXXXG motif is present in one *S. pneumoniae* clade, the SXXXG motif in one *Staphylococcus aureus* clade, and the TXXXG motif in several *Staphylococcus* species. The *Listeria* clade is the most divergent of the Fst/Ldr family and has three glycines, one within a PKN(L/I)GF motif which fits in the PXXXG motif clade. Interestingly, all of the *Staphylococcus* isolates, but none of the isolates from other species, contain a cysteine residue at the C-terminal end of the motif and this residue has been postulated to play an important role in the function of this class (see below). While the Ldr and Fst toxins are clearly related, their mechanisms of regulation are distinct; with *ldr/rdl* loci regulated more like the *hok/sok* system of *Escherichia coli* [17].

NMR structures of two Fst family members in membrane mimetics have been determined: the Fst_{pAD1} prototype [18] and the PepA1 toxin of the *S. aureus* SprA1-$SprA1_{AS}$ locus [19] (Figure 2). Fst_{pAD1} forms a TM α-helix with the first ~two and last ~seven amino acids protruding. The α-helix structure is maintained at the proline residue in spite of its potential destabilizing effect, but is slightly bent at the glycine residue. The highly charged C-terminal seven amino acids are intrinsically unstructured. Molecular dynamic simulations suggested that the TM α-helix extends from and includes aspartic acid residues at positions 3 and 26. Since the C-terminus contains a stretch of five negatively charged amino acids at the membrane exit point and the external side of Gram-positive membranes is known to be negatively charged, the authors postulated that the C-terminus is located within the cytosol. The TM helix prediction tool TMpred [20] also predicted this orientation. A similar orientation was also proposed for LdrA by molecular modeling [21]. NMR analysis of PepA1 revealed a discontinuous α-helical TM domain with a flexible hinge near the cysteine residue that is unique to the staphylococcal toxins. Molecular dynamic simulations predicted that this α-helix would condense into an extended straight helix when inserted into the cell membrane. Unlike Fst_{pAD1}, it was predicted that the C-terminal domain of PepA1 was folded while the N-terminal domain was unstructured. It was also proposed that the arginine-rich C-terminus interacts with the anionic head groups of membrane phospholipids to lock the TM domain in place. PepA1 lacks the multiple acidic amino acids present in the Fst_{pAD1} C-terminus, so the possibility that the two toxins insert into the membrane in opposite orientations must be considered. These structural differences might account for the differences in the behavior of the two toxins. Fst_{pAD1}, and also the closely related *fst-Sm* from *Streptococcus mutans*, functions strictly from the inside of the cell and neither lyses bacterial cells nor red blood cells when added extracellularly [18,22,23]. It was proposed that, rather than forming oligomeric pores, the unstructured C-terminus of Fst_{pAD1} might interact with a specific intracellular target [18]. Conversely, PepA1 is capable of lysing both bacterial cells and red blood cells when added extracellularly [24]. It was proposed that the cysteine residue might promote intermolecular disulfide bonds facilitating oligomerization and pore formation [19]. Curiously, PepA2, a *S. aureus* TA-1 toxin 50% identical to PepA1 and containing the same PXXXGC motif, does not lyse bacterial cells extracellularly but is approximately 10X more effective in lysing red blood cells [25]. Thus far, only PepA1 and LdrA have been experimentally determined to be located in the cytoplasmic membrane of their native hosts [19,21].

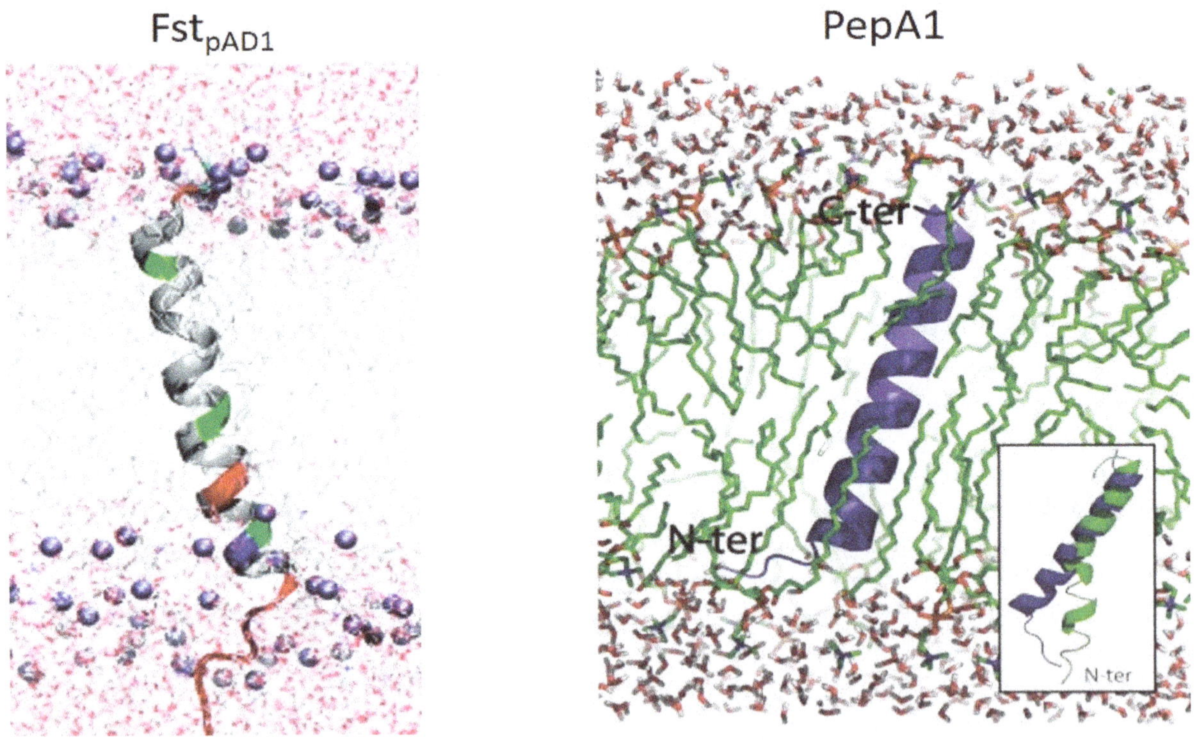

Figure 2. NMR-derived structure of two Fst-family proteins in membrane mimetics. Fst_{pAD1} presents a linear helix postulated to have the N-terminus outside the cell (top) and the C-terminus inside the cell (bottom). Polar amino acids are shown in green, negatively charged residues in red and positively charged residues in blue. Reprinted with permission from [18]. PepA1 forms a similar linear helix spanning the membrane. Membrane interaction appears to straighten a kink produced at the cysteine residue in all staphylococcal Fst-family toxins (inset). Reprinted with permission from [19].

An extensive mutagenesis survey of Fst_{pAD1} revealed several sequence features affecting toxicity in *E. faecalis* [15]. (1) Maintaining the hydrophobicity of the putative TM domain was essential for toxin function. In most cases isoleucine, leucine and valine could be interchanged for one another but alanine substitutions were not tolerated. (2) The conserved proline and glycine residues in the PXXXG motif were essential for toxicity. (3) The two charged amino acids at the N-terminus (lysine and aspartic acid) could be substituted with either negatively or positively charged, but not uncharged, amino acids without disrupting function. (4) Truncation of the charged C-terminus retained toxicity, although recent data suggests that toxicity is reduced (K. Weaver, unpublished results). In the only other mutagenic analysis of an Fst/Ldr family member published to date, the proline residue of the PXXXG motif of the *Lactobacillus rhamnosus* Lpt toxin was shown to be essential for toxicity when expressed in *E. coli* as for Fst_{pAD1} [26]. Unlike Fst_{pAD1}, however, the charged C-terminus was also required.

3. Effects of Toxin Over-Expression

As with most TA toxins, the effects of Fst/Ldr superfamily toxins have been determined primarily in toxin over-producing strains. Thus far, overproduction of Fst_{pAD1} [1] and Fst_{EF0409} [27] of *E. faecalis*, PepA1 [19] and PepA2 [25] of *S. aureus*, *fst-Sm* of *S. mutans* [22], and LdrA [21] and LdrD [16] of *E. coli* have all been shown to cause cell death of their native hosts and, in the case of Fst_{pAD1} [15,28], PepA1 [24] and PepA2 [25], several heterologous hosts as well. In addition, the Lpt toxin from a *par* homolog on a *L. rhamnosus* plasmid was demonstrated to be toxic when over-produced in *E. coli* [26]. Disruption of the cell membrane was demonstrated directly for Fst_{pAD1} [23,28], PepA1 [19] and Lpt [26] by DNA staining with membrane impermeant dyes and inferred for LdrA by simultaneous inhibition of macromolecular synthesis (which was also observed for Fst_{pAD1} [23]) and inhibition of ATP production [21]. Lpt was shown to form pores in the outer membrane of *E. coli* cells by atomic force microscopy, but the relevance of this observation to function in a native Gram-positive host without an outer membrane is uncertain [26]. It was also demonstrated that Fst_{pAD1} and nisin, a lantibiotic that disrupts both peptidoglycan biosynthesis and membrane integrity, had synergistic effects on *E. faecalis* cells, strongly supporting a cell envelope effect of Fst_{pAD1} [23]. However, membrane disruption is most likely a secondary effect as time course experiments revealed that nucleoid condensation along with aberrant chromosomal segregation and cell division occurred prior to permeation to DNA staining dyes in both *E. faecalis* and heterologous hosts [28]. Nucleoid condensation was also observed with both LdrD and Lpt [16,26]. A primary role for membrane perturbation has yet to be demonstrated for PepA1 as well, with possible effects on membrane associated functions and/or nucleoid condensation still under consideration [29].

Microarray and RNA-seq analyses revealed the induction of a preponderance of transporters, particularly ATP-utilizing transporters, in response to Fst_{pAD1} expression in *E. faecalis* cells [27,30]. A spontaneous Fst_{pAD1}-resistant mutant with a single base change in the *rpoC* gene, encoding the β′ subunit of RNA polymerase, showed reduced transporter induction, suggesting that over-expression of one or more transporters might be responsible for growth inhibition, perhaps by depleting ATP pools. In support of this hypothesis, the inhibition of transporter function with the broad-spectrum translocase inhibitor reserpine had a protective effect against Fst_{pAD1} over-expression [30]. A microarray analysis of cells overproducing LdrD showed the upregulation of genes involved in purine metabolism [16].

Other than the loss of plasmid stability due to the deletion of plasmid-encoded systems, no phenotype has been associated with the deletion of any Fst/Ldr TA-1 system. Plasmid-encoded ectopic over-expression of the complete *S. mutans* Fst-Sm/srSm system led to a dramatic decrease in the number of oxacillin, cefotaxime and vancomycin tolerant persister cells [22]. Since the plasmid-encoded expression of the locus would be expected to proportionately increase both toxin mRNA and antitoxin regulatory RNA, it is not clear how this effect is established and no molecular mechanism has been suggested. Somewhat surprisingly, Michaux et al. were able to construct a mutant in the antitoxin gene of the chromosomal *par*$_{EF0409}$ locus in *E. faecalis* strain V583 [31]. Such a deletion would be expected to be lethal as a result of the loss of the repression of Fst_{Ef0409} expression. However, the transcription of the toxin message from this locus is quite low and, given the intramolecular translational inhibitory structures, expression of the toxin might be below that required to induce cell death even in the absence of the antitoxin. The mutant strain showed increased virulence in a *G. mellonella* larval model and a mouse urinary tract infection model and showed increased resistance to oxidative stress, bile salts, and acidity. The authors suggested that these results implicated the par_{EF0409} locus in colonization rather than virulence. Proteomic analysis was also performed to determine the response of the mutant to the absence of the antitoxin RNA. Numerous changes potentially related to the observed increased resistance to stress were identified, but curiously the gene expression changes were distinct from those observed by RNA-seq during mild over-expression of Fst_{EF0409} in *E. faecalis* strain OG1RF [27,31]. Indeed, in one case, the expression of the *arc* operon, the effects were opposite. The reason for these differences is currently unknown, but could relate to differences in toxin expression levels, strain differences, differences between effects on transcription and translation, or the presence of second site suppressor mutations.

Given the various effects of over-production of Fst/Ldr family members, it seems unlikely that they have a generally similar mechanism of action, e.g., forming membrane pores. To examine the degree of variability directly, the transcriptomic response of *E. faecalis* OG1RF cells to Fst_{pAD1} and Fst_{EF0409}, plasmid- and chromosomally encoded toxins that presumably have distinct functions, was determined [27]. Results showed substantial differences in response to the two toxins with 113 genes showing higher expression when exposed to Fst_{pAD1} and 90 showing higher expression when exposed to Fst_{EF0409}. For example, OG1RF_RS02610, annotated as a copper-translocating P-type ATPase, is induced greater than 100-fold in response to Fst_{pAD1} but not induced significantly by Fst_{EF0409}. Conversely, OG1RF_RS01655, encoding an ABC-transporter closely linked genetically to the par_{EF0409} locus, is induced 16-fold by Fst_{EF0409} but only eightfold by Fst_{pAD1}. These results suggest that, rather than simply poking holes in the membrane, the toxin sequences may be optimized for specific functions. Interestingly, TA-1 toxins in *B. subtilis* unrelated to the Fst/Ldr family have been observed to have subtle effects that similarly bring into question a role in pore formation [32]. Given the small size of the toxins, it should be possible to identify specific amino acids responsible for distinct responses. Such results will pave the way for more detailed structure/function analyses.

At this time, it is unclear how small, membrane-localized proteins could have such disparate effects on nucleoid structure, cell division, stress response and gene expression. It seems unlikely that the proteins act directly with the transcription apparatus, so effects on gene expression are probably an indirect effect of the perturbation of the cell envelope structure or interference with membrane protein function. Discerning the mechanistic aspects of toxin function is a focus of ongoing research.

4. Regulation of Toxin Expression

The performance of a post-segregational killing (PSK) function by plasmid-encoded TA-1 systems imposes certain requirements for proper regulation. First, and most obvious, the toxin mRNA must be more stable than the antitoxin so that it can persist in cells that lose the plasmid. Second, the antitoxin must be transcribed at a level high enough to prevent translation from all available active toxin mRNA, but not so high that it allows several generations to pass without killing after plasmid loss. Ideally, the level of antitoxin to translatable toxin mRNA should be close to 1:1. Third, the antitoxin cannot immediately bind and degrade the toxin mRNA, as occurs in most negatively regulated antisense systems, because then no mRNA would remain for translation upon plasmid loss. In the case of the prototypical TA-1 system *E. coli hok/sok*, this is accomplished by the initial adoption of a conformation of *hok* toxin mRNA that can neither be translated nor interact with the *sok* antitoxin, allowing a pool of inactive mRNA to accumulate in the cell. The *hok* mRNA is then slowly degraded from the 3′ end, triggering a conformational change that allows *sok* binding and degradation if the plasmid is retained, or translation if it has been lost (for a review, see [33]). Regulation of par_{pAD1} adopts a different strategy with the formation of a mRNA:antitoxin complex that is more stable than either RNA is alone [8]. A pool of the complex is then maintained in plasmid-containing cells, but with continual replacement of the antitoxin RNA removed from the complex. The 5′-SL sequesters the Fst_{pAD1} ribosome binding site to prevent translation of transiently antitoxin-free mRNA [7]. Only after the plasmid is lost is enough antitoxin removed from the complex to free sufficient toxin mRNA for translation. It seems likely that other plasmid-encoded *par* homologs are regulated in a similar manner. One possible exception is the *L. rhamnosus* plasmid-encoded Lpt TA system which shows elevated levels of both toxin and antitoxin RNA under conditions mimicking those that occur during cheese ripening [34]

The regulation of chromosomally encoded TA systems is most likely tailored to their specific functions, so examining the regulatory features of these systems might shed light on their physiological roles. Unlike its plasmid-encoded paralog, the *E. faecalis* antitoxin RNA of chromosomally encoded par_{EF0409} is transcribed in large excess over the Fst_{EF0409} toxin mRNA. Indeed, toxin mRNA is barely detectable on Northern blots and is barely above background in qRT-PCR [27,31]. Furthermore, induced expression of the toxin mRNA ectopically reduced levels of the antitoxin RNA suggesting that, unlike in par_{pAD1}, complex formation destabilizes the antitoxin [27]. Michaux et al. reported that the antitoxin promoter showed threefold decreased activity in the presence of bile salts, a 13-fold decrease in the presence of H_2O_2 and a twofold decrease during growth in glycerol as compared to glucose [31]. However, because the antitoxin is produced in such molar excess over toxin mRNA, it seems unlikely that such small stressor-induced decreases in antitoxin RNA would liberate sufficient toxin mRNA for translation. Levels of the antitoxin RNA have also been shown to decrease in the stationary phase, but remain in excess over the toxin message [27]. The plasmid-encoded par_{pAD1} and the chromosomally encoded par_{EF0409} systems can co-exist in the same cell, apparently without interfering with each other's function. Thus, RNA II_{pAD1} protects cells from the over-expression of RNA I_{pAD1} but not RNA I_{EF0409} [27]. This is presumably due to variations in the sequence of the DRa and DRb RNA–RNA interaction sites which alignments of *par*-family systems show are poorly conserved.

The regulation of expression of two Fst homologs in *S. aureus*, PepA1 and PepA2, have been examined in detail. As in par_{EF0409}, SprA1 mRNA, encoding PepA1, is constitutively expressed at low levels and expression of the antitoxin, $SprA_{AS}$, is in large molar excess (35-90-fold) over the toxin mRNA and peaks in mid-exponential phase [24]. PepA1 production was increased by acid stress (twofold) and oxidative stress (threefold) and levels of the antitoxin were decreased by 25% and 50% by those two stresses, respectively. This is consistent with the reported decrease in par_{EF0409} antitoxin promoter activity under similar stress conditions [31]. As with Fst_{EF0409}, however, it is not clear how a fractional drop in antitoxin production would lead to toxin expression. Nonetheless, the similar expression levels and responses to stress in the two systems from divergent hosts is intriguing.

The second *S. aureus par*-homolog, SprA2/$SprA2_{AS}$, produces both toxin mRNA and regulatory antisense RNA at easily detectable levels, though the relative quantities of the toxin mRNA and antitoxin RNA were not determined [25]. Unlike par_{EF0409} but similarly to par_{pAD1}, overexpression of the toxin mRNA stabilized the antitoxin RNA. This is curious given that the SprA2 system is part of the core genome and therefore likely does not play a PSK role. Antitoxin expression levels were reduced under osmotic shock and stringent conditions, suggesting that it may play a role in stress response, but to different stresses than the SprA1/$SprA1_{AS}$ system.

In summary, while the mechanism and rationale for the regulation of plasmid-encoded PSK systems is well understood, that of the chromosomal systems, as with their functions, remains obscure. Some appear to produce toxin mRNA and antitoxin sRNA at relatively equal levels, while others produce the antitoxin sRNA in high molar excess. In the latter cases, it remains unclear how or when antitoxin levels are reduced sufficiently to allow toxin expression. Further studies on promoter structure and function would certainly be helpful, as would a broader examination of related systems in different species. An examination of this expression under a greater variety of growth conditions, including biofilm and in vivo infection models, would also be helpful.

5. Speculations on Function

Except for the well-established role of plasmid-encoded TA systems in maintaining stable inheritance, TA system functions remain largely mysterious and the case of Fst/Ldr systems is no exception. Indeed, it is even tempting to question the canonical PSK function of the plasmid-encoded systems. For example, although the par_{pAD1} locus clearly stabilizes heterologous, artificially destabilized plasmids at the expense of host cell growth rate [10], it is reasonable to ask whether this is the most efficient way to stabilize the native pAD1 plasmid. Since the plasmid has a highly efficient conjugation system and enterococci grow in chains, would it not be more efficient to just slow the growth of the plasmid-free segregant and reacquire the plasmid from a neighboring cell in the chain? This possibility certainly raises many questions about how the pheromone-response system required for conjugation would function within an enterococcal chain, but, to the best of our knowledge, no efforts have been directed toward answering these questions.

The functioning of the chromosomally encoded Fst/Ldr systems is even more speculative. As is common for TA systems in general, the deletion of the Fst/Ldr TA systems has not been associated with any phenotypic effect on host cells; so either their functions are redundant with other bacterial genes or conditions have not yet been found under which their function can be observed. As described above, overexpression of the toxins has been demonstrated to have numerous effects on the host cells, but it is unclear whether such high expression levels are ever obtained under normal physiological conditions. A variety of stress conditions have been shown to modestly reduce the expression of antitoxin RNA, which could conceivably increase toxin levels and affect growth, but this observation comes with a number of caveats. First, in the SprA1 and $\mathrm{Fst}_{\mathrm{EF0409}}$ cases, the antitoxin is produced in such excess over the toxin mRNA levels that it is hard to see how modest changes in antitoxin levels could lead to sufficient toxin expression to have an effect [24,31]. Increased toxin expression was observed in the SprA1 system, but only when the entire locus was present ectopically on a multicopy plasmid [19]. With $\mathrm{Fst}_{\mathrm{EF0409}}$, improved growth was observed in an antitoxin knockout under the same conditions that showed reduced antitoxin expression in the wild type, which provides some circumstantial evidence that the system may be involved in stress survival [31]. However, genomic sequencing should be performed to ensure that this mutant does not have compensatory mutations that allow the cell to grow in the presence of the toxin. The SprA2 antitoxin RNA also appears to be regulated by stress conditions [25], but whether this affects toxin expression has not been addressed. Overall, the evidence supporting a role for these TA loci in stress response is not strong.

The *S. aureus* SprA1-$\mathrm{SprA1}_{\mathrm{AS}}$ system is located on a pathogenicity island [24] and therefore could conceivably function as a PSK to ensure maintenance of that MGE. However, the vast excess of antitoxin RNA under the conditions examined thus far makes such a role unlikely. Both PepA1 and PepA2 are cytolytic for human cells and therefore could be considered as virulence factors during *S. aureus* infection [24,25]. However, no evidence from animal models has been presented supporting such a role. Furthermore, it is not clear how the peptide toxin would be released from the membranes of the producing bacterium to attack red blood cells. Synthetic SprA1 also lyses Gram-positive and Gram-negative bacterial cells [24], but, again, problems with release from the producing cell's membrane would limit its usefulness in microbial competition, though it does provide potential as an antibacterial agent [35].

Circumstantial evidence based on gene location suggests that some of the chromosomal *par* homologs may be integrated with core metabolic functions. For example, par_{EF0409} is located in all sequenced *E. faecalis* strains between two paralogous mannitol class phosphotransferase systems [15] (see Figure 3). Our unpublished results suggest that the downstream operon plays an essential role in mannitol transport, while the upstream operon plays a regulatory role, particularly the *mtlR* gene, which is homologous to mannitol regulators in other Gram-positive bacteria [36]. Interestingly, there is no transcriptional terminator between the *mtlF* gene and the antitoxin RNA, suggesting that readthrough transcription could modulate expression of the toxin by altering antitoxin levels. We have also observed that mannitol-grown cells are approximately 10-fold more sensitive to $\mathrm{Fst}_{\mathrm{EF0409}}$ than glucose-grown cells. Finally, a mutation of a genetically linked ABC transporter increased toxicity in both glucose and mannitol, suggesting that this transporter was important for recovery from toxin expression. How exactly Fst_{EF0409} might fit into the regulation of mannitol transport and/or metabolism is not clear, but it is worth noting that small peptides have been shown to directly modulate the transport of other sugars [37]. Alternatively, we also note that there are significant regions of DNA sequence homology between the *mtlA* and *mtlA2* genes that could result in the recombination and excision of the intervening DNA. The par_{EF0409} locus could conceivably perform a PSK role to keep the excised circle of DNA from being lost, although the ratio of antitoxin to toxin mRNA would not be ideal for this function. Several other chromosomally encoded *par* homologs show a similarly integrative association with core metabolism genes, but little is known about how these systems are regulated and function [14,15].

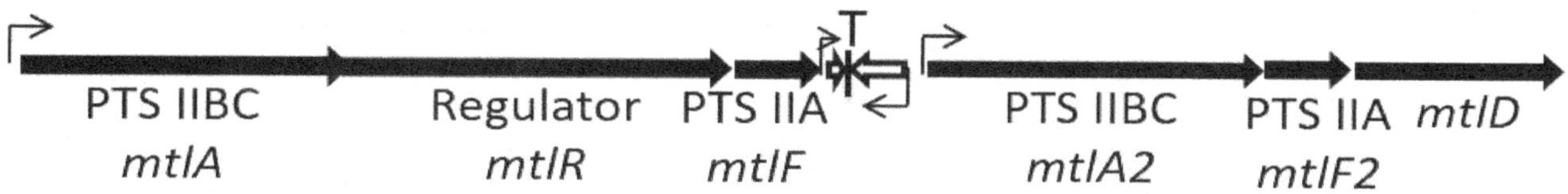

Figure 3. Genetic context of par_{EF0409}. The "T" marks the position of the bidirectional terminator shown in Figure 1A with the shorter white arrow on the left depicting the position of RNA II and the longer white arrow on the right depicting RNA I. The locus is situated between two paralogous mannitol-family phosphotransferase systems. The upstream operon includes *mtlR* encoding a putative positive regulator of the downstream operon.

6. Conclusions

The Fst/Ldr proteins are a widespread family of apparently membrane active peptides regulated by small RNAs in TA-1 systems. The Fst subfamily is present in the Firmicutes, including in pathogenic species of *Staphylococcus*, *Streptococcus*, *Enterococcus* and *Listeria* as well as important commensal species of *Clostridia* and *Lactobacillus*. The Ldr subfamily is present in the enterobacteria, including pathogenic species of *Escherichia*, *Shigella*, and *Salmonella*. While the toxins of the two subfamilies clearly share conserved motifs, their regulatory mechanisms are distinct, and it is unclear whether they share a common ancestor or evolved convergently. Their taxonomic coherence with the phylogenetic trees of their host bacterial species suggest that they have not been spread broadly by horizontal genetic transfer despite their presence on MGE [13].

Although widespread in numerous bacterial pathogens, the demonstration of a specific role in pathogenesis for the Fst/Ldr family has been elusive. The prototype Fst-encoding TA-1 system, par_{pAD1}, has been demonstrated to perform a stabilization function for its host plasmid and related systems are likely to perform similar functions for other MGE. The locus is highly conserved on pheromone-responsive conjugative plasmids of *E. faecalis*, which are known to encode both antibiotic resistance and virulence factors [38–40], and therefore likely plays a critical role in the maintenance and dissemination of these determinants in this important pathogen. Other putative roles in pathogenesis are more speculative. Limited toxin expression could conceivably function to slow bacterial growth and lead to the accumulation of persisters, but no evidence of such an effect has been observed and some evidence actually contradicts such a role [22]. Expression analysis suggests that several of the Fst-related toxins may be involved in stress response [24,25,31], but the mechanisms have yet to be elucidated. An *E. faecalis* strain over-producing the Fst_{EF0409} was found to increase virulence in two model systems [31], but the conditions that would lead to such levels of toxin production have not been defined. Synthetic derivatives of two *S. aureus* Fst-related toxins have been demonstrated to be cytolytic for both human and bacterial cells [24,25], but whether these toxins are actually secreted from natural producers has yet to be demonstrated. Circumstantial evidence suggests that some related loci are integrated with basic sugar metabolism [14,15], but evidence for a specific role is lacking.

Thus, like most other TA system toxins, the role of the Fst/Ldr toxins in pathogenesis, and indeed in metabolism in general, remains frustratingly obscure. A clear demonstration of function remains tantalizingly just beyond our reach. Once a function has been defined for a specific system, it will be important not to jump to the conclusion that all of the related systems perform the same function. The possibility that their functions may be tailored to the needs of specific species must always be considered. Clearly, more work is justified and required to identify the role(s) of these ubiquitous systems.

References

1. Weaver, K.E.; Clewell, D.B. Construction of Enterococcus faecalis pAD1 miniplasmids: Identification of a minimal pheromone response regulatory region and evaluation of a novel pheromone-dependent growth inhibition. *Plasmid* **1989**, *22*, 106–119. [CrossRef]
2. Weaver, K.E.; Jensen, K.D.; Colwell, A.; Sriram, S. Functional analysis of the *Enterococcus faecalis* plasmid pAD1-encoded stability determinant *par*. *Mol. Microbiol.* **1996**, *20*, 53–63. [CrossRef] [PubMed]
3. Weaver, K.E.; Tritle, D.J. Identification and characterization of an *Enterococcus faecalis* plasmid pAD1-encoded stability determinant which produces two small RNA molecules necessary for its function. *Plasmid* **1994**, *32*, 168–181. [CrossRef]
4. Greenfield, T.J.; Franch, T.; Gerdes, K.; Weaver, K.E. Antisense RNA regulation of the *par* post-segregational killing system: Structural analysis and mechanism of binding of the antisense RNA, RNAII and its target, RNAI. *Mol. Microbiol.* **2001**, *42*, 527–537. [CrossRef] [PubMed]
5. Greenfield, T.J.; Ehli, E.; Kirshenmann, T.; Franch, T.; Gerdes, K.; Weaver, K.E. The antisense RNA of the *par* locus of pAD1 regulates the expression of a 33-amino-acid toxic peptide by an unusual mechanism. *Mol. Microbiol.* **2000**, *37*, 652–660. [CrossRef] [PubMed]
6. Greenfield, T.J.; Weaver, K.E. Antisense RNA regulation of the pAD1 *par* post-segregational killing system requires interaction at the 5′ and 3′ ends of the RNAs. *Mol. Microbiol.* **2000**, *37*, 661–670. [CrossRef]
7. Shokeen, S.; Patel, S.; Greenfield, T.J.; Brinkman, C.; Weaver, K.E. Translational regulation by an intramolecular stem-loop is required for intermolecular RNA regulation of the *par* addiction module. *J. Bacteriol.* **2008**, *190*, 6076–6083. [CrossRef]
8. Weaver, K.E.; Ehli, E.A.; Nelson, J.S.; Patel, S. Antisense RNA regulation by stable complex formation in the *Enterococcus faecalis* plasmid pAD1 *par* addiction system. *J. Bacteriol.* **2004**, *186*, 6400–6408. [CrossRef]
9. Shokeen, S.; Greenfield, T.J.; Ehli, E.A.; Rasmussen, J.; Perrault, B.E.; Weaver, K.E. An intramolecular upstream helix ensures the stability of a toxin-encoding RNA in *Enterococcus faecalis*. *J. Bacteriol.* **2009**, *191*, 1528–1536. [CrossRef]
10. Weaver, K.E.; Walz, K.D.; Heine, M.S. Isolation of a derivative of *Escherichia coli–Enterococcus faecalis* shuttle vector pAM401 temperature sensitive for maintenance in *E. faecalis* and its use in evaluating the mechanism of pAD1 *par*-dependent plasmid stabilization. *Plasmid* **1998**, *40*, 225–232. [CrossRef]
11. Hayes, F. Toxins-antitoxins: Plasmid maintenance, programmed cell death, and cell cycle arrest. *Science* **2003**, *301*, 1496–1499. [CrossRef] [PubMed]
12. Weaver, K.E. The *par* toxin-antitoxin system from *Enterococcus faecalis* plasmid pAD1 and its chromosomal homologs. *RNA Biol.* **2012**, *9*, 1498–1503. [CrossRef] [PubMed]
13. Fozo, E.M.; Makarova, K.S.; Shabalina, S.A.; Yutin, N.; Koonin, E.V.; Storz, G. Abundance of type I toxin–antitoxin systems in bacteria: Searches for new candidates and discovery of novel families. *Nucleic Acids Res.* **2010**, *38*, 3743–3759. [CrossRef]
14. Kwong, S.M.; Jensen, S.O.; Firth, N. Prevalence of Fst-like toxin–antitoxin systems. *Microbiology* **2010**, *156*, 975–977. [CrossRef] [PubMed]
15. Weaver, K.E.; Reddy, S.G.; Brinkman, C.L.; Patel, S.; Bayles, K.W.; Endres, J.L. Identification and characterization of a family of toxin–antitoxin systems related to the *Enterococcus faecalis* plasmid pAD1 *par* addiction module. *Microbiology* **2009**, *155*, 2930–2940. [CrossRef]

16. Kawano, M.; Oshima, T.; Kasai, H.; Mori, H. Molecular characterization of long direct repeat (LDR) sequences expressing a stable mRNA encoding for a 35-amino-acid cell-killing peptide and a cis-encoded small antisense RNA in *Escherichia coli*. *Mol. Microbiol.* **2002**, *45*, 333–349. [CrossRef] [PubMed]
17. Kawano, M. Divergently overlapping *cis*-encoded antisense RNA regulating toxin-antitoxin systems from *E. coli*. *RNA Biol.* **2012**, *9*, 1520–1527. [CrossRef]
18. Göbl, C.; Kosol, S.; Stockner, T.; Rückert, H.M.; Zangger, K. Solution structure and membrane binding of the toxin Fst of the *par* addiction module. *Biochemistry* **2010**, *49*, 6567–6575. [CrossRef]
19. Sayed, N.; Nonin-Lecomte, S.; Réty, S.; Felden, B. Functional and structural insights of a *Staphylococcus aureus* apoptotic-like membrane peptide from a toxin-antitoxin module. *J. Biol. Chem.* **2012**, *287*, 43454–43463. [CrossRef]
20. Hofmann, K.; Stoffl, W. TMbase—A database of membrane spanning proteins segments. *Biol. Chem. Hoppe-Seyler* **1993**, *374*, 166.
21. Yamaguchi, Y.; Tokunaga, N.; Inouye, M.; Phadtare, S. Characterization of LdrA (Long direct repeat A) protein of *Escherichia coli*. *J. Mol. Microbiol. Biotechnol.* **2014**, *24*, 91–97. [CrossRef] [PubMed]
22. Koyanagi, S.; Lévesque, C.M. Characterization of a *Streptococcus mutans* intergenic region containing a small toxic peptide and its *cis*-encoded antisense small RNA antitoxin. *PLoS ONE* **2013**, *8*, e54291. [CrossRef] [PubMed]
23. Weaver, K.E.; Weaver, D.M.; Wells, C.L.; Waters, C.M.; Gardner, M.E.; Ehli, E.A. *Enterococcus faecalis* plasmid pAD1-encoded Fst toxin affects membrane permeability and alters cellular responses to lantibiotics. *J. Bacteriol.* **2003**, *185*, 2169–2177. [CrossRef]
24. Sayed, N.; Jousselin, A.; Felden, B. A *cis*-antisense RNA acts in *trans* in *Staphylococcus aureus* to control translation of a human cytolytic peptide. *Nat. Struct. Mol. Biol.* **2012**, *19*, 105–112. [CrossRef] [PubMed]
25. Germain-Amiot, N.; Augagneur, Y.; Camberlein, E.; Nicolas, I.; Lecureur, V.; Rouillon, A.; Felden, B. A novel *Staphylococcus aureus cis–trans* type I toxin–antitoxin module with dual effects on bacteria and host cells. *Nucleic Acids Res.* **2018**, *47*, 1759–1773. [CrossRef]
26. Maggi, S.; Yabre, K.; Ferrari, A.; Lazzi, C.; Kawano, M.; Rivetti, C.; Folli, C. Functional characterization of the type I toxin Lpt from *Lactobacillus rhamnosus* by fluorescence and atomic force microscopy. *Sci. Rep.* **2019**, *9*, 15208. [CrossRef]
27. Weaver, K.E.; Chen, Y.; Miiller, E.M.; Johnson, J.N.; Dangler, A.A.; Manias, D.A.; Clem, A.M.; Schjodt, D.J.; Dunny, G.M. Examination of *Enterococcus faecalis* toxin-antitoxin system toxin Fst function utilizing a pheromone-inucible expression vector with tight repression and broad dynamic range. *J. Bacteriol.* **2017**, *199*, e00065-17. [CrossRef]
28. Patel, S.; Weaver, K.E. Addiction toxin Fst has unique effects on chromosome segregation and cell division in *Enterococcus faecalis* and *Bacillus subtilis*. *J. Bacteriol.* **2006**, *188*, 5374–5384. [CrossRef]
29. Brielle, R.; Pinel-Marie, M.L.; Felden, B. Linking bacterial type I toxins with their actions. *Curr. Opin. Microbiol.* **2016**, *30*, 114–121. [CrossRef]
30. Brinkman, C.L.; Bumgarner, R.; Kittichotirat, W.; Dunman, P.M.; Kuechenmeister, L.J.; Weaver, K.E. Characterization of the effects of an *rpoC* mutation that confers resistance to the Fst peptide toxin-antitoxin system toxin. *J. Bacteriol.* **2013**, *195*, 156–166. [CrossRef]
31. Michaux, C.; Hartke, A.; Martini, C.; Reiss, S.; Albrecht, D.; Budin-Verneuil, A.; Sanguinetti, M.; Engelmann, S.; Hain, T.; Verneuil, N.; et al. Involvement of *Enterococcus faecalis* small RNAs in stress response and virulence. *Infect. Immun.* **2014**, *82*, 3599–3611. [CrossRef] [PubMed]
32. Brantl, S.; Muller, P. Toxin-antitoxin systems in *Bacillus subtilis*. *Toxins* **2019**, *11*, 262. [CrossRef] [PubMed]
33. Gerdes, K.; Wagner, E.G.H. RNA antitoxins. *Curr. Opin. Microbiol.* **2007**, *10*, 117–124. [CrossRef] [PubMed]
34. Folli, C.; Levante, A.; Percudani, R.; Amidani, D.; Bottazzi, S.; Ferrari, A.; Rivetti, C.; Neviani, E.; Lazzi, C. Toward the identification of a type I toxin-antitoxin system in the plasmid DNA of dairy *Lactobacillus rhamnosus*. *Sci. Rep.* **2017**, *7*, 12051. [CrossRef] [PubMed]
35. Solecki, O.; Mosbah, A.; Baudy Floc'h, M.; Felden, B. Converting a *Staphylococcus aureus* toxin into effective cyclic pseudopeptide antibiotics. *Chem. Biol.* **2015**, *22*, 329–335. [CrossRef]
36. Deutscher, J.; Aké, F.M.D.; Derkaoui, M.; Zébré, A.C.; Cao, T.N.; Bouraoui, H.; Kentache, T.; Mokhtari, A.; Milohanic, E.; Joyet, P. The bacterial phosphoenolpyruvate:carbohydrate phosphotransferase system: Regulation by protein phosphorylation and phosphorylation-dependent protein-protein interactions. *Microbiol. Mol. Biol. Rev.* **2014**, *78*, 231–256. [CrossRef]

37. Lloyd, C.R.; Park, S.; Fei, J.; Vanderpool, C.K. The small protein SgrT controls transport activity of the glucose-specific phosphotransferase system. *J. Bacteriol.* **2017**, *199*, e00869-16. [CrossRef]
38. Clewell, D.B. Properties of *Enterococcus faecalis* plasmid pAD1, a member of a widely disseminated family of pheromone-responding, conjugative, virulence elements encoding cytolysin. *Plasmid* **2007**, *58*, 205–227. [CrossRef]
39. Weaver, K.E. Enterococcal genetics. *Microbiol. Spectr.* **2019**, *7*. [CrossRef]
40. Zheng, B.; Tomita, H.; Inoue, T.; Ike, Y. Isolation of VanB-type *Enterococcus faecalis* strains from nosocomial infections: First report of the isolation and identification of the pheromone-responsive plasmids pMG2200, encoding VanB-type vancomycin resistance and a Bac41-type bacteriocin, and pMG2201, encoding erythromycin resistance and cytolysin (Hly/Bac). *Antimicrob. Agents Chemother.* **2009**, *53*, 735–747. [CrossRef]

9

Semicarbazone EGA Inhibits Uptake of Diphtheria Toxin into Human Cells and Protects Cells from Intoxication

Leonie Schnell [1], Ann-Katrin Mittler [1], Andrea Mattarei [2], Domenico Azarnia Tehran [3], Cesare Montecucco [3] and Holger Barth [1,*]

1 Institute of Pharmacology and Toxicology, University of Ulm Medical Center, Albert-Einstein-Allee 11, 89081 Ulm, Germany; leonie.schnell@uni-ulm.de (L.S.); ann-katrin.mittler@uni-ulm.de (A.-K.M.)
2 Department of Chemical Sciences, University of Padova, 35121 Padova, Italy; andrea.mattarei@unipd.it
3 Department of Biomedical Sciences, University of Padova, 35131 Padova, Italy; doazte@gmail.com (D.A.T.); cesare.montecucco@gmail.com (C.M.)
* Correspondence: holger.barth@uni-ulm.de.

Abstract: Diphtheria toxin is a single-chain protein toxin that invades human cells by receptor-mediated endocytosis. In acidic endosomes, its translocation domain inserts into endosomal membranes and facilitates the transport of the catalytic domain (DTA) from endosomal lumen into the host cell cytosol. Here, DTA ADP-ribosylates elongation factor 2 inhibits protein synthesis and leads to cell death. The compound 4-bromobenzaldehyde *N*-(2,6-dimethylphenyl)semicarbazone (EGA) has been previously shown to protect cells from various bacterial protein toxins which deliver their enzymatic subunits from acidic endosomes to the cytosol, including *Bacillus anthracis* lethal toxin and the binary clostridial actin ADP-ribosylating toxins C2, iota and *Clostridium difficile* binary toxin (CDT). Here, we demonstrate that EGA also protects human cells from diphtheria toxin by inhibiting the pH-dependent translocation of DTA across cell membranes. The results suggest that EGA might serve for treatment and/or prevention of the severe disease diphtheria.

Keywords: diphtheria; diphtheria toxin; cellular uptake; membrane transport; EGA

1. Introduction

Corynebacterium diphtheriae produces the single-chain diphtheria toxin (DT, 58 kDa), which is the causative agent of diphtheria [1]. DT is efficiently taken up into human cells and its catalytic domain (DTA, 21 kDa) acts as an extremely potent enzyme in the cytosol. DTA covalently transfers ADP-ribose from cellular NAD^+ onto a modified histidine residue (diphthamide) of the elongation factor 2 (EF-2) thereby inhibiting protein synthesis and causing cell death [2,3], which can be monitored in terms of cell-rounding using HeLa cells [4,5]. DTA is located in the N-terminal domain of DT [6] while the C-terminal part (DTB, 37 kDa) mediates binding of the toxin to susceptible cells and the subsequent transport of DTA into the cytosol. DTB contains a receptor-binding (B) domain, which binds to the heparin-binding epidermal growth factor-like growth factor precursor (HB-EGF) [7,8] and a translocation (T) domain [9], which inserts into the membranes of acidified endosomes [10,11] allowing the membrane translocation of DTA from the endosomal lumen into the cytosol [12–18]. This process is prevented by bafilomycin A1, an inhibitor of endosomal acidification [19] and can be experimentally mimicked on the surface of cultured cells by exposure of cell-bound DT to an acidic pulse [20]. This triggers the insertion of DTB directly into the plasma membrane and the translocation of DTA into the cytosol where it modifies its substrate [21–23].

Translocation of DTA across endosomal membranes is facilitated by host cell factors including the chaperone heat shock protein (Hsp) 90 [24,25] and thioredoxin reductase [5,24,26]. DTA is separated from DTB by cleavage prior or during DT uptake [27] but these two subunits remain linked via an interchain disulfide between Cys-186 of DTA and Cys-201 of DTB [28]. The integrity of the interchain disulfide bond is essential during toxin uptake into endosomes, as well as DTA translocation across the membranes [27,29] but its reduction is necessary for the subsequent release of DTA on the cytosolic side [23] and this process is the rate-limiting step during DT uptake [30]. Reduction of the disulfide bond likely happens after membrane insertion of the T-domain [30] during or after DTA translocation to the cytosol [31]. Thioredoxin 1 reduces this disulfide under acidic conditions in vitro [32] and we recently demonstrated that pharmacological inhibition of thioredoxin reductase prevents DTA transport across cell membranes and protects cells from intoxication [5], implicating that this enzyme is crucial for the reduction of the disulfide bond and the subsequent release of DTA in the cell cytosol of living cells.

The compound 4-bromobenzaldehyde *N*-(2,6-dimethylphenyl)semicarbazone (EGA) was previously identified as a potent inhibitor that protects cells from toxins and viruses which enter the cytosol of cells via acidified vesicles [33,34], such as the *Bacillus anthracis* lethal toxin and DT [34], as well as the binary actin ADP-ribosylating toxins C2 from *Clostridium* (*C.*) *botulinum*, iota from *C. perfringens* and CDT from *C. difficile* [35]. EGA also protects neuronal cells from *C. botulinum* neurotoxins [36] and it was suggested that this compound might modulate intracellular toxin trafficking [34–36]. Prompted by these findings, we analyzed the effect of EGA on the intoxication of HeLa cells with DT in more detail. Here, we demonstrate that EGA significantly delays intoxication of cells with DT in a time- and concentration-dependent manner and analyzed the underlying molecular mechanism.

2. Results and Discussion

EGA protects HeLa cells from intoxication with DT. In a first set of experiments the possible inhibitory effect of EGA on the intoxication of HeLa cells by DT was investigated. To this end, cells were pre-incubated for 1 h with increasing concentrations of EGA and then challenged with DT. After different incubation periods, the number of round cells was determined because this is an established, highly specific and sensitive endpoint to monitor the intoxication process [5]. As shown in Figure 1, EGA significantly delayed the DT-induced cell-rounding in a time- and concentration-dependent manner, indicating that EGA interferes with the mode of action of DT in these cells. EGA delayed intoxication with DT even when cells were not grown to confluence and therefore more susceptible to DT. Importantly, EGA alone had no effects on the cells under such conditions (Figure 1A). Adverse effects on the cells were observed at concentrations of 500 μM EGA and above (not shown). The usage of 100 μM instead of 50 μM was still tolerated by the cells but did not result in a relevant improvement of the inhibitory effect (Figure 1B). Therefore, for the following investigation of the underlying molecular mechanism of the EGA-mediated protection of cells from DT, an EGA concentration of 50 μM was used. A prolonged pre-incubation period with EGA (up to 6 h) did not significantly enhance the inhibitory effect of EGA towards DT compared to an 1 h pre-incubation period. Moreover, EGA showed some protective effect towards DT, even when applied 15 min after DT (not shown).

EGA did not completely inhibit intoxication with DT but delayed it, which was expected from our earlier results with EGA and binary clostridial toxins [35] and with DT and other pharmacological inhibitors of toxin uptake [5]. The reason is most likely related to the extreme potency of AB-toxins in cells, as few molecules of their catalytic moieties in the host cell cytosol is usually sufficient to exhibit the full cytotoxic effects [37]. A pharmacological inhibition might therefore not be able to completely prevent the uptake of some DTA molecules into the cytosol over longer incubation periods.

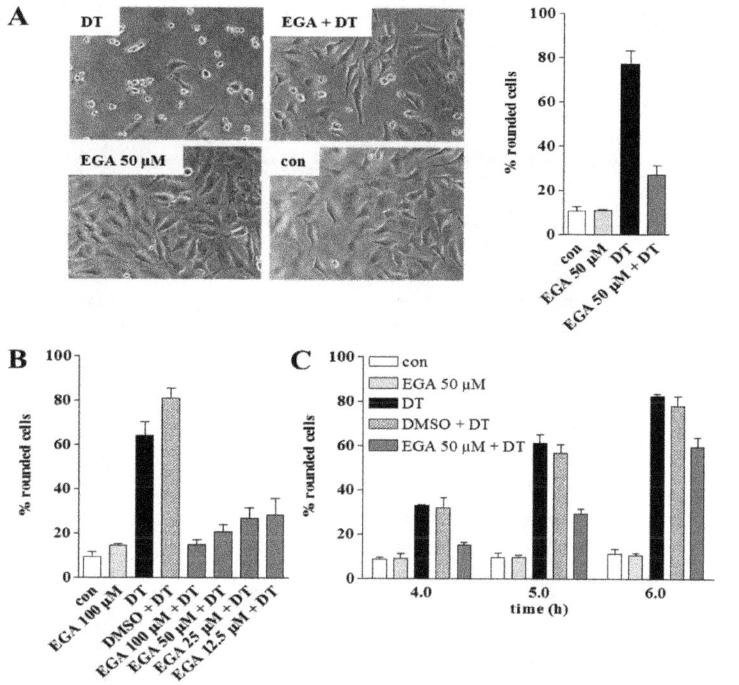

Figure 1. EGA protects HeLa cells from intoxication with DT. Cells were pre-incubated for 1 h at 37 °C with EGA or the solvent DMSO for control. Thereafter, DT (0.86 nM) was added and cells were further incubated at 37 °C. For control, cells were left untreated (con) or incubated with DT in the absence of EGA (DT). At the indicated time points, phase contrast pictures were taken and for quantitative analysis, the percentage of rounded cells was determined. Values are gives as mean ± SD (n = 3). (**A**) Protective effect of 50 µM EGA on the intoxication of cells with DT. Representative pictures after 4 h are shown; (**B**) Concentration-dependent inhibition of DT-intoxication after 4 h by EGA, determined in a separate experiment. The maximum dosage of EGA (100 µM) was tested on cells without DT to exclude any morphological alteration of the cells in response to EGA alone. In addition, a DMSO solvent control of the maximum dosage was performed; (**C**) Time-dependent inhibitory effect mediated by 50 µM EGA, determined in a separate experiment with confluent cells.

EGA inhibits the membrane transport of DTA in cells. The effect of EGA on the membrane transport of DTA was investigated by using a well-established assay where the endosomes acidic environment is mimicked on the surface of cultured HeLa cells. DT was allowed to bind to the receptor at 4 °C where no endocytosis occurs and the normal toxin uptake via acidic endosomes was inhibited by bafilomycin A1 treatment. When such cells are exposed to warm acidic medium, DTB inserts into the plasma membrane and mediates the translocation of DTA into the cytosol where it ADP-ribosylates EF-2 leading to cell-rounding, as shown in Figure 2. No cell-rounding was observed under neutral conditions, by incubation of cells in the acidic medium in the absence of DT, or in the presence of EGA without DT (Figure 2), indicating that cell-rounding is indeed specifically induced by the action of DTA, as described earlier [5,20]. Importantly, the DT-induced cell-rounding was significantly reduced when cells were pre-treated with EGA (Figure 2). Given that this inhibitor did not inhibit the in vitro enzymatic activity of DTA (Figure 3A) nor the binding of DT to HeLa cells (Figure 3B) and it leads to inhibition of intoxication also when applied after receptor-binding of nicked DT (i.e., DT that was proteolytically activated in vitro prior to its application to the cells) (Figure 3C), our results strongly suggest that EGA interferes with the toxin pH-dependent transport. In addition, the results obtained by performing the translocation assay with nicked DT also exclude an effect of EGA on the proteolytic activation of DT during cellular uptake on the cell surface and/or in endosomal vesicles. To demonstrate that the observed inhibitory effect of EGA was mediated by the molecule itself, in each experiment a vehicle control (DMSO) was performed in parallel to EGA treatment.

In contrast to EGA, DMSO did not inhibit the intoxication of cells with DT (see Figure 1B,C) or the translocation of cell-bound DT (Figure 2).

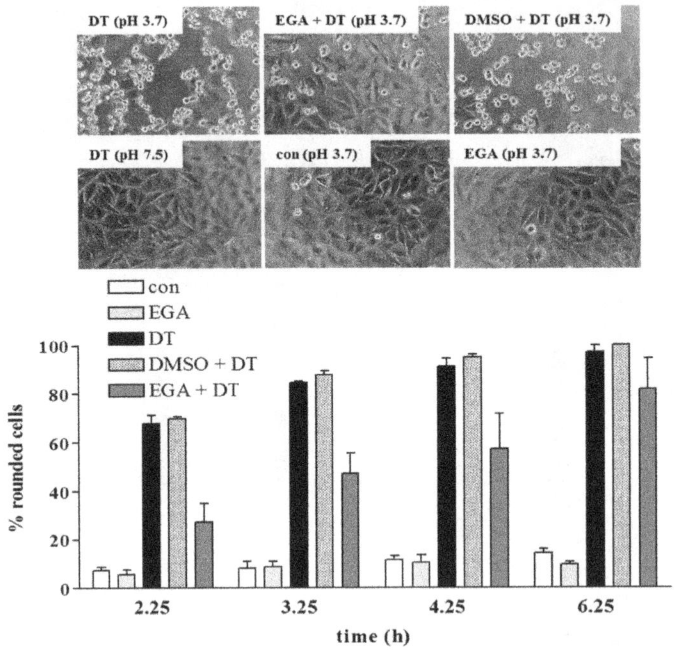

Figure 2. EGA inhibits the pH-dependent transport of the enzyme component DTA of DT across the cytoplasmic membranes in living cells. HeLa cells were pre-incubated for 1 h at 37 °C in serum-free medium with 100 nM bafilomycin (Baf) A1. In addition, some cells were also treated with 50 μM EGA or the corresponding solvent DMSO for control. Subsequently, cells were kept on ice for 10 min. Thereafter, nicked DT (13 nM) was added and cells were incubated for 40 min at 4 °C to enable toxin-binding to the receptors on the cell surface. Cells were further incubated for 15 min at 37 °C and pH 3.7 to trigger the pH-driven membrane transport of DTA. For control (con), cells were incubated at pH 7.5 (not shown) or at pH 3.5 (shown). During this step, there was no EGA present in the medium. Thereupon, all cells were further incubated at 37 °C in neutral medium (pH 7.5) containing serum, Baf A1 ± EGA. At the indicated time points, phase contrast pictures were taken and for quantitative analysis, the percentage of rounded cells was determined. Representative pictures are shown after 3.25 h. Values are gives as mean ± SD (n = 3).

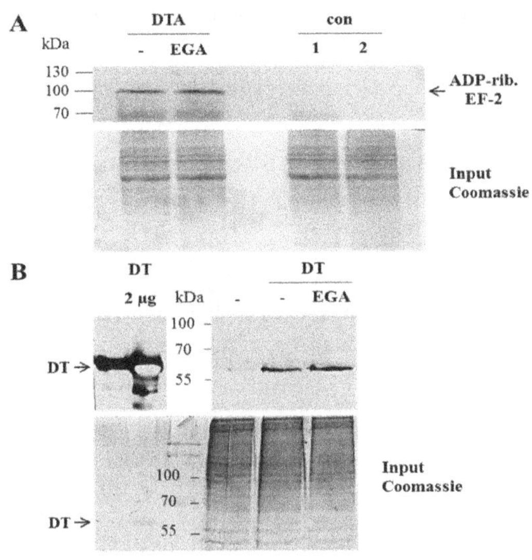

Figure 3. *Cont.*

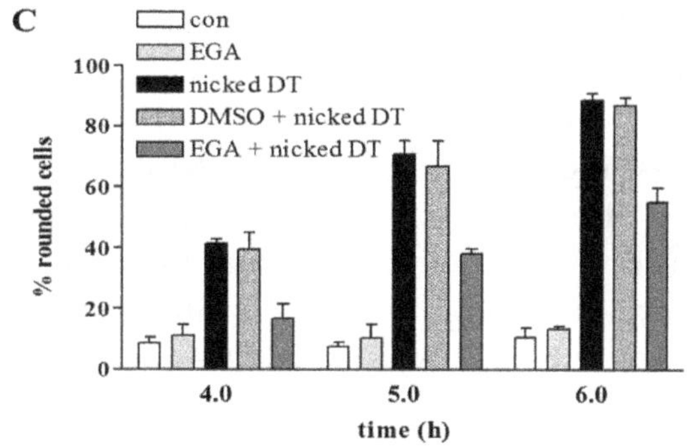

Figure 3. Treatment with EGA has no effect on the in vitro enzymatic activity of DTA nor on the receptor-binding of DT to cells. (**A**) DTA (100 ng) was pre-incubated with 50 μM EGA for 10 min at 37 °C. For control (con), a sample of DTA was left untreated. Thereupon, HeLa lysate protein (20 μg) as well as biotin-labelled NAD^+ were added followed by an incubation of the samples for 10 min at 37 °C. Thereafter, samples were subjected to SDS-PAGE, blotted and biotinylated (i.e., ADP-ribosylated, indicated as ADP-rib.) EF-2 was detected (upper panel). A sample of HeLa lysate with biotin-labelled NAD^+ (1) and one of only HeLa lysate (2) were additionally analyzed. As control of equal protein loading, the SDS-gel was stained with Coomassie after the blotting process (lower panel). Noteworthy, there was no obvious difference in the amount of ADP-ribosylated EF-2 when lysate from EGA-treated cells was used (not shown), indicating that EGA-treatment does not interfere with the generation of the diphthamide in EF-2; (**B**) HeLa cells were pre-treated with EGA (50 μM) for 1 h at 37 °C in serum-free medium or left untreated for control. Thereupon, cells were kept on ice for 10 min. Then, DT (34.4 nM) was added followed by further incubation for 1 h 10 min at 4 °C to enable binding of DT to the cell surface. Thereafter, three washing steps with ice-cold PBS were performed to remove unbound toxin and cells were scraped off in 2.5-fold SDS sample buffer containing 10% DTT. Proteins were separated by SDS-PAGE, blotted and bound DT was detected using a specific primary and a peroxidase-coupled secondary antibody and the ECL system (upper panel). For loading control, two dilutions of DT were analyzed on the left side of the same blot. Comparable amounts of total protein were confirmed by Coomassie staining of the proteins in the SDS-gel after the blotting process (lower panel); (**C**) Cells were kept on ice for 10 min before addition of nicked DT (3.4 nM) and further incubation at 4 °C for 30 min. Thereafter, the medium was removed and warm, serum-free medium containing EGA (50 μM) or DMSO was added followed by further incubation at 37 °C. For control (con), cells were left untreated or treated only with EGA without toxin. At the indicated time points, phase contrast pictures were taken and for quantitative analysis, the percentage of rounded cells was determined. Values are gives as mean ± SD ($n = 3$).

3. Conclusions

In conclusion, our results demonstrate that EGA protects cultured human cells from intoxication with DT because it interferes with the uptake of DTA into the host cell cytosol. Moreover, our findings suggest that EGA inhibits the pH-dependent transport of DTA across membranes into the cytosol. This step was also inhibited by EGA in the case of clostridial binary toxins, which also deliver their A subunit from acidic endosomes to the cytosol [35]. All toxins which were inhibited by EGA, exploit acidic endosomal vesicles for cellular trafficking, suggesting a common inhibitory mode of action for EGA in cells. The binary anthrax toxins and the binary clostridial actin ADP-ribosylating toxins share an overall comparable mechanism to deliver their enzymatically active components through pores formed by their separate heptameric transport components across endosomal membranes into the cytosol. Thus, a common mechanism by which EGA inhibits the membrane transport of these binary toxins seems plausible. However, EGA inhibits the membrane transport of DT, as shown in this present study, but it did not inhibit the translocation of botulinum neurotoxins across the plasma membrane [36], suggesting that the protective effect of EGA towards both single-chain toxins might not be mediated via the same molecular mechanism. Thus far, the molecular mechanisms by which

EGA interferes with the membrane transport of DT are not known and it can be speculated that EGA might block the transmembrane pore formed by the translocation domain of DT or might inhibit host cell factors which are involved in translocation of DTA across endosomal membranes.

However, although the precise mode of action how EGA inhibits the transport of the enzymatically active moieties of bacterial toxins into the cytosol requires further elucidation, this compound inhibits a series of the most powerful and medically most relevant bacterial toxins including DT, botulinum neurotoxins and anthrax toxin, and thus might have very attractive clinical applications. In previous studies [36] the treatment of mice with EGA by multiple intraperitoneally injections were well tolerated. It should be taken into account that higher concentrations of DT might be used in the experimental approach compared to the situation in patients. Moreover, EGA could represent a lead compound and chemical modifications may result in even more efficient compounds for future clinical applications.

4. Experimental Section

4.1. Materials and Reagents

Cell culture media (MEM) and fetal calf serum (FCS) were obtained from Invitrogen (Karlsruhe, Germany) and cell culture materials from TPP (Trasadingen, Switzerland). The protein molecular weight marker Page Ruler prestained Protein ladder® was purchased from Thermo Fisher Scientific Inc. (Waltham, MA, USA). Complete® protease inhibitor was supplied by Roche (Mannheim, Germany). Biotinylated NAD^+ was from R&D Systems GmbH (Wiesbaden-Nordenstadt, Germany). Bafilomycin (Baf) A1 was purchased from Calbiochem (Bad Soden, Germany), 2-(4-bromobenzylidene)-*N*-(2,6-dimethylphenyl)hydrazinecarboxamide (EGA) was synthesized as described [36]. The quality of EGA was tested by HPLC-MS analysis. The purity is >95% [36].

Streptavidin-peroxidase was obtained from Roche (Mannheim, Germany), the enhanced chemiluminescence (ECL) system from Millipore (Schwalbach, Germany) and the nitrocellulose blotting membrane from Whatman® (Dassel, Germany). Diphtheria toxin antibody was obtained from GeneTex Inc. (Irvine, CA, USA) and goat anti-mouse IgG-HRP secondary antibody from Santa-Cruz Biotechnology Inc. (Heidelberg, Germany). Diphtheria toxin (DT) was supplied by Calbiochem (Bad Soden, Germany). DT was proteolytically activated in vitro with trypsin to yield nicked DT as described in Section 4.4. and DTA was expressed and purified as described earlier [5].

4.2. Cell Culture and Cytotoxicity Assays

HeLa cells were cultivated at 37 °C and 5% CO_2 in MEM containing 10% heat-inactivated FCS, 1 mM sodium-pyruvate, 2 mM L-glutamine, 0.1 mM non-essential amino acids and 1% penicillin-streptomycin. Cells were trypsinized and reseeded for at most 30 times.

For cytotoxicity experiments, cells were grown in 96-well plates and pre-incubated with EGA or the vehicle DMSO in serum-free medium for 1 h at 37 °C. For control, cells were incubated without inhibitor. Thereafter, DT (0.86 nM) was added and cells were further incubated with toxin plus inhibitor at 37 °C. For analysis of the cellular effect of EGA when added after DT-binding to cells, cells grown in 96-well plates were kept on ice for 10 min in serum-free medium before addition of nicked DT (3.4 nM). After incubation for 30 min at 4 °C, medium was removed and warm, serum-free medium containing EGA (50 µM) or the vehicle DMSO was added followed by further incubation at 37 °C. After the indicated incubation periods, a Zeiss Axiovert 40 CFL microscope (Oberkochen, Germany) with a Jenoptik progress C10 CCD camera (Jena, Germany) was used to visualize the cells to analyze the DT-induced morphological changes. The characteristic rounding up of the cells was taken as specific endpoint to monitor the intoxication process. For quantitative analysis, cells were counted per picture (ImageJ software; NIH, Bethesda, MD, USA) and the amount of rounded cells was determined in percent.

4.3. SDS-PAGE and Western Blotting

For immunoblot analysis, according to the method of Laemmli [38], equal amounts of protein were subjected to SDS-PAGE. Afterwards, the proteins were transferred to a nitrocellulose membrane which was then blocked for 1 h at RT with 5% dry milk powder in PBS containing 0.1% Tween-20 (PBS-T) or alternatively overnight at 4 °C. For detection of the biotin-labelled EF-2 or DT, the samples were either probed with streptavidin-peroxidase or a specific antibody against DTA, respectively, for 1 h followed by washing steps with PBS-T. Subsequently, in case of DT, the membrane was probed with anti-mouse secondary antibody coupled to horseradish peroxidase for 1 h followed by further washing steps with PBS-T. Thereafter, the proteins were visualized using the ECL system according to the manufacturer's instructions. Ponceau S staining of the membrane and Coomassie staining of the gel were used to confirm equal amounts of protein.

4.4. Proteolytic Activation (Nicking) of DT

To activate DT, the toxin was treated with trypsin (3 μg/mL) for 3 h at 37 °C and then kept on ice. Following this, trypsin was neutralized by incubation with trypsin inhibitor (30 μg/mL) for 80 min at 4 °C and the concentration of nicked DT was determined using SDS-PAGE.

4.5. ADP-Ribosylation of EF-2 by DTA in a Cell-Free System

DTA (100 ng) was pre-incubated for 10 min at 37 °C with EGA (50 μM) or left untreated for control followed by the addition of HeLa lysate protein (20 μg) and biotin-labelled NAD^+ (10 μM). After further incubation of the samples for 10 min at 37 °C, the proteins were subjected to SDS-PAGE, blotted onto a nitrocellulose membrane. Finally, biotin-labelled (i.e., ADP-ribosylated) EF-2 was detected using Western blotting.

4.6. Binding of DT to Its Cell Surface Receptor

HeLa cells were pre-incubated in serum-free medium with EGA (50 μM) for 1 h at 37 °C. For control, cells were left untreated. Following this, cells were kept at 4 °C for 10 min. Then, DT (34.4 nM) was added and cells were further incubated at 4 °C for 1 h and 10 min to enable binding of DT to its specific cell surface receptor. Three washing steps with ice-cold PBS were used to remove unbound toxin and cells were subsequently scraped in 2.5-fold concentrated sample buffer containing 10% DTT [38]. SDS-PAGE was used for protein separation and after blotting, bound DT was detected using a specific antibody against DT and peroxidase-coupled secondary anti-mouse antibody.

4.7. DTA Translocation Assay across the Cytoplasmic Membrane of Living Cells

The pH-dependent translocation of DT across endosomal membranes was experimentally mimicked on the cytoplasmic membranes of intact cells and performed as described earlier [5]. In brief, HeLa cells were pre-incubated in serum-free medium with Baf A1 (100 nM) plus EGA (50 μM) or the vehicle DMSO for 1 h at 37 °C. Then, the cells were kept on ice for 10 min followed by an incubation with nicked DT (13 nM) for 40 min at 4 °C to allow binding of the toxin to the cell surface. Thereafter, to trigger pH-driven toxin-translocation across the surface membrane, cells were exposed to an acidic pulse (pH 3.7) for 15 min at 37 °C. Additional wells with toxin-treated cells were treated with neutral medium (pH 7.5) for control. Subsequently, all cells were further incubated at 37 °C in neutral medium containing FCS, Baf A1 (100 nM) ± EGA (50 μM). After the given incubation periods, DT-induced cell-rounding was monitored using photography. From these, the amount of rounded cells was determined in per cent of total cell count per picture.

4.8. Reproducibility of the Experiments

All experiments were independently performed at least two times and in the figures, results from representative experiments are shown. Quantification was performed by calculating the values

($n = 3$; $n = 3$ refers to three fields of view in one experiment and several independent experiments were performed, all with comparable results) as the means ± standard deviation (S.D.) using the Prism4 Software (Version 4.0, GraphPad Software Inc., La Jolla, CA, USA, 2003).

Acknowledgments: The work was financially supported by the Deutsche Forschungsgemeinschaft DFG (grant BA 2087/2-2). Leonie Schnell is a member of the International School in Molecular Medicine Ulm (IGradU) and thanks IGradU for the support.

Author Contributions: Conceived and designed the experiments: H.B., L.S.; Performed the experiments: L.S., A.-K.M.; Analyzed the data: H.B., L.S.; Wrote the paper: H.B., L.S.; Contributed essential materials, reagents, tools: A.M., D.A.T., C.M.

References

1. Murphy, J.R. Mechanism of diphtheria toxin catalytic domain delivery to the eukaryotic cell cytosol and the cellular factors that directly participate in the process. *Toxins* **2011**, *3*, 294–308. [CrossRef] [PubMed]
2. Pappenheimer, A.M. Diphtheria toxin. *Annu. Rev. Biochem.* **1977**, *46*, 69–94. [CrossRef] [PubMed]
3. Collier, R.J.; Cole, H.A. Diphtheria toxin subunit active in vitro. *Science* **1969**, *164*, 1179–1181. [CrossRef] [PubMed]
4. Blumenthal, B.; Hoffmann, C.; Aktories, K.; Backert, S.; Schmidt, G. The cytotoxic necrotizing factors from Yersinia pseudotuberculosis and from Escherichia coli bind to different cellular receptors but take the same route to the cytosol. *Infect. Immun.* **2007**, *75*, 3344–3353. [CrossRef] [PubMed]
5. Schnell, L.; Dmochewitz-Kück, L.; Feigl, P.; Montecucco, C.; Barth, H. Thioredoxin reductase inhibitor auranofin prevents membrane transport of diphtheria toxin into the cytosol and protects human cells from intoxication. *Toxicon* **2016**, *116*, 23–28. [CrossRef] [PubMed]
6. Collier, R.J.; Kandel, J. Structure and activity of diphtheria toxin. I. Thiol-dependent dissociation of a fraction of toxin into enzymically active and inactive fragments. *J. Biol. Chem.* **1971**, *246*, 1496–1503. [PubMed]
7. Iwamoto, R.; Higashiyama, S.; Mitamura, T.; Taniguchi, N.; Klagsbrun, M.; Mekada, E. Heparin-binding EGF-like growth factor, which acts as the diphtheria toxin receptor, forms a complex with membrane protein DRAP27/CD9, which up-regulates functional receptors and diphtheria toxin sensitivity. *EMBO J.* **1994**, *13*, 2322–2330. [PubMed]
8. Naglich, J.G.; Metherall, J.E.; Russell, D.W.; Eidels, L. Expression cloning of a diphtheria toxin receptor: Identity with a heparin-binding EGF-like growth factor precursor. *Cell* **1992**, *69*, 1051–1061. [CrossRef]
9. Choe, S.; Bennett, M.J.; Fujii, G.; Curmi, P.M.; Kantardjieff, K.A.; Collier, R.J.; Eisenberg, D. The crystal structure of diphtheria toxin. *Nature* **1992**, *357*, 216–222. [CrossRef] [PubMed]
10. Johnson, V.G.; Nicholls, P.J.; Habig, W.H.; Youle, R.J. The role of proline 345 in diphtheria toxin translocation. *J. Biol. Chem.* **1993**, *268*, 3514–3519. [PubMed]
11. Perier, A.; Chassaing, A.; Raffestin, S.; Pichard, S.; Masella, M.; Ménez, A.; Forge, V.; Chenal, A.; Gillet, D. Concerted protonation of key histidines triggers membrane interaction of the diphtheria toxin T domain. *J. Biol. Chem.* **2007**, *282*, 24239–24245. [CrossRef] [PubMed]
12. Drazin, R.; Kandel, J.; Collier, R.J. Structure and activity of DT II. Attack by trypsin at a specific site within the intact molecule. *J. Biol. Chem.* **1971**, *246*, 1504–1510. [PubMed]
13. Papini, E.; Sandoná, D.; Rappuoli, R.; Montecucco, C. On the membrane translocation of diphtheria toxin: At low pH the toxin induces ion channels on cells. *EMBO J.* **1988**, *7*, 3353–3359. [PubMed]
14. Lemichez, E.; Bomsel, M.; Devilliers, G.; vanderSpek, J.; Murphy, J.R.; Lukianov, E.V.; Olsnes, S.; Boquet, P. Membrane translocation of diphtheria toxin fragment A exploits early to late endosome trafficking machinery. *Mol. Microbiol.* **1997**, *23*, 445–457. [CrossRef] [PubMed]
15. Wiedlocha, A.; Madshus, I.H.; Mach, H.; Middaugh, C.R.; Olsnes, S. Tight folding of acidic fibroblast growth factor prevents its translocation to the cytosol with diphtheria toxin as vector. *EMBO J.* **1992**, *11*, 4835–4842. [PubMed]
16. Falnes, P.O.; Choe, S.; Madshus, I.H.; Wilson, B.A.; Olsnes, S. Inhibition of membrane translocation of diphtheria toxin A-fragment by internal disulfide bridges. *J. Biol. Chem.* **1994**, *269*, 8402–8407. [PubMed]

17. Stenmark, H.; Olsnes, S.; Madshus, I.H. Elimination of the disulphide bridge in fragment B of diphtheria toxin: Effect on membrane insertion, channel formation, and ATP binding. *Mol. Microbiol.* **1991**, *5*, 595–606. [CrossRef] [PubMed]
18. Leka, O.; Vallese, F.; Pirazzini, M.; Berto, P.; Montecucco, C.; Zanotti, G. Diphtheria toxin conformational switching at acidic pH. *FEBS J.* **2014**, *281*, 2115–2122. [CrossRef] [PubMed]
19. Umata, T.; Moriyama, Y.; Futai, M.; Mekada, E. The cytotoxic action of diphtheria toxin and its degradation in intact Vero cells are inhibited by bafilomycin A1, a specific inhibitor of vacuolar-type H(+)-ATPase. *J. Biol. Chem.* **1990**, *265*, 21940–21945. [PubMed]
20. Sandvig, K.; Olsnes, S. Rapid entry of nicked diphtheria toxin into cells at low pH. Characterization of the entry process and effects of low pH on the toxin molecule. *J. Biol. Chem.* **1981**, *256*, 9068–9076. [PubMed]
21. Sandvig, K.; Olsnes, S. Diphtheria toxin-induced channels in Vero cells selective for monovalent cations. *J. Biol. Chem.* **1988**, *263*, 12352–12359. [PubMed]
22. Moskaug, J.O.; Stenmark, H.; Olsnes, S. Insertion of diphtheria toxin B-fragment into the plasma membrane at low pH. Characterization and topology of inserted regions. *J. Biol. Chem.* **1991**, *266*, 2652–2659. [PubMed]
23. Falnes, P.O.; Madshus, I.H.; Sandvig, K.; Olsnes, S. Replacement of negative by positive charges in the presumed membrane-inserted part of diphtheria toxin B fragment. Effect on membrane translocation and on formation of cation channels. *J. Biol. Chem.* **1992**, *267*, 12284–12290. [PubMed]
24. Ratts, R.; Zeng, H.; Berg, E.A.; Blue, C.; McComb, M.E.; Costello, C.E.; vanderSpek, J.C.; Murphy, R. The cytosolic entry of diphtheria toxin catalytic domain requires a host cell cytosolic translocation factor complex. *J. Cell Biol.* **2003**, *160*, 1139–1150. [CrossRef] [PubMed]
25. Dmochewitz, L.; Lillich, M.; Kaiser, E.; Jennings, L.D.; Lang, A.E.; Buchner, J.; Fischer, G.; Aktories, K.; Collier, R.J.; Barth, H. Role of CypA and Hsp90 in membrane translocation mediated by anthrax protective antigen. *Cell. Microbiol.* **2011**, *13*, 359–373. [CrossRef] [PubMed]
26. Mustacich, D.; Powis, G. Thioredoxin reductase. *Biochem. J.* **2000**, *346*, 1–8. [CrossRef] [PubMed]
27. Tsuneoka, M.; Nakayama, K.; Hatsuzawa, K.; Komada, M.; Kitamura, N.; Mekada, E. Evidence for the involvement of furin in cleavage and activation of diphtheria toxin. *J. Biol. Chem.* **1993**, *268*, 26461–26465. [PubMed]
28. Gill, D.M.; Pappenheimer, A.M. Structure activity relationships in diphtheria toxin. *J. Biol. Chem.* **1971**, *246*, 1485–1491. [PubMed]
29. Ariansen, S.; Afanasiev, B.N.; Moskaug, J.O.; Stenmark, H.; Madhaus, I.H.; Olsnes, S. Membrane translocation of diphtheria toxin A-fragment: Role of carboxy-terminal region. *Biochemistry* **1993**, *32*, 83–90. [CrossRef] [PubMed]
30. Papini, E.; Rappuoli, R.; Murgia, M.; Montecucco, C. Cell penetration of diphtheria toxin. Reduction of the interchain disulfide bridge is the rate-limiting step of translocation in the cytosol. *J. Biol. Chem.* **1993**, *268*, 1567–1574. [PubMed]
31. Madshus, I.H.; Wiedlocha, A.; Sandvig, K. Intermediates in translocation of diphtheria toxin across the plasma membrane. *J. Biol. Chem.* **1994**, *269*, 4648–4652. [PubMed]
32. Moskaug, J.O.; Sandvig, K.; Olsnes, S. Cell-mediated reduction of the interfragment disulfide in nicked DT. A new system to study toxin entry at low pH. *J. Biol. Chem.* **1987**, *262*, 10339–10345. [PubMed]
33. Young, J.A.; Collier, R.J. Anthrax toxin: Receptor binding, internalization, pore formation, and translocation. *Annu. Rev. Biochem.* **2007**, *76*, 243–265. [CrossRef] [PubMed]
34. Gillespie, E.J.; Ho, C.L.; Balaji, K.; Clemens, D.L.; Deng, G.; Wang, Y.E.; Elsaesser, H.J.; Tamilselvam, B.; Gargi, A.; Dixon, S.D.; et al. Selective inhibitor of endosomal trafficking pathways exploited by multiple toxins and viruses. *Proc. Natl. Acad. Sci. USA* **2013**, *110*, E4904–E4912. [CrossRef] [PubMed]
35. Schnell, L.; Mittler, A.K.; Sadi, M.; Popoff, M.R.; Schwan, C.; Aktories, K.; Mattarei, A.; Tehran, D.A.; Montecucco, C.; Barth, H. EGA protects mammalian cells from *Clostridium difficile* CDT, *Clostridium perfringens* iota toxin, and *Clostridium botulinum* C2 toxin. *Toxins* **2016**, *8*, 101. [CrossRef] [PubMed]
36. Tehran, D.A.; Zanetti, G.; Leka, O.; Lista, F.; Fillo, S.; Binz, T.; Shone, C.C.; Rossetto, O.; Montecucco, C.; Paradisi, C.; et al. A Novel inhibitor prevents the peripheral neuroparalysis of botulinum neurotoxins. *Sci. Rep.* **2015**, *5*, 17513. [CrossRef] [PubMed]

37. Yamaizumi, M.; Mekada, E.; Uchida, T.; Okada, Y. One molecule of diphtheria toxin fragment A introduced into a cell can kill the cell. *Cell* **1978**, *15*, 245–250. [CrossRef]
38. Laemmli, U.K. Cleavage of structural proteins during the assembly of the head of bacteriophage T4. *Nature* **1970**, *227*, 680–685. [CrossRef] [PubMed]

10

Multi-Stress Induction of the *Mycobacterium tuberculosis* MbcTA Bactericidal Toxin-Antitoxin System

Kanchiyaphat Ariyachaokun [1,2,†], Anna D. Grabowska [1,3,†], Claude Gutierrez [1,*] and Olivier Neyrolles [1]

1 Institut de Pharmacologie et de Biologie Structurale, IPBS, Université de Toulouse, CNRS, UPS, 205 route de Narbonne, 31400 Toulouse, France; kanchiyaphat.a@ubu.ac.th (K.A.); dr.anna.grabowska@gmail.com (A.D.G.); olivier.neyrolles@ipbs.fr (O.N.)

2 Department of Biological Science, Ubon Ratchathani University, 85 Sathollmark Rd., Warin Chamrab, Ubon Ratchathani 34190, Thailand

3 Department of Biophysics and Human Physiology, Medical University of Warsaw, 02-091 Warsaw, Poland

* Correspondence: claude.gutierrez@ipbs.fr

† K.A. and A.D.G. contributed equally to this work.

Abstract: MbcTA is a type II toxin/antitoxin (TA) system of *Mycobacterium tuberculosis*. The MbcT toxin triggers mycobacterial cell death in vitro and in vivo through the phosphorolysis of the essential metabolite NAD^+ and its bactericidal activity is neutralized by physical interaction with its cognate antitoxin MbcA. Therefore, the MbcTA system appears as a promising target for the development of novel therapies against tuberculosis, through the identification of compounds able to antagonize or destabilize the MbcA antitoxin. Here, the expression of the *mbcAT* operon and its regulation were investigated. A dual fluorescent reporter system was developed, based on an integrative mycobacterial plasmid that encodes a constitutively expressed reporter, serving as an internal standard for monitoring mycobacterial gene expression, and an additional reporter, dependent on the promoter under investigation. This system was used both in *M. tuberculosis* and in the fast growing model species *Mycobacterium smegmatis* to: (i) assess the autoregulation of *mbcAT*; (ii) perform a genetic dissection of the *mbcA* promoter/operator region; and (iii) explore the regulation of *mbcAT* transcription from the *mbcA* promoter (P_{mbcA}) in a variety of stress conditions, including in vivo in mice and in macrophages.

Keywords: tuberculosis; toxin-antitoxin systems; bacterial cell death; NAD^+; stress-response

Key Contribution: A dual fluorescent reporter system, designed to monitor gene expression and regulation in mycobacteria, was constructed and used to address the multi-stress regulation of the MbcTA bactericidal toxin/antitoxin system of *Mycobacterium tuberculosis*.

---

1. Introduction

Mycobacterium tuberculosis (*Mtb*), the etiological agent of tuberculosis (TB), is a strictly human-adapted obligate pathogen of major public health importance (WHO, Global Tuberculosis report, 2019). Indeed, TB was responsible for 1.5 million deaths and about ten million new TB cases were reported in 2018. Of particular concern, an estimated 484,000 new infections were due to *Mtb* strains resistant to one or more frontline antibiotics in 2018, and 80% of these *Mtb* strains were multi-drug resistant (MDR). Therefore, the development of novel therapies against TB is urgently needed. In line with this, we recently started to explore the potential utilization of intrinsic bactericidal toxin/antitoxin (TA) systems as new therapeutic tools [1]. TA systems are genetic modules encoding an endogenous protein toxin, which targets an essential process in the bacterial cell, and a toxin-neutralizing antidote, the so-called antitoxin [2–4]. Under favorable conditions, the antitoxin is stable and prevents auto-intoxication of the TA-producing cell. Although the precise mechanisms remain to be characterized for most TA systems, it has been proposed that when cells are facing environmental stresses, the less stable antitoxin is degraded, leading to an increased production of toxin, resulting in growth arrest of the bacterial cells [5,6]. TA systems are classified into six families, depending on the nature of the antitoxin, i.e., ribonucleic acid (RNA) or protein, and its mechanism of toxin neutralization. For instance, in the most studied family of type II TA systems, the antitoxin is a protein that neutralizes the toxin through direct protein-protein interaction [2,4]. The *Mtb* genome harbors more than 80 TA systems, which were proposed to contribute to virulence and persistence in the infected host [7–11]. Among those, the recently characterized MbcTA system encodes an NAD^+ phosphorylase toxin, able to trigger bacterial cells death through thorough depletion of the NAD^+ pool [1]. In the present study, we investigated the expression and regulation of the *mbcAT* operon. We developed a dual fluorescent reporter system, which we used to dissect the operon promoter/operator region at the genetic level. Using this system, we demonstrate that transcription from P_{mbcA} is induced by a range of stress conditions, reflecting those encountered inside the infected host.

2. Results

2.1. MbcT Toxin Depletes NAD^+ and Is Bactericidal in Mycobacterium smegmatis

In order to express the MbcT toxin in *M. smegmatis*, we used an expression vector, pGMCS-TetR-P1-*mbcT*, that can stably integrate into the mycobacterial chromosome, within the *glyV* tRNA gene, through site-specific recombination catalyzed by the L5 mycobacteriophage integrase [12]. In this plasmid, *mbcT* is expressed under the tetracycline-inducible promoter $P_{myc1\ tetO}$ [13], hereafter abbreviated P1. Upon transformation into a *Mtb* strain deleted for the endogenous *mbcAT* operon, pGMCS-TetR-P1-*mbcT* conferred anhydro-tetracycline (Atc)-dependent depletion of the essential metabolite NAD^+, resulting in bacterial cell death [1]. In contrast, growth of a derivative of wild-type *M. smegmatis* strain mc^2 155 carrying this plasmid was not affected by addition of Atc to its growth medium (Figure 1a), presumably because the amount of MbcT produced in such conditions was not sufficient to deplete NAD^+ from the *M. smegmatis* cells (Figure 1d). To verify this hypothesis, we constructed plasmids expressing *mbcT* under the control of stronger Atc-inducible promoters, namely P606Pld and P606 (obtained from D. Schnappinger, Weill Cornell Medical College, NY, USA). Induction of P606Pld-*mbcT* had no effect on mc^2 155 growth (Figure 1b), although it reduced the NAD^+ content to 8% of the amount measured in the uninduced strain carrying pGMCS-TetR-P1-*mbcT*. In contrast, *mbcT* expression from the Atc-induced P606 promoter abolished mycobacterial growth (Figure 1c) and, accordingly, reduced the amount of NAD^+ to less than 2%, compared to that in control cells (Figure 1d). Furthermore, it resulted in an approximately 99% loss of colony forming units (CFU) within few hours following induction (Figure 1e), and assessment of cell viability using the LIVE/DEAD BacLight stains followed by flow cytometry analysis demonstrated that the proportion of propidium iodide-permeable cells, which was very low in uninduced exponentially growing cells (Figure 1e), increased markedly by 8 or 24 h after the addition of Atc (Figure 1f–i). This confirmed the bactericidal effect of the NAD^+ depletion following the induction of P606-*mbcT*. In conclusion, providing that it is driven by a relatively strong promoter, production of MbcT in *M. smegmatis* can recapitulate the bactericidal effect of the toxin previously observed in *Mtb*.

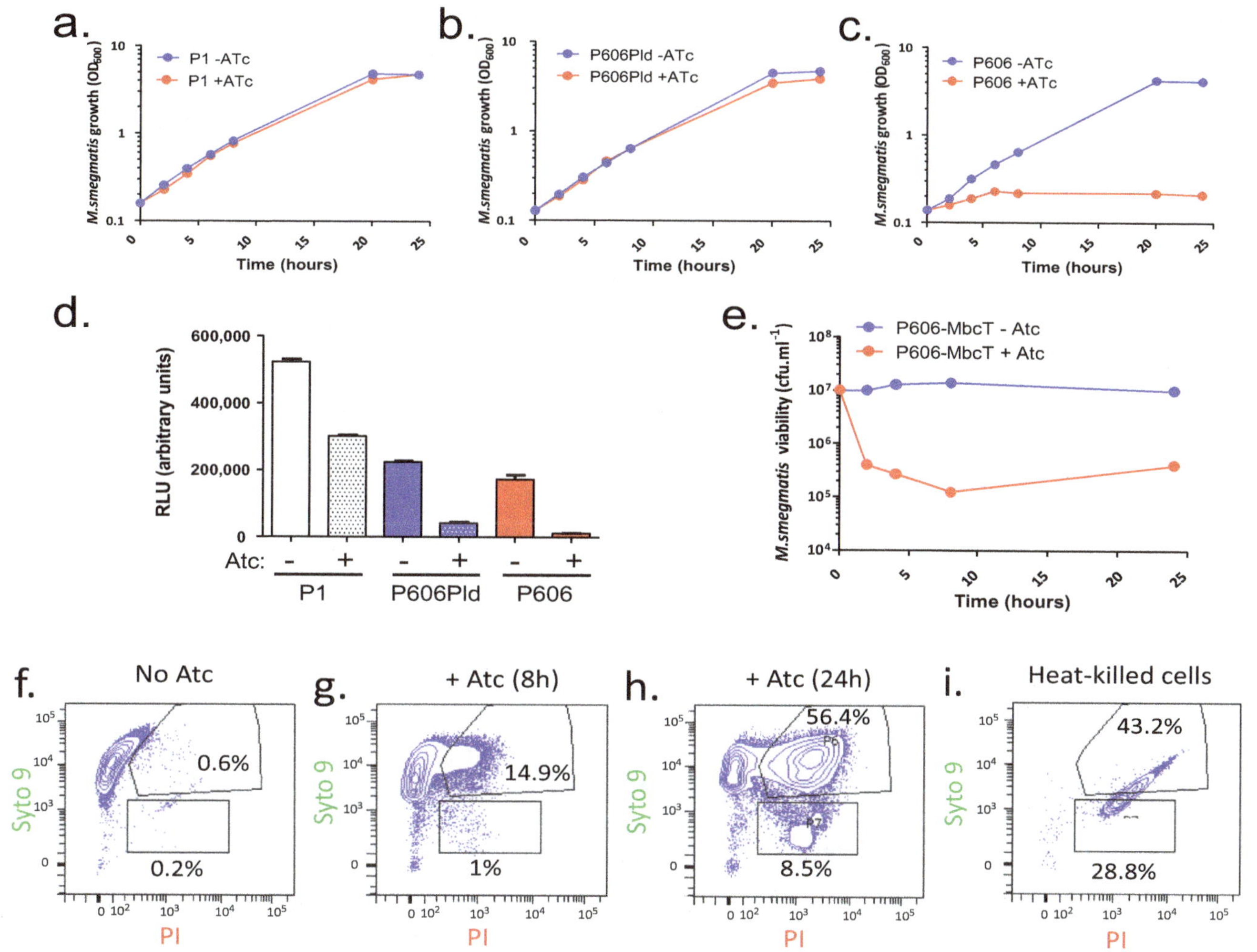

Figure 1. The toxin MbcT depletes NAD^+ and is bactericidal in *Mycobacterium smegmatis*. (**a–c**) Cultures of *M. smegmatis* strain mc^2 155 transformed with plasmids producing MbcT under the control of the indicated promoter were diluted at time 0 in fresh medium supplemented with streptomycin alone (blue) or streptomycin and 200 $ng.mL^{-1}$ anhydro-tetracycline (Atc) (red) and bacterial growth was followed over up to 24 h. (**d**) Samples of the *M. smegmatis* cultures were harvested after 8 h of treatment and relative content of NAD^+ was measured in cell extracts. Values are representative of two independent replicates of the same experiment. (**e**) Samples of the cultures of *M. smegmatis* mc^2 155/pGMCS-TetR-P606-mbcT were harvested at the indicated times, diluted, and plated on L agar. Colonies were counted after 3 days of growth at 37 °C. (**f–h**) Samples of the cultures of mc^2 155/pGMCS-TetR-P606-mbcT were harvested after 8 or 24 h of treatment with Atc, labeled with the LIVE/DEAD BacLight dyes (Syto9 and Propidium Iodide (PI)) and analyzed by flow cytometry (FACS). (**i**) Cells grown without Atc were heat-killed at 98 °C for 30 min before LIVE/DEAD labeling and FACS analysis. Data are representative of two experiments with similar results.

2.2. Construction of a Dual Fluorescent Reporter System to Assess Gene Expression in Mycobacteria

In order to monitor the expression of mycobacterial genes of interest in in vitro cultures or during infection, we developed a dual fluorescent reporter system in which an integrative plasmid carries both (i) a constitutive reporter, serving as an internal standard or probe for the detection of the bacterial cells and (ii) an additional reporter, expressed from the promoter/operator region of the gene under investigation. This reporter system was constructed by multisite gateway recombination [14], using plasmid pDE43-MCS as backbone, and entry vectors carrying either a 5′, a middle or a 3′ module, as depicted in Figure 2a. The 5′ module consists of a fluorescent reporter, which can be either green fluorescent protein (GFP), mCherry, mTurquoise, or mVenus, expressed from the constitutive P1 promoter. The middle entry clone introduces a second fluorescent reporter gene, from the same list but different from the first one, fused to the promoter/operator region of the *mbcAT* operon. Finally, the 3′ entry clone introduces either nothing, using an empty entry clone (pEN23A-MluI), or any gene(s) of interest, which effect on P_{mbcA} is to be tested. Figure 2b depicts an example of such construct, pGMCS-P1-mCherry-P_{mbcA}-GFP-*mbcA*, in which the *mbcA* gene is inserted to assess the effect of MbcA on its own transcription.

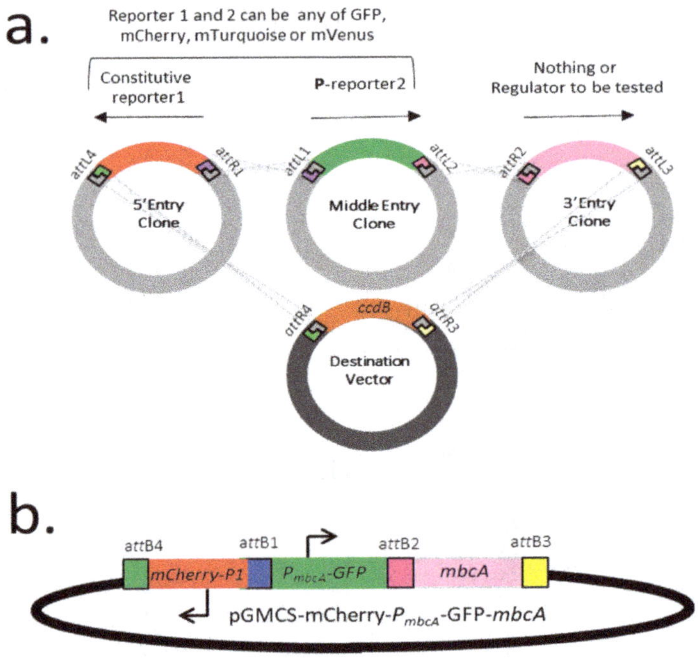

Figure 2. A dual florescent protein reporter system developed for studying gene expression in mycobacteria. (**a**). Gateway cloning strategy used to construct the required plasmids. (**b**). Example of plasmid construction designed to analyze the expression and regulation of *mbcAT*. GFP = green fluorescent protein.

2.3. Use of a Dual Fluorescent Reporter System to Study the Expression of mbcAT

2.3.1. Auto-Repression of *mbcAT*

A set of three plasmids, namely pGMCS-P1-mCherry-P_{mbcA}-GFP, pGMCS-P1-mCherry-P_{mbcA}-GFP-*mbcA*, and pGMCS-P1-mCherry-P_{mbcA}-GFP-*mbcAT*, was constructed and transformed into *M. smegmatis* mc^2 155. The resulting strains were grown until exponential phase in complete 7H9 medium and the fluorescence of mCherry and GFP were measured by flow cytometry. The untransformed mc^2 155 strain exhibited a very low background level of fluorescence (Figure 3a). In contrast, the three transformed strains exhibited both red and green fluorescence (Figure 3b–d). Comparison of the mean fluorescence intensities showed a similar level of fluorescence for mCherry in the three constructs, whereas the fluorescence for GFP was approximately 5-fold lower when the MbcA antitoxin was expressed, regardless of MbcT toxin co-expression (Figure 3h). Therefore, we conclude that the transcription from P_{mbcA} is auto-repressed by MbcA, independently of the presence of the toxin. We confirmed these observations in a qualitative assay, by observation of the bacterial cells in fluorescence microscopy (Figure S1). In addition, we also introduced the same set of plasmids into *M. tuberculosis* H37Rv and observed the cells in fluorescence microscopy after fixation with *p*-formaldehyde. As expected, in the wild-type H37Rv, expression of GFP remained very low in all our constructs because of the MbcA repressor produced from the chromosomal copy of the *mbcAT* operon. In contrast, in H37Rv Δ*mbcAT*::Kan^R, the pattern of mCherry and GFP expression was comparable to that observed in *M. smegmatis*, demonstrating that autoregulation occurs similarly in the two mycobacterial species.

Analysis of the MbcA sequence predicts a helix-turn-helix motif between amino-acids 20 and 40, forming a putative DNA binding domain. In order to test whether this motif is important for repression by MbcA, the Arg^{33} and Arg^{37} residues were substituted with glutamic acid by site-directed mutagenesis of pGMCS-P1-mCherry-P_{mbcA}-GFP-*mbcA*, and the resulting plasmids were transformed into *M. smegmatis* mc^2 155. Both modifications abolished the repression by the MbcA variants, leading to an expression level comparable to that observed in the absence of MbcA (Figure 3h). When the $R^{33}E$ substitution was introduced in the plasmid pGMCS-P1-mCherry-P_{mbcA}-GFP-*mbcAT*, the co-expression of the toxin MbcT with the $R^{33}E$ variant of MbcA still resulted in absence of repression.

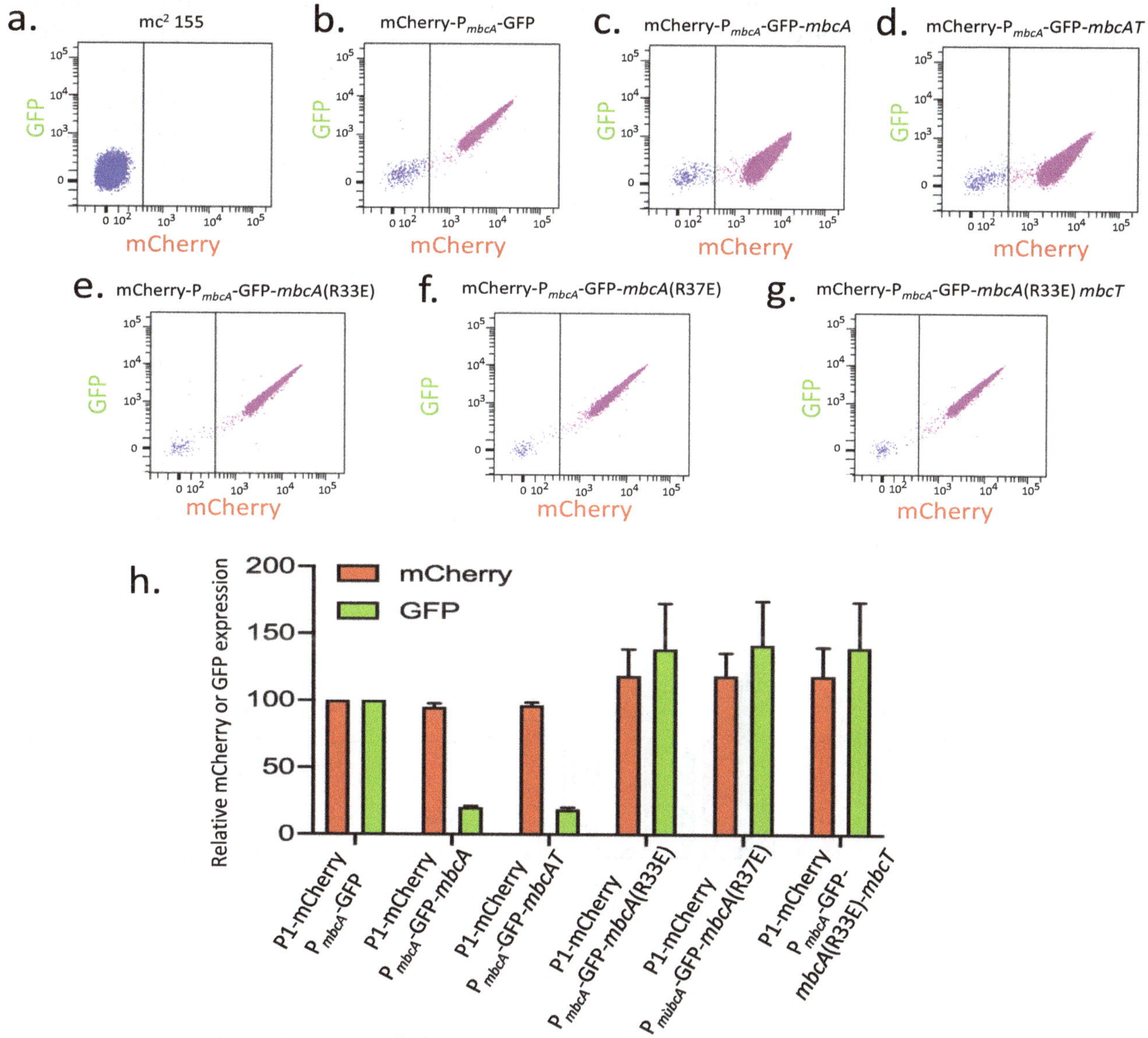

Figure 3. Autorepression of the *mbcAT* operon. (**a**)–(**g**) *M. smegmatis* mc^2 155 wild-type and transformed with the indicated pGMCS-mCherry-P_{mbcA}-GFP derivatives were grown to exponential phase in complete 7H9 medium and analyzed by flow cytometry. Blue and pink points separated by a vertical line indicate the cells negative or positive for mCherry fluorescence, respectively. (**h**) Mean fluorescence intensities (MFI) of the fluorescence positive cells (pink points) relative to that of the strain harboring pGMCS-mCherry-P_{mbcA}-GFP (a) are reported for mCherry (red bars) and for GFP (green bars). Values are the average of two biological replicates with standard deviation.

2.3.2. Genetic Dissection of the *mbcAT* Promoter/Operator Region

In the plasmids described above (Figure 3), the *mbcAT* promoter is located on a 261-bp fragment ($P_{mbcA(261)}$-GFP, labeled in red letters in Figure 4a) amplified with the oligonucleotides Fw1/Rv1 (Table S1). A shorter 103-bp fragment, extending between oligonucleotide Fw2 and Rv1, also contains the entire promoter and operator, because it produced the same pattern of expression ($P_{mbcA(103)}$-GFP, labeled in green letters in Figure 4b). A putative -10 element of σ^{A}-dependent promoters is present upstream from the *mbcA* transcription start site identified by RNA-seq experiments [15]. We confirmed this promoter location by generating two 239-bp DNA fragments amplified with the oligonucleotides Fw1/Rv2, with the wild-type sequence and Fw1/Rv3, with a mutation changing a consensus T into a non-consensus G in the -10 element. Whereas the first fragment still allowed the expression of GFP ($P_{mbcA(239)}$-GFP, labeled in blue letters in Figure 4b), the T=>G substitution almost completely abolished the production of the downstream encoded GFP ($P_{mbcA(239\text{-}G)}$-GFP, labeled in brown letters in Figure 4b). In addition, we observed that the Fw1/Rv2 fragment was much less sensitive to repression by MbcA. To further analyze the *mbcA* operator, we constructed mutations of 4 (Mut1; changing CAAA into TCGT) or 3 nucleotides (Mut2; changing ACA into GTG), and we observed that Mut1 resulted in a high expression in the presence of MbcA, whereas Mut 2 did not abolish repression by MbcA.

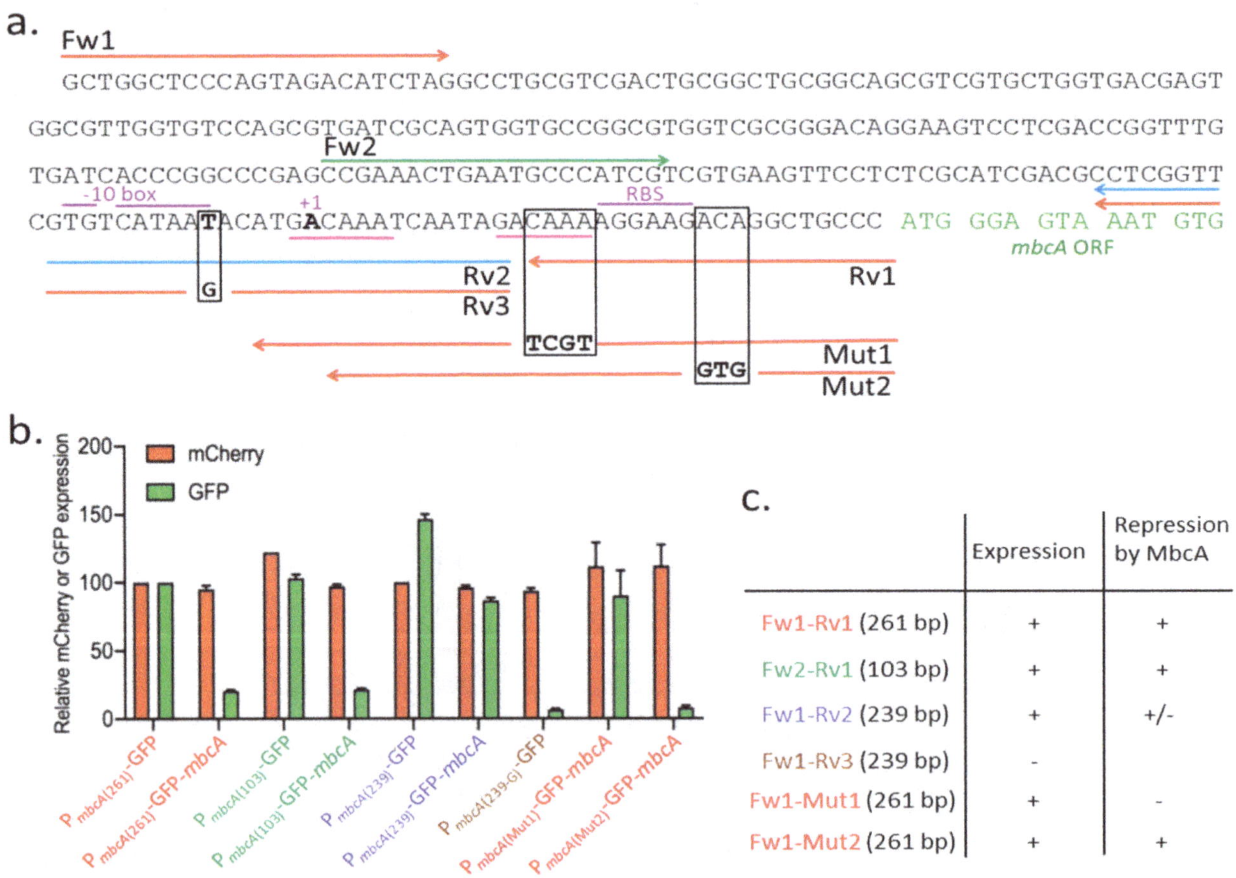

Figure 4. Genetic analysis of the promoter/operator region of the *mbcAT* operon. (**a**) DNA sequence of the *mbcA* promoter/operator region. Arrows indicate oligonucleotides used to amplify various fragments encompassing the promoter region. Black squares surround the nucleotides modified by mutations introduced in the oligonucleotides Rv3 (T=>G), Mut1 (CAAA=>TCGT), and Mut2 (ACA=>GTG). +1 points to the transcription start site. (**b**) Mean fluorescence intensities (MFI) were determined by flow cytometry with strains harboring the different constructs. Red and green bars indicate MFI relative to that of strain harboring pGMCS-mCherry-P_{MbcA}-GFP. Values are the average of two biological replicates with standard deviation. (**c**) Summary of the results obtained with the different constructs.

2.4. Multistress Induction of mbcAT Expression

To assess the regulation of *mbcAT* expression, the plasmid pGMCS-P1-mTurquoise-P_{mbcA}-mVenus-*mbcA* was introduced into wild-type Mtb H37Rv, and the resulting strain was grown to mid-exponential phase and subjected to various stress conditions. As shown in Figure 5, we could observe qualitatively that the expression from P_{mbcA} was increased upon starvation following incubation of the cells in phosphate saline buffer (PBS), or upon treatment with H_2O_2 or with the NO-generating reagent diethylenetriamine/nitric oxide adduct (DETA/NO).

Finally, *Mtb* strain H37Rv/pGMCS-P1-mTurquoise-P_{mbcA}-mVenus-*mbcA* was used to infect mouse or human macrophages, and fluorescence microscopy of fixed infected macrophages demonstrated that transcription from P_{mbcA} is stimulated upon *Mtb* phagocytosis by macrophages (Figure 6).

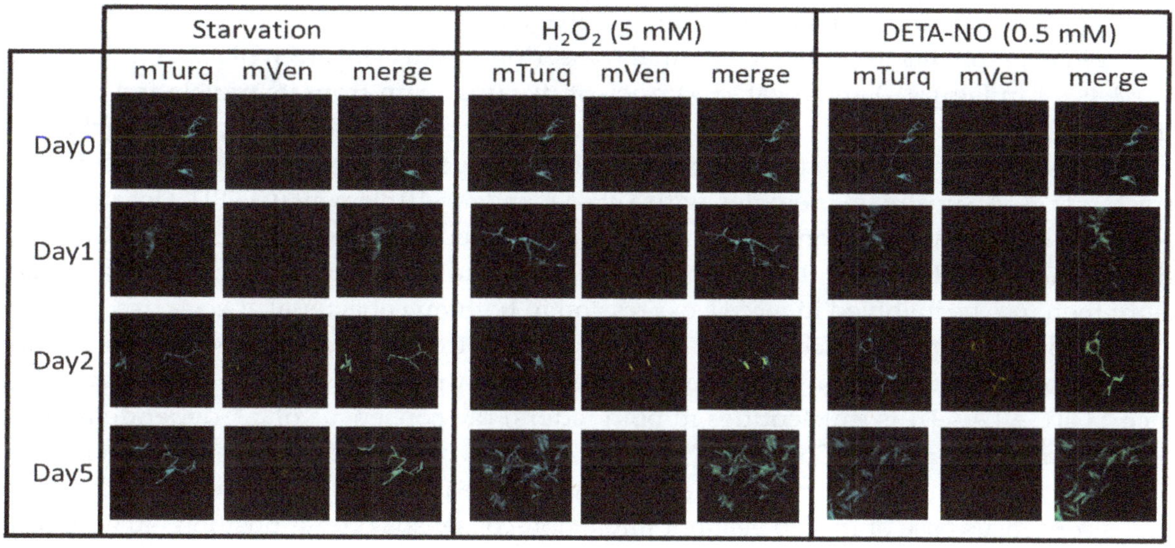

Figure 5. Effect of stress conditions on expression of the *mbcAT* operon in *M. tuberculosis.* Wild-type H37Rv transformed with plasmid pGMCS-P1-mTurq-P_{mbcA}-mVenus-*mbcA* was grown in complete 7H9 medium. At day 0, cells were harvested and resuspended in phosphate saline buffer (PBS) buffer (starvation), or 7H9 medium supplemented with 5 mM H_2O_2 or 0.5 mM DETA/NO. At the indicated times, samples were harvested, fixed with 4% *p*-formaldehyde and observed by fluorescent microscopy using large field Leica DMIRB, magnification 630×.

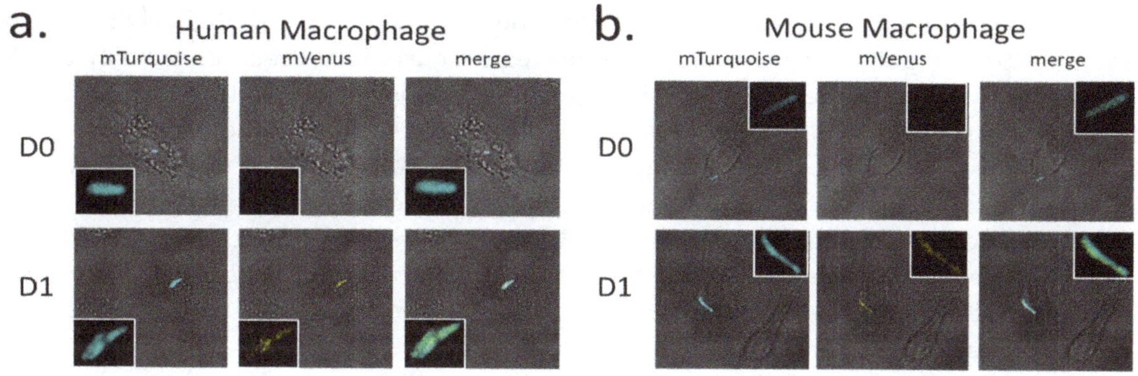

Figure 6. Induction of the *mbcAT* operon expression in *M. tuberculosis* upon infection of macrophages. *Mtb* H37Rv/pGMCS-P1-mTurq-P_{mbcA}-mVenus-*mbcA* was used to infect human (**a**) or mouse (**b**) macrophages at multiplicity of infection of 0.1. Infected cells were harvested at day 0 (D0) or 24 h post-infection (D1), fixed with 4% *p*-formaldehyde and observed by fluorescent microscopy using large field Leica DMIRB, magnification 630×. Images show the merged bright field and either the blue, yellow, or merged fluorescence. Insets show enlarged view of the phagocytized bacteria.

3. Discussion

The dual fluorescent reporter system presented here, constructed using the Gateway cloning technology [12], allows for quantification and/or visualization of bacterial cells, and in parallel for following expression of a gene of interest (Figure 2). As shown here, this system is stable and robust, since it is based on an integrative vector and exhibited only minor variations of the expression of the internal standard reporter (mCherry in Figures 3 and 4). Allowing both quantitative (Figures 3 and 4) and qualitative measurements (Figures 5 and 6), this system can be used to follow gene expression not only in bacterial cultures, or inside infected macrophages, as shown in this work, but also inside host tissues during infection of animal models [16]. Finally, it is also very versatile, since it offers a straightforward way to construct bacterial variants carrying a set of different reporters, as exemplified here with a variety of fluorescent proteins (Figure 2). We believe that it can be of broad interest to investigate gene expression and regulation in mycobacteria.

RNA-seq experiments indicated a potential transcription start site (TSS) of *mbcA* in *M. tuberculosis* (indicated as +1 in Figure 4) and located 34 nucleotides upstream from its translation initiation site (TIS) [15]. Based on bioinformatics analysis, it has been proposed that transcription of the *mbcAT* operon occurs from two overlapping promoters recognized by RNA polymerase holoenzyme using alternative sigma factors [17]. An upstream promoter, with a σ^L-dependent -10 element (GGTTC) located at position -54 to -50 from the TIS, and a second promoter, using a σ^H-dependent -10 element (CGTGTC), located at positions -50 to -45. The σ^L and σ^H alternative sigma factors are both present in *M. smegmatis* [18] and might therefore be involved in *mbcAT* expression in both mycobacterial species used in this study. Our data do not rule out the possibility of *mbcAT* expression from σ^L- or σ^H-dependent promoters, but they demonstrate that yet another -10 element is essential for the expression of *mbcAT* (Figure 4), which sequence **TG**TC**A**T**AA**T fits (nucleotides in bold letters) the consensus of σ^A-dependent promoters with an extended -10 motif (TGNTANNNT [15]). The involvement of this -10 element in *mbcAT* expression has been confirmed by our genetic analysis since its activity was abolished by a mutation changing a consensus base of the -10 element into a non-consensus one, at position -40 with respect to the TIS (Figure 4).

Many type II TA systems are regulated by auto-repression, with the antitoxin binding to an operator site in the vicinity of the operon promoter [2,4]. It is also the case for the *mbcAT* operon (Figures 3 and 4). In solution, MbcA and MbcT form a heterododecameric complex (3 × [MbcT-MbcA]$_2$) arranged around a 3-fold symmetry axis in which MbcA folds into a single structured domain consisting of seven α helices [1]. Helices $\alpha2$ and $\alpha3$ form a putative helix-turn-helix motif, and Arg^{33} and Arg^{37}, which are pointing out of helix 3, could be important for the recognition of the operator site. Consistently, substitutions of any of these two arginine residues totally abolished the transcription repression by MbcA (Figure 3). The *mbcA* promoter region harbors a direct repeat of two GACAAA motifs, separated by 12 bp (underlined with a magenta line in Figure 4a). Deletion of the downstream motif on promoter fragment Fw1-Rv2 abolishes the repression by MbcA and mutation Mut1, which disrupts this downstream motif, increases transcription in the presence of MbcA to the level observed in absence of the repressor. In contrast, modification of nucleotides downstream from this motif (mutation Mut2) did not relieve the repression by MbcA (Figure 4b). Although additional biochemical experiments are needed to determine the precise interaction of MbcA with its operator site, these observations suggest that MbcA could bind the GACAAA direct repeats and that both repeats are indispensable to constitute a functional operator.

Several autoregulated type II TA systems are subject to the so-called conditional cooperativity, i.e., a repression mechanism in which the antitoxin-repressor can participate in different types of complexes, alone, or with its cognate toxin, exhibiting different repression efficacies [2,4,19]. Conditional cooperativity has been proposed to contribute to several aspects of the TA systems biology [2]. In such systems, the antitoxin alone binds its operator site with low affinity, resulting in poor repression. When toxin and antitoxin are present in stoichiometric amounts, binding of the toxin to the antitoxin increases its affinity for the operator, leading to a more potent repression of transcription than is observed for the antitoxin alone. Finally, high amounts of toxin can result in a different heteromeric complex exhibiting low affinity for the operator, thereby relieving repression in excess of toxin. As already shown for a number of TA systems [20], our data suggest that the MbcTA system is not subject to conditional cooperativity, since the level of transcription repression by the antitoxin appeared independent on the production of the toxin (Figure 3). We note that in the dodecameric MbcTA complex, the helix-turn-helix motif of MbcA, necessary for the repression (Figure 3), is masked by an interaction with the $\alpha6$ helix of MbcT [1]. In consequence, the dodecameric complex is most likely unable to participate in repression. In the wild-type H37Rv *M. tuberculosis* strain, the ectopic expression of MbcT under the control of the P1 promoter revealed not to be toxic [1], indicating that expression of the wild-type *mbcAT* operon leads to higher amounts of MbcA antitoxin than of MbcT toxin. Therefore, only a fraction of MbcA must be is involved in MbcT-neutralizing complexes, and it is likely that the remaining excess of MbcA can function as auto-repressor.

Previous work has already reported increased expression of *mbcAT* in stress conditions. The MbcT toxin (Rv1989c) has been identified by proteomic analysis in culture filtrates of PBS-starved *M. tuberculosis* cells, whereas it is not detectable during exponential growth [21]. Transcriptomic studies demonstrated increased amounts of *mbcA* and/or *mbcT* mRNA in stationary compared to exponential phase [15], during the so-called enduring hypoxic response [22], or in *M. tuberculosis* persisters surviving a treatment with D-cycloserine [8]. Our data are in agreement with these observations, and further demonstrate that transcription of *mbcAT* increases upon exposure to oxidative stress caused by H_2O_2 or NO-generating compounds (Figure 5), conditions mimicking those encountered inside macrophages [23–25]. In agreement, we also observed increased transcription from the *mbcAT* promoter within infected macrophages (Figure 6). To date, the physiological role of the MbcTA system has not been firmly established. However, the existence of a gene mimicking the antitoxin MbcA in the genome of several myco-bacteriophages [1] strongly suggests that this system could belong to the family of TA systems used to fight against phage infections [26,27]. The multi-stress induction of *mbcAT*, leading to accumulation of both the toxin and the antitoxin, should not be toxic for the *M. tuberculosis* cells. However, it is reminiscent of the so-called general stress response observed in many bacterial species, which results in the expression of a variety of adaptive responses to stresses not yet encountered by bacterial cells but that contribute to reinforce their resistance if they eventually happen to deal with one of these stresses [28].

The *M. tuberculosis* genome is particularly rich in TA systems [7–11], and, as already proposed for other bacterial pathogens [29,30], it is tempting to speculate that at least some of these systems could be used as novel tools in adjunct therapies with classical antibiotics. However, a number of studies have proposed that activation of TA systems may trigger a non-growing physiological state, or so-called persistence, in which the bacterial cells become tolerant to antibiotics [31–34]. Although the role of TA systems in the development of tolerance is still a matter of debate [35], this raises a serious drawback to their potential therapeutic use. The MbcTA system attracted our attention because it is one of the only two TA systems of *M. tuberculosis* that harbor an essential antitoxin gene, i.e., which cannot be genetically disrupted [36]. We reasoned that this could be the consequence of an atypical and particularly toxic activity of MbcT. Indeed, we have demonstrated that MbcT activity, when expressed in recombinant conditions generating physiological expression levels of the toxin, is bactericidal for *M. tuberculosis* through depletion of NAD^+ [1], a metabolite known to be essential for mycobacteria survival, even in a dormant state [37–39]. In addition, we have shown that production of the MbcT toxin during infection could act in synergy with isoniazid treatment to reduce the bacterial load in mice infected by *M tuberculosis*, supporting the idea that MbcTA could be a novel therapeutic target [1]. The observations reported here strengthen this proposal, given that the induction of *mbcAT* during infection ensures the presence of the toxin inside the infecting *M. tuberculosis* cells, where targeting its activation could help fighting against TB.

4. Materials and Methods

4.1. Bacterial Strains and Cultures

M. tuberculosis strains H37Rv (ATCC27294) and H37Rv Δ*mbcAT*::Kan^R [1] and *M. smegmatis* wild-type strain mc^2 155 (ATCC700084) were routinely grown aerobically at 37 °C in complete 7H9 medium, i.e., Middlebrook 7H9 medium (Difco, Franklin Lakes, NJ, USA) supplemented with 10% albumin-dextrose-catalase (ADC, Difco) and 0.05% Tween 80 (Sigma-Aldrich, Saint-Louis, MI, USA) or on Middlebrook 7H11 agar medium (Difco) supplemented with 10% oleic acid-albumin-dextrose-catalase (OADC, Difco). When required, streptomycin (25 μg mL^{-1}), kanamycin (50 μg mL^{-1}), or anhydrotetracycline (Atc; 200 ng mL^{-1}) were added to the culture media. *E. coli* Stellar strain (Clontech Laboratories, Inc., Mountain View, CA, USA) was grown aerobically at 37 °C in L Broth medium or on L Agar, supplemented with streptomycin (25 μg mL^{-1}).

4.2. Oligonucleotides and Plasmids Used in This Work

Oligonucleotides and plasmids used in this work are listed in Tables S1 and S2, respectively. Plasmids pEN- and pGMCS- were constructed by multisite gateway recombination [14] following the manufacturer's instructions (Life Technologies, Carlsbad, CA, USA). DNA fragments were amplified by PCR, using Phusion DNA polymerase (Thermo-Fisher Scientific, Waltham, MA, USA) with the templates and oligonucleotide pairs indicated in Table S1. Plasmids pEN41- and pEN12- were generated by "BP" cloning, using pDO41A- or pDO221A- as destination vectors, and DNA fragments flanked by *attB4/attB1* or *attB1/attB2*, respectively. Plasmids pGMCS-TetR-P1-*mbcT*, pGMCS-TetR-P606Pld-*mbcT* and pGMCS-TetR-P606-*mbcT* were constructed by "LR" cloning reactions, using plasmid pDE43-MCS as destination vector, plasmid pEN41-TetR as 5′ entry clone, plasmids pEN12-P1, pEN12-P606Pld, or pEN12-P606 as middle entry clone, and plasmid pEN23-*mbcT* as 3′ entry clone. Plasmids pGMCS-P1-reporter1-P_{mbcA}-Reporter2 were constructed by "LR" cloning reactions, using plasmid pDE43-MCS as destination vector, a plasmid pEN41-P1-Reporter as 5′ entry clone, a plasmids pEN12-P_{mbcA}-reporter as middle entry clone, and plasmids pEN23-MluI, pEN23-*mbcA*, or pEN23-*mbcAT* as 3′ entry clone. Site directed mutagenesis of plasmids was performed by amplification by PCR of the plasmid to be mutated with a pair of overlapping oligonucleotides carrying the mutation to be introduced (Table S1). The linear plasmid DNA was purified on agarose gels and circularized using HD In-Fusion cloning reactions (Takara, Kusatsu, Japan). All plasmid constructs were verified by DNA sequencing.

4.3. M. Smegmatis Viability Assays and NAD$^+$ Measurement

Exponentially growing cultures (OD_{600} between 0.05 and 0.2) of *M. smegmatis* mc^2 155 transformed with the desired plasmids were divided in two: half was left in the same medium (uninduced cultures) and the other half was treated with 200 ng mL^{-1} of Atc to induce expression from the tetracycline-inducible promoters. After various time post-induction, samples were harvested and centrifuged to eliminate residual Atc. Cells were resuspended in PBS buffer and dilutions were plated on L Agar supplemented with streptomycin, colonies were counted after 3 days at 37 °C. For labeling with LIVE/DEAD® BacLight dyes (Molecular Probes, Eugene, OR, USA), cells were harvested 8 or 24 h post-Atc induction. Cells were centrifuged, resuspended in PBS buffer and stained as recommended by the manufacturer. Labeled cells were analyzed by fluorescence activated cell sorting using a BD FACS LSRFortessa X20 flow cytometer. Flow cytometry data analysis were performed using FlowJo software (Version 10; Becton, Dickinson and Company, Ashland, OR, USA).

Samples of the above cultures were harvested 8 h post-Atc induction and centrifuged. Cells were resuspended in PBS, adjusting the OD_{600} to 0.2. 0.1 μm-diameter glass beads were added to tubes containing 500 μL of each cell suspensions, and cells were lysed by four 60-s pulses at full speed in a bead-beater device. The samples were centrifuged for 1 min at 20,200× *g* and 50 μL of the lysates were mixed with an equal volume of NAD/NADH-Glo™ Detection Reagent (Promega). Luciferin bioluminescence was measured after 30 min of incubation at room temperature, using a CLARIOstar® plate reader (BMG LABTECH, Champigny s/Marne, France) and normalized to background signal (PBS-only) to monitor the relative amounts of NAD^+ present in the cell extracts.

4.4. *Effect of Stress Conditions on P_{mbcA}*

Exponential cultures in complete 7H9 medium of *M. tuberculosis* H37Rv transformed with plasmid pGMCS-P1-mTurq-P_{mbcA}-mVenus-*mbcA* were harvested and centrifuged. Cell pellets were and resuspended in PBS buffer (for starvation), or complete 7H9 medium supplemented with 5 mM H_2O_2 or 0.5 mM DETA/NO. At different times, samples were harvested, fixed with 4% *p*-formaldehyde and observed by fluorescent microscopy using large field Leica DMIRB, magnification 630×.

4.5. *Human and Mice Macrophage Cultures and Infection*

Human monocytes were obtained from healthy blood donors by Etablissement Français du Sang, EFS, Toulouse, France). Written informed consents were obtained from the donors before sample collection (under EFS Contract n°21PLER2017-0035 valid until July 02 2020, which was approved by the French Ministry of Science and Technology, agreement nr. AC2009-921, following articles L1243-4 and R1243-61 of the French Public Health Code). Monocytes were prepared following a previously published procedure [40]. Cells were purified using CD14 microbead positive selection and MACS separation columns (Miltenyi Biotec, Bergisch Gladbach, Germany), according to manufacturer's instructions. For differentiation of monocyte-derived macrophages, monocytes were allowed to adhere to glass coverslips (VWR international, Radnor, PA, USA) in 6-well plates (Thermo-Fisher Scientific), at 1.5×10^6 cells/well, for 1 h at 37 °C in pre-warmed RPMI-1640 medium (GIBCO, Thermo-Fisher Scientific, Waltham, MA, USA). The medium was then supplemented with the following additives: 10% Fetal Bovine Serum (Sigma-Aldrich), 1 mM sodium pyruvate (GIBCO), 50 μM β-mercaptoethanol (GIBCO), and 20 ng mL^{-1} human Macrophage Colony-Stimulating Factor (Miltenyi Biotec, Bergisch Gladbach, Germany). Cells were allowed to differentiate for seven days at 37 °C under 5% CO_2 atmosphere. Murine bone-marrow derived macrophages were prepared as described [41]. Bone-marrow cells were flushed out of the femurs and tibias of 6- to 8-week-old female C57BL/6 mice and cultured in Petri dishes (2×10^6 cells per dish) in pre-warmed RPMI-1640 GlutaMax medium (GIBCO or Life Technologies, Carlsbad, CA, USA), supplemented with 10% FBS (Sigma-Aldrich), 1 mM sodium pyruvate (GIBCO or Life Technologies), 50 μM β-mercaptoethanol (GIBCO or Life Technologies), and 20 ng mL^{-1} M-CSF (PeproTech, Rocky Hill, NJ, USA) at 37 °C under 5% CO_2 atmosphere. Medium was refreshed every three days of culture.

For infection, *M. tuberculosis* strain H37Rv/pGMCS-P1-mTurq-P_{mbcA}-mVenus-*mbcA* was grown to exponential phase in complete 7H9 medium. Mycobacterial clumps were disaggregated by at least 20 passages through a 25G needle, and macrophages were infected at MOI of 0.1 in complete RPMI medium for 4 h at 37 °C under 5% CO_2 atmosphere. Cells were then washed with RPMI and further incubated at 37 °C in RPMI supplemented medium. Infected cells were harvested at day 0 or 24 h post-infection, fixed with 4% *p*-formaldehyde and observed by fluorescent microscopy using large field Leica DMIRB, at a magnification of 630X.

Author Contributions: Conceptualization, K.A., A.D.G., and C.G., and O.N.; funding acquisition, C.G. and O.N.; methodology and investigation, K.A., A.D.G., and C.G.; writing—original draft preparation, C.G. and A.D.G.; writing—review and editing, C.G., A.D.G., K.A., and O.N. All authors have read and agreed to the published version of the manuscript.

Acknowledgments: We thank D. Schnappinger (Weill Cornell Medical College; New York, USA) for the gift of plasmids for Gateway cloning; Jérôme Rech and Jean-Yves Bouet (LMGM-CNRS; Toulouse, France) for the gift of plasmid carrying the mTurquoise and mVenus reporter; and A. Peixoto and E. Näser (IPBS Genotoul-TRI imaging facility) for help with microscopy and flow cytometry analysis.

References

1. Freire, D.M.; Gutierrez, C.; Garza-Garcia, A.; Grabowska, A.D.; Sala, A.J.; Ariyachaokun, K.; Panikova, T.; Beckham, K.S.H.; Colom, A.; Pogenberg, V.; et al. An NAD(+) Phosphorylase Toxin Triggers Mycobacterium tuberculosis Cell Death. *Mol. Cell* **2019**, *73*, 1282–1291.e8. [CrossRef]
2. Harms, A.; Brodersen, D.E.; Mitarai, N.; Gerdes, K. Toxins, Targets, and Triggers: An Overview of Toxin-Antitoxin Biology. *Mol. Cell* **2018**, *70*, 768–784. [CrossRef] [PubMed]
3. Lobato-Marquez, D.; Diaz-Orejas, R.; Garcia-Del Portillo, F. Toxin-antitoxins and bacterial virulence. *FEMS Microbiol. Rev.* **2016**, *40*, 592–609. [CrossRef] [PubMed]
4. Page, R.; Peti, W. Toxin-antitoxin systems in bacterial growth arrest and persistence. *Nat. Chem. Biol.* **2016**, *12*, 208–214. [CrossRef] [PubMed]
5. Deter, H.S.; Jensen, R.V.; Mather, W.H.; Butzin, N.C. Mechanisms for Differential Protein Production in Toxin-Antitoxin Systems. *Toxins* **2017**, *9*, 211. [CrossRef] [PubMed]
6. Song, S.; Wood, T.K. Toxin/Antitoxin System Paradigms: Toxins Bound to Antitoxins Are Not Likely Activated by Preferential Antitoxin Degradation. *Adv. Biosyst.* **2020**, *4*, e1900290. [CrossRef] [PubMed]
7. Akarsu, H.; Bordes, P.; Mansour, M.; Bigot, D.J.; Genevaux, P.; Falquet, L. TASmania: A bacterial Toxin-Antitoxin Systems database. *PLoS Comput. Biol.* **2019**, *15*, e1006946. [CrossRef]
8. Keren, I.; Minami, S.; Rubin, E.; Lewis, K. Characterization and transcriptome analysis of Mycobacterium tuberculosis persisters. *MBio* **2011**, *2*, e00100–e00111. [CrossRef]
9. Ramage, H.R.; Connolly, L.E.; Cox, J.S. Comprehensive functional analysis of Mycobacterium tuberculosis toxin-antitoxin systems: Implications for pathogenesis, stress responses, and evolution. *PLoS Genet.* **2009**, *5*, e1000767. [CrossRef]
10. Sala, A.; Bordes, P.; Genevaux, P. Multiple toxin-antitoxin systems in Mycobacterium tuberculosis. *Toxins* **2014**, *6*, 1002–1020. [CrossRef]
11. Slayden, R.A.; Dawson, C.C.; Cummings, J.E. Toxin-antitoxin systems and regulatory mechanisms in Mycobacterium tuberculosis. *Pathog. Dis.* **2018**, *76*, fty039. [CrossRef] [PubMed]
12. Lee, M.H.; Pascopella, L.; Jacobs, W.R., Jr.; Hatfull, G.F. Site-specific integration of mycobacteriophage L5: Integration-proficient vectors for Mycobacterium smegmatis, Mycobacterium tuberculosis, and bacille Calmette-Guerin. *Proc. Natl. Acad. Sci. USA* **1991**, *88*, 3111–3115. [CrossRef] [PubMed]
13. Ehrt, S.; Guo, X.V.; Hickey, C.M.; Ryou, M.; Monteleone, M.; Riley, L.W.; Schnappinger, D. Controlling gene expression in mycobacteria with anhydrotetracycline and Tet repressor. *Nucleic Acids Res.* **2005**, *33*, e21. [CrossRef] [PubMed]
14. Schnappinger, D.; Ehrt, S. Regulated Expression Systems for Mycobacteria and Their Applications. *Microbiol. Spectr.* **2014**, *2*, MGM2-0018-2013. [CrossRef] [PubMed]
15. Cortes, T.; Schubert, O.T.; Rose, G.; Arnvig, K.B.; Comas, I.; Aebersold, R.; Young, D.B. Genome-wide mapping of transcriptional start sites defines an extensive leaderless transcriptome in Mycobacterium tuberculosis. *Cell Rep.* **2013**, *5*, 1121–1131. [CrossRef] [PubMed]
16. Levillain, F.; Poquet, Y.; Mallet, L.; Mazeres, S.; Marceau, M.; Brosch, R.; Bange, F.C.; Supply, P.; Magalon, A.; Neyrolles, O. Horizontal acquisition of a hypoxia-responsive molybdenum cofactor biosynthesis pathway contributed to Mycobacterium tuberculosis pathoadaptation. *PLoS Pathog.* **2017**, *13*, e1006752. [CrossRef]
17. Thakur, Z.; Saini, V.; Arya, P.; Kumar, A.; Mehta, P.K. Computational insights into promoter architecture of toxin-antitoxin systems of Mycobacterium tuberculosis. *Gene* **2018**, *641*, 161–171. [CrossRef]
18. Waagmeester, A.; Thompson, J.; Reyrat, J.M. Identifying sigma factors in Mycobacterium smegmatis by comparative genomic analysis. *Trends Microbiol.* **2005'** *13*, 505–509. [CrossRef]

19. Garcia-Pino, A.; Balasubramanian, S.; Wyns, L.; Gazit, E.; De Greve, H.; Magnuson, R.D.; Charlier, D.; Van Nuland, N.A.; Loris, R. Allostery and intrinsic disorder mediate transcription regulation by conditional cooperativity. *Cell* **2010**, *142*, 101–111. [CrossRef]
20. Song, S.; Wood, T.K. Post-segregational Killing and Phage Inhibition Are Not Mediated by Cell Death Through Toxin/Antitoxin Systems. *Front. Microbiol.* **2018**, *9*, 814. [CrossRef]
21. Albrethsen, J.; Agner, J.; Piersma, S.R.; Hojrup, P.; Pham, T.V.; Weldingh, K.; Jimenez, C.R.; Andersen, P.; Rosenkrands, I. Proteomic profiling of Mycobacterium tuberculosis identifies nutrient-starvation-responsive toxin-antitoxin systems. *Mol. Cell. Proteom.* **2013**, *12*, 1180–1191. [CrossRef] [PubMed]
22. Rustad, T.R.; Harrell, M.I.; Liao, R.; Sherman, D.R. The enduring hypoxic response of Mycobacterium tuberculosis. *PLoS ONE* **2008**, *3*, e1502. [CrossRef] [PubMed]
23. Flannagan, R.S.; Cosio, G.; Grinstein, S. Antimicrobial mechanisms of phagocytes and bacterial evasion strategies. *Nat. Rev. Microbiol.* **2009**, *7*, 355–366. [CrossRef] [PubMed]
24. Russell, D.G. Mycobacterium tuberculosis and the intimate discourse of a chronic infection. *Immunol. Rev.* **2011**, *240*, 252–268. [CrossRef] [PubMed]
25. Tsunawaki, S.; Nathan, C.F. Enzymatic basis of macrophage activation. Kinetic analysis of superoxide production in lysates of resident and activated mouse peritoneal macrophages and granulocytes. *J. Biol. Chem.* **1984**, *259*, 4305–4312. [PubMed]
26. Dy, R.L.; Przybilski, R.; Semeijn, K.; Salmond, G.P.; Fineran, P.C. A widespread bacteriophage abortive infection system functions through a Type IV toxin-antitoxin mechanism. *Nucleic Acids Res.* **2014**, *42*, 4590–4605. [CrossRef]
27. Pecota, D.C.; Wood, T.K. Exclusion of T4 phage by the hok/sok killer locus from plasmid R1. *J. Bacteriol.* **1996**, *178*, 2044–2050. [CrossRef]
28. Gottesman, S. Trouble is coming: Signaling pathways that regulate general stress responses in bacteria. *J. Biol. Chem.* **2019**, *294*, 11685–11700. [CrossRef]
29. Lee, K.Y.; Lee, B.J. Structure, Biology, and Therapeutic Application of Toxin-Antitoxin Systems in Pathogenic Bacteria. *Toxins* **2016**, *8*, 305. [CrossRef]
30. Williams, J.J.; Hergenrother, P.J. Artificial activation of toxin-antitoxin systems as an antibacterial strategy. *Trends Microbiol.* **2012**, *20*, 291–298. [CrossRef]
31. Helaine, S.; Cheverton, A.M.; Watson, K.G.; Faure, L.M.; Matthews, S.A.; Holden, D.W. Internalization of Salmonella by macrophages induces formation of nonreplicating persisters. *Science* **2014**, *343*, 204–208. [CrossRef]
32. Keren, I.; Kaldalu, N.; Spoering, A.; Wang, Y.; Lewis, K. Persister cells and tolerance to antimicrobials. *FEMS Microbiol. Lett.* **2004**, *230*, 13–18. [CrossRef]
33. Ronneau, S.; Helaine, S. Clarifying the Link between Toxin-Antitoxin Modules and Bacterial Persistence. *J. Mol. Biol.* **2019**, *431*, 3462–3471. [CrossRef] [PubMed]
34. Wang, X.; Wood, T.K. Toxin-antitoxin systems influence biofilm and persister cell formation and the general stress response. *Appl. Environ. Microbiol.* **2011**, *77*, 5577–5583. [CrossRef] [PubMed]
35. Fraikin, N.; Goormaghtigh, F.; Van Melderen, L. Type II Toxin-Antitoxin Systems: Evolution and Revolutions. *J. Bacteriol.* **2020**, *202*, e00763-19. [CrossRef]
36. DeJesus, M.A.; Gerrick, E.R.; Xu, W.; Park, S.W.; Long, J.E.; Boutte, C.C.; Rubin, E.J.; Schnappinger, D.; Ehrt, S.; Fortune, S.M.; et al. Comprehensive Essentiality Analysis of the Mycobacterium tuberculosis Genome via Saturating Transposon Mutagenesis. *MBio* **2017**, *8*, e02133-16. [CrossRef] [PubMed]
37. Kim, J.H.; O'Brien, K.M.; Sharma, R.; Boshoff, H.I.; Rehren, G.; Chakraborty, S.; Wallach, J.B.; Monteleone, M.; Wilson, D.J.; Aldrich, C.C.; et al. A genetic strategy to identify targets for the development of drugs that prevent bacterial persistence. *Proc. Natl. Acad Sci. USA* **2013**, *110*, 19095–19100. [CrossRef]
38. Rodionova, I.A.; Schuster, B.M.; Guinn, K.M.; Sorci, L.; Scott, D.A.; Li, X.; Kheterpal, I.; Shoen, C.; Cynamon, M.; Locher, C.; et al. Metabolic and bactericidal effects of targeted suppression of NadD and NadE enzymes in mycobacteria. *MBio* **2014**, *5*, e00747-13. [CrossRef]
39. Vilcheze, C.; Weinrick, B.; Wong, K.W.; Chen, B.; Jacobs, W.R., Jr. NAD+ auxotrophy is bacteriocidal for the tubercle bacilli. *Mol. Microbiol.* **2010**, *76*, 365–377. [CrossRef]

40. Troegeler, A.; Lastrucci, C.; Duval, C.; Tanne, A.; Cougoule, C.; Maridonneau-Parini, I.; Neyrolles, O.; Lugo-Villarino, G. An efficient siRNA-mediated gene silencing in primary human monocytes, dendritic cells and macrophages. *Immunol. Cell Biol.* **2014**, *92*, 699–708. [CrossRef]
41. Benard, A.; Sakwa, I.; Schierloh, P.; Colom, A.; Mercier, I.; Tailleux, L.; Jouneau, L.; Boudinot, P.; Al-Saati, T.; Lang, R.; et al. B Cells Producing Type I IFN Modulate Macrophage Polarization in Tuberculosis. *Am. J. Respir. Crit. Care Med.* **2018**, *197*, 801–813. [CrossRef] [PubMed]

11

Protection against Shiga Toxins

Simona Kavaliauskiene [1,2], Anne Berit Dyve Lingelem [1,2], Tore Skotland [1,2] and Kirsten Sandvig [1,2,3,*]

1 Department of Molecular Cell Biology, Institute for Cancer Research, Oslo University Hospital, N-0379 Oslo, Norway; simona.kavaliauskiene@rr-research.no (S.K.); Anne.Berit.Dyve@rr-research.no (A.B.D.L.) Tore.Skotland@rr-research.no (T.S.)

2 Center for Cancer Biomedicine, Faculty of Medicine, Oslo University Hospital, N-0379 Oslo, Norway

3 Department of Biosciences, University of Oslo, N-0316 Oslo, Norway

* Correspondence: kirsten.sandvig@ibv.uio.no.

Abstract: Shiga toxins consist of an A-moiety and five B-moieties able to bind the neutral glycosphingolipid globotriaosylceramide (Gb3) on the cell surface. To intoxicate cells efficiently, the toxin A-moiety has to be cleaved by furin and transported retrogradely to the Golgi apparatus and to the endoplasmic reticulum. The enzymatically active part of the A-moiety is then translocated to the cytosol, where it inhibits protein synthesis and in some cell types induces apoptosis. Protection of cells can be provided either by inhibiting binding of the toxin to cells or by interfering with any of the subsequent steps required for its toxic effect. In this article we provide a brief overview of the interaction of Shiga toxins with cells, describe some compounds and conditions found to protect cells against Shiga toxins, and discuss whether they might also provide protection in animals and humans.

Keywords: Shiga toxin; Stx1; Stx2; hemolytic uremic syndrome; inhibitors; chloroquine; fluorodeoxyglucose; Mn^{2+}

---

1. Introduction

Shiga toxins (Stxs) comprise a family of related bacterial protein toxins that are similar in structure and mechanism of action, but are produced by different types of bacteria. Shiga toxin is secreted by *Shigella dysenteriae*, whereas Shiga-like toxin 1 (Stx1) and Shiga-like toxin 2 (Stx2) are produced by certain strains of *Escherichia coli* (Shiga toxin-producing *E. coli* (STEC)) and some other bacteria [1]. Prototypic Stx1 (Stx1a) differs from Shiga toxin only in one amino acid residue in the catalytic A-moiety of the toxin, whereas Stx2 shares only ~60% sequence similarity with Shiga toxin and defines an immunologically distinct subgroup comprised of at least seven subtypes of Stx2 [2]. Stx2 is more lethal than Stx1 in animal models [3,4] and is thought to be the main cause of life-threatening infections in humans. Some STEC produce only one toxin type, either Stx1 or Stx2, while others express a combination of both types and different subtypes [5]. For simplicity, we will use the abbreviation Stx to refer to the whole family of Shiga toxins when discussing general facts about the toxin and/or where the exact type or variant is not known.

Infection with enterohemorrhagic STEC may cause hemorrhagic colitis, hemolytic uremic syndrome (HUS), and death [6]. There is no approved treatment of STEC-induced HUS, and the use of antibiotics may worsen the disease by increasing toxin formation and release by the bacteria [7]. In general, HUS occurs in 5%–15% of cases with STEC infection, with children having the highest risk [8], although the large outbreak with a Stx2a-producing enteroaggregative STEC strain in Northern

Europe in 2011 demonstrated that there are bacterium-toxin combinations that can be as dangerous to adults as to children [9]. HUS will most often occur 5–13 days after the onset of diarrhea, with a mortality of 3%–5% [10,11]. In addition to direct renal damage, neurological complications may also occur in HUS patients and are important determinants of severity of the condition and mortality rate [12–15]. Neurological symptoms may be caused by fatigue, cerebral microvascular thrombi, ischemia-hypoxia, or the direct neuronal effects of Stxs [12,14,16].

One of the first specific therapeutic approaches against infections with Stxs was the idea of sequestering the toxin once it is released in the gut. In this regard, a novel agent composed of silicon dioxide particles covalently linked to the trisaccharide moiety of the globotriaosylceramide molecule that mediates Stx binding (Synsorb® Pk, Synsorb Biotech) was developed. However, although Synsorb® Pk was shown to bind and neutralize Stx1 (and Stx2, but less efficiently) in vitro [17], it failed to improve the clinical course of diarrhea-associated HUS in pediatric patients when tested in a randomized clinical trial [18]. The main drawback of neutralization of Stxs in the intestine for the prevention of HUS is that only trace amounts of the toxin reaching circulation are sufficient to induce HUS, and thus a more systemic treatment is required. Taking this into account, analogues of the globotriaosylceramide (Gb3) receptor and Stx antibodies for systematic administration have been developed and proven promising in in vivo models [19–21]. In addition, human serum amyloid component P (HuSAP) has been found to neutralize Stx2, but not Stx1, in vitro [22], and to protect mice against a lethal dose of Stx2 [23]. Moreover, eculizumab, an antibody directed against the complement protein C5, was used in patients with HUS during the outbreak in Northern Europe in 2011 [24] in order to counteract the activation of complement by the toxin [25]. These novel strategies based on direct neutralization of Stx in the intestine and/or circulation and the inhibition of complement have been well described in a recent review by Melton-Celsa and O'Brien [26] and thus are not further discussed here. In this review we will first provide a short overview of the toxin structure, toxin binding to the glycosphingolipid Gb3, and the intracellular transport, before we focus on the potential therapeutic agents for treatment of STEC infections and HUS that target specific cellular functions and protect cells against Stx by inhibiting toxin binding and/or intracellular trafficking.

1.1. Stx Structure

Stxs belong to the AB_5 class of protein toxins and consist of an A-moiety (~32 kDa), which is non-covalently attached to a homo-pentameric B-moiety (7.7 kDa per monomer) (Figure 1) [27,28]. Nearly all Stxs bind exclusively to the globotriaosylceramide Gb3 [29–31] with the exception of one Stx2 subtype, Stx2e, which has been shown to bind to Gb4 [32]. Each B subunit harbors three Gb3 binding sites [33], making the toxin capable of binding up to 15 Gb3 molecules on the cell surface (Figure 1C). However, not all binding sites have equal affinity for the carbohydrates of Gb3 [34,35] and, therefore, not all sites might be required for binding to the cell surface, but might rather mediate additional recognition. The B-moiety alone is not toxic to cells (with the exception of B cells, where it may induce apoptosis [36]) and functions as a delivery tool for the enzymatically active A-moiety. It is still not clear why Stx2 is more lethal to humans than Stx1, but crystallographic studies and investigations of deletion mutants reveal important differences when it comes to the role of the C-terminal end of the A-subunit for retrograde transport and complex stability [28,37,38].

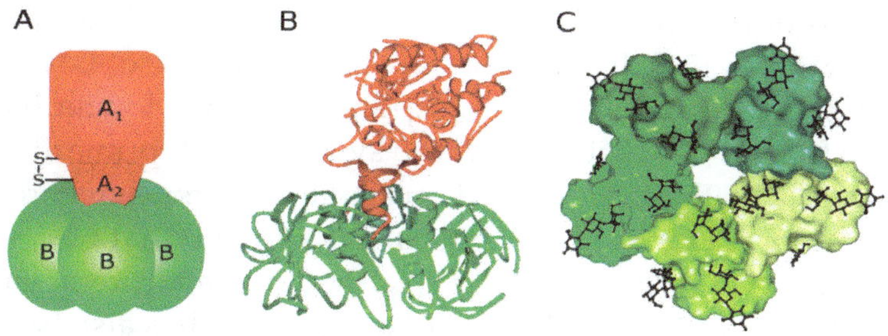

Figure 1. The schematic and structural models of Shiga toxins (Stxs). (**A**) Stxs consist of two non-covalently linked moieties: an A-moiety of ~32 kDa (shown in red), and a B-moiety (shown in green), comprised of five 7.7 kDa B-chains [27,28]. During intracellular toxin transport, the A-moiety is cleaved by the protease furin [39] into two fragments: an enzymatically active A_1 fragment (~27 kDa) and a carboxyl terminal A_2 fragment, which remain linked by a disulfide bond until arrival to the endoplasmic reticulum (ER) [40]. (**B**) The structure of the holotoxin as determined by X-ray crystallography [28] (PDB ID:1DM0); (**C**) The receptor-binding surface of the B-pentamer based on the structure of Stx1 complexed with the Gb3 analogue MCO-PK (methoxycarbonyloctyl glycoside of P^k trisaccharide) [33] (PDB ID:1BOS); the sugar moieties of MCO-PK are shown in black. Structure images were prepared using PDB ProteinWorkshop 4.2.

1.2. Gb3 and Its Interaction with Stx

Globotriaosylceramide (Gal-α1→4Gal-β1→4Glc-β1→Cer, Gb3; Figure 2) is a glycosphingolipid expressed on the surface of certain cell types. Gb3 is formed by the addition of one galactose residue to lactosylceramide (LacCer), which is a common precursor for different classes of glycosphingolipids, and the reaction is catalyzed by Gb3 synthase (lactosylceramide α-1,4-galactosyltransferase). Gb3 is the first glycosphingolipid in the globo-series and thus serves as a precursor for the synthesis of more complex globo-series glycosphingolipids, such as globotetraosylceramide (Gb4). Gb4 is formed after addition of *N*-acetylgalactosamine (GalNAc) to the terminal galactose of Gb3. Although Gb3 is the primary receptor for all Stxs, it has been suggested that Gb4 might facilitate Stx2 binding to colon epithelium cells, which normally have no Gb3 or very low levels of Gb3 [41].

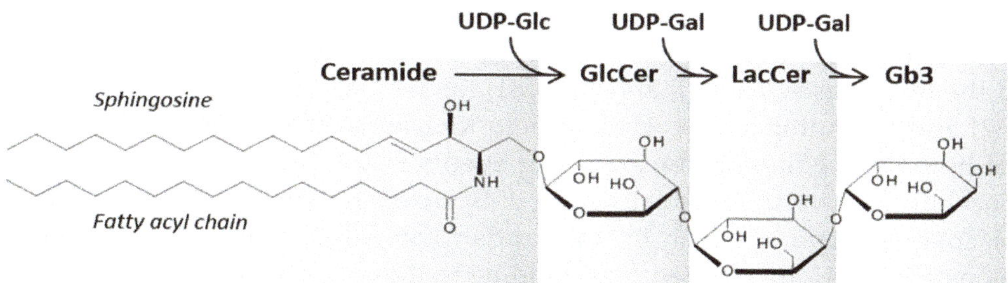

Figure 2. Chemical structure and biosynthesis of the Stx receptor globotriaosylceramide (Gb3). Sphingosine most often contains 18 carbon atoms, whereas the fatty acyl chain of ceramide varies both in length and saturation (here shown as C16:0). Gb3 is synthesized from LacCer by the addition of one galactose, and the reaction is catalyzed by Gb3 synthase (lactosylceramide α-1,4-galactosyltransferase). The sugar chain for Gb3 is: Gal-α1→4Gal-β1→4Glc-β1→Ceramide.

The sphingosine chain in the ceramide part of Gb3 is relatively invariable (most often it is monounsaturated with 18 carbon atoms, i.e., d18:1), but the *N*-amidated fatty acyl chain varies both in length (most common are 16–24 carbon atoms) and saturation resulting in multiple Gb3 species present in cells. Importantly, the receptor function of Gb3 has been shown to depend on its species composition [42–45], which in turn depends on cell type [46] and growth conditions [47], and might change in response to certain treatments, like exposure to butyric acid and cytokines [43,48–52]. It has been suggested that the production of butyric acid by the bacterial flora in the intestine may affect the expression and composition of Gb3 in the target cells and in turn lead to different susceptibility to the toxin between individuals [53]. In addition, the turnover time of Gb3 in the cells depends on its species, with longer fatty acyl chain-containing species having a longer half-life than the species with short fatty acyl chain [54]. Thus, inhibition of Gb3 synthesis will primarily lead to changes in Gb3 species composition in cells [43,50]. Studies based on artificial systems, where Gb3 was immobilized on thin layer chromatography (TLC) or ELISA plates, have shown that Stx1 and Stx2 have different binding preferences for different Gb3 species [44,55], although a mixture of various Gb3 species was required for the highest binding affinity [45]. Stx1 binding to Gb3 has also been shown to depend on cholesterol levels in the membrane [56,57]. Stx2 has been shown to be more potent in mice [3] and is more often associated with disease in humans [58], although the binding affinity of Stx2 to Gb3 is lower than that of Stx1, when measured using Gb3 adsorbed on a microtiter plate [59], and Stx2 is less toxic to Vero cells than Stx1 [60]. The different pathology observed for Stx1 and Stx2 might be caused by differences in receptor binding and thus differential targeting to susceptible tissues [61–63], as well as differences in intracellular transport of the toxins [37,64–66].

In the human body, the expression of Gb3 is restricted to only certain tissues. Normally, the highest Gb3 content is found in the microvascular glomeruli and proximal tubule cells of the kidney, consistent with the renal pathology of HUS [63,67–69]. Gb3 is also found in microvascular endothelial cells, and during infection with STEC, main Stx-target sites are the vascular endothelium of the colon [70,71] and the central nervous system [16,72]. Moreover, Gb3 is expressed in platelets [73,74], and in the carbohydrate defined P histo-blood group system, Gb3 constitutes the rare P^k antigen present on erythrocytes [75]. In the immune system, Gb3 represents a lymphocyte differentiation antigen, termed CD77, which is expressed in a subset of germinal center B lymphocytes [76]. It should also be noted that Gb3 expression is frequently increased in cancer cells [77]. However, the physiological role of Gb3 is still unclear, and it is not known why Gb3 expression is restricted to certain tissues. In vivo studies of Gb3 synthase knock-out mice, which displayed a total loss of Gb3 and other globo-series glycosphingolipids, showed no changes in birth-rates and no apparent abnormalities over a year of nurturing, with the exception of total loss of sensitivity to Stx1 and Stx2 as compared to wild-type mice [78].

1.3. Intracellular Transport of Stx

Upon binding to cells, the toxin has been found to activate a number of tyrosine kinases, including Syk [79], and the Src kinases Yes [80] and Lyn [81], as well as the serine/threonine protein kinase Cδ (PKCδ) [82] and the mitogen-activated protein kinase (MAPK) p38α [83]. Although the exact mechanism of how Stx mediates these signaling events is not yet fully understood, a recent study from our group have shown that the activation of Syk depends on the multivalent cross-linking of Gb3 at the plasma membrane, which in turn leads to an increase in cytosolic calcium levels and phosphorylation of Syk [84]. In addition, StxB binding to the cells has been shown to induce the release of cytoplasmic phospholipase A2 (cPLA2) from a cPLA2-annexin A2 complex, thereby facilitating Golgi transport, which has been found to be dependent on cPLA2 [85]. Furthermore, binding of the Stx B-moiety has been reported to stimulate remodeling of cytoskeleton components, such as actin, ezrin and dynein [86–88]. Thus, Stx is able to induce cell signaling and to modulate various cellular components to favor its uptake and intracellular transport.

Receptor-bound Stx becomes internalized by different endocytic mechanisms, including both clathrin- and dynamin-dependent and independent pathways [1]. After internalization, the toxin is transported from early/recycling endosomes directly to the Golgi apparatus [89] and then further to the endoplasmic reticulum (ER) [90–93]. During the transport, the A-moiety is cleaved by the protease furin, leaving two fragments, A_1 and A_2 (Figure 1), which remain linked to each other by a disulfide bridge [39]. Cleavage of Stx is optimal at low pH [94], indicating that it can occur early in the transport pathway. However, cells that lack furin can also cleave Stx, but less efficiently and at a later stage of the transport [39,95]. In the ER, the disulfide bond between the A_1 and A_2 subunits is reduced and the

A_1-subunit is released from the toxin. Finally, the A_1-subunit is translocated across the ER membrane and inhibits protein synthesis by cleaving one adenine residue from the 28S RNA of the 60S ribosomal subunit [1]. However, the action of Stx in the cells is not limited to the inhibition of protein synthesis, and other cellular responses, such as cytokine expression and apoptosis, are triggered by the toxin (for review see [96,97]). Thus, efficient intracellular Stx transport depends on various cellular proteins and kinases, ER chaperones and other factors (for review see [1]). Drugs that affect any of these factors might interfere with proper intracellular transport of Stx and protect cells against the cytotoxic action of the toxin. In the next sections we will give an overview of compounds shown to protect cells against Stx, and we will discuss mechanisms responsible for the protection against the toxin, as well as the potential applicability of these drugs in vivo. An overview of the compounds discussed here and how they might act against Stx is given in Table 1. The intracellular transport of Stx and which steps of the transport are affected by the different compounds are shown in Figure 3.

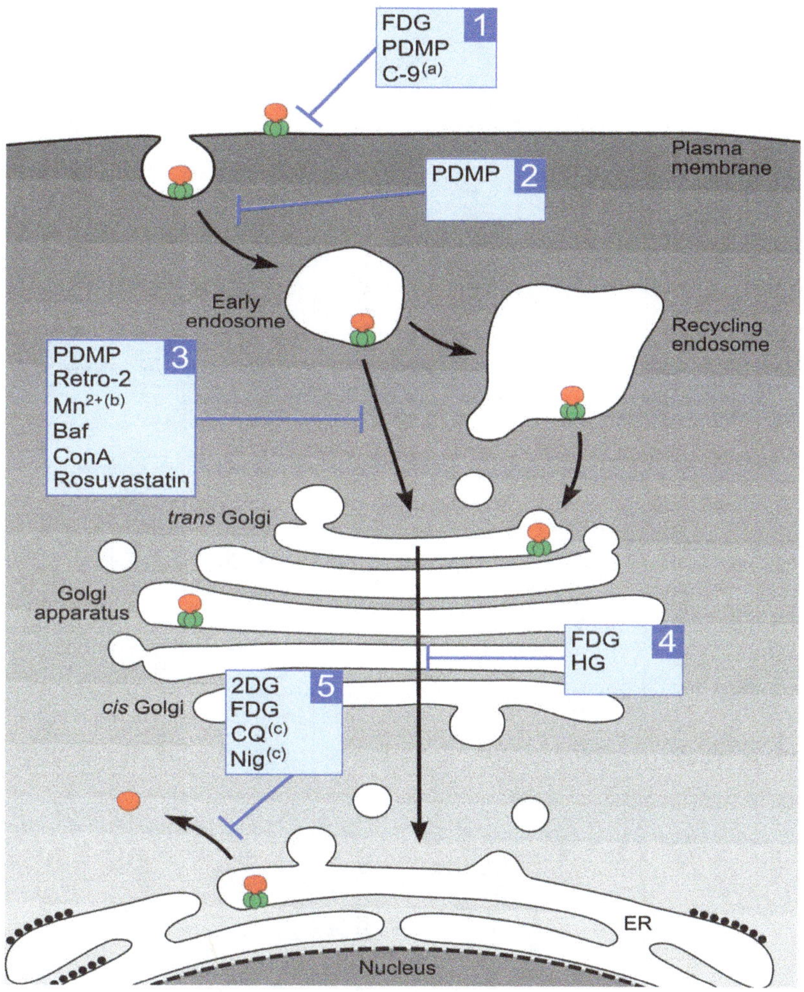

Figure 3. Stx uptake and intracellular transport, and the steps affected by different compounds. Stx binds to Gb3 on the cell surface and is taken up by various endocytic mechanisms. Following endocytosis, the toxin is transported through early endosomes and recycling endosomes and to the Golgi apparatus. From the Golgi, Stx is transported retrogradely to the ER, where its catalytically active A_1-subunit is released and translocated into the cytosol. The different compounds discussed in this review are shown with their suggested action on different steps of Stx intoxication: 1—Stx binding; 2—Stx endocytosis; 3—Stx sorting to the Golgi; 4—Stx transport via Golgi to ER; 5—release and translocation of $StxA_1$; (a) predicted effect for Stx2; (b) no effect for Stx2; (c) predicted effect for Shiga toxin.

Table 1. Compounds that protect cells against Stx.

Compound	Cellular Action	Targeted Step of Stx Intoxication	Cell Lines Tested	In Vivo Studies	Reference(s)
CQ	Elevation of pH in acidified organelles	Translocation of A_1-moiety to cytosol (predicted)	HEp-2	-	[98]
Baf	V-ATPase inhibitor	Transport to the Golgi	HEp-2	-	[98]
ConA	V-ATPase inhibitor	Transport to the Golgi	HEp-2	-	[98]
Nig	Ionophore that exchanges H^+ for monovalent cations	Not determined; later than Golgi	HEp-2	-	[98]
2DG	Inhibition of glycolysis and protein *N*-glycosylation; Ca^{2+} release from the endoplasmic reticulum (ER); Inhibition of Gb3 synthesis	Release of A_1-moiety	HEp-2, HT-29, SW480, HeLa	-	[99]
FDG	Inhibition of glycolysis and protein *N*-glycosylation; Ca^{2+} release from the ER; Inhibition of GlcCer synthesis	Binding; Transport from Golgi to ER; Release of A_1-moiety	HEp-2, HT-29, MCF-7, HBMEC	-	[48]
Retro-2 substances	Relocalization of Syntaxins 5 and 6	Transport from endosomes to the Golgi	A459, HeLa	Reduction in mortality rate from 70% to 40% in mice infected with *E. coli* O104:H4	[100,101]
Mn^{2+}	Induction of GPP130 oligomerization and its sorting to lysosomes for degradation	Transport from endosomes to the Golgi (no effect on Stx2 transport)	HeLa, Vero	Protection against lethal doses of Stx1 in BALB/c mice; No protection against either Stx1 or Stx2 in CD-1 mice	[65,102–105]
PDMP	Inhibition of GlcCer synthesis	Binding and endocytosis; Transport from endosomes to the Golgi	HEp-2	-	[43]
C-9	Inhibition of GlcCer synthesis	Not investigated	Human renal tubular epithelial cells, Human glomerular endothelial cells	50% reduction in mortality rate in rats injected with supernatant from *E. coli* expressing Stx2	[106–108]
HG	Ether lipid precursor	Transport from Golgi to ER	HEp-2, HMEC-1, HBMEC	-	[109]
Rosuvastatin	Inhibition of cholesterol biosynthesis and protein prenylation	Transport to the Golgi	ACHN	-	[110]
Furin inhibitors	Inhibition of furin	Proteolytic cleavage of the A-moiety	HEp-2	-	[111,112]

2. Compounds that Protect Cells against Stx

2.1. Chloroquine

Chloroquine (CQ; *N′*-(7-chloroquinolin-4-yl)-*N,N*-dietyl-pentane-1,4-diamine) is a weak base that in its unprotonated form easily can diffuse across membranes and accumulate in acidic compartments of the cell. There, CQ becomes protonated and trapped, leading to an elevated pH and swelling of the compartments. CQ (first named resochin) was developed in 1934 by Bayer laboratories as a synthetic antimalarial drug [113]. CQ was FDA-approved in 1949 and has proven to be one of the most effective and best tolerated agents against malaria [113,114]. It was widely used throughout the world in the 1950s and 1960s, and also later, but due to the emergence of the CQ-resistant malaria parasites *Plasmodium falciparum* and *Plasmodium vivax*, CQ has been abandoned as a prophylactic drug in most countries [115]. CQ and its analogues are also FDA-approved for the treatment of systemic lupus erythematosus and rheumatoid arthritis, and are in clinical trials as an adjuvant for anti-cancer chemotherapy and radiotherapy [115,116].

In short-term 3 h toxicity experiments in HEp-2 cells, a 2 h preincubation with 100 μM CQ gave a 15-fold increase in the Shiga toxin concentration needed to inhibit protein synthesis by 50% (IC50) [98]. In the retrograde pathway, the acidification decreases in the direction from early endosomes to the Golgi and the ER, and one might therefore expect CQ to have the most prominent effect early in the retrograde pathway. However, Shiga toxin transport to the Golgi apparatus was not affected by CQ treatment, neither was release of the A_1-moiety of the toxin in the ER, suggesting that transport to the ER was also normal [98]. Thus, CQ might interfere with $StxA_1$ translocation into the cytosol. Other compounds that disrupt the pH gradient, such as the V-ATPase inhibitors bafilomycin A1 (Baf) and concanamycin A (ConA) and the ionophore nigericin (Nig), also protect against Shiga toxin [98]. In contrast to CQ, the V-ATPase inhibitors interfere with Shiga toxin transport to the Golgi [98]. Even though all the compounds inhibit acidification, they interfere with different toxin transport steps, suggesting that also pH-independent processes are affected. The V-ATPase inhibitors show poor selectivity in vivo [117,118], and are therefore not suitable for clinical application.

Shiga toxin is shown to interact with the Sec61 channel [119,120], and the knockdown of Sec61B has been shown to protect cells against Shiga toxin [119], suggesting that the translocation of $StxA_1$ into the cytosol is dependent on the Sec61 channel. Interestingly, CQ has previously been shown to block several channels, such as the inward rectifier potassium channel Kir2.1 [121], and the translocation pores of the pore-forming toxins anthrax and C2 toxin [122,123]. One can speculate that CQ might also interact with the Sec61 channel to block the translocation of Shiga toxin. Importantly, as shown here, CQ also protects against Stx2 (see below).

For acute treatment of malaria, the recommended dose of CQ is 10 mg/kg/day. CQ is rapidly distributed in the body and reaches a peak plasma concentration within 3–12 h after an oral dose [124]. In most cases, CQ is slowly eliminated with a half-life median value of 40 days, but there are big differences between individuals and the half-life of CQ has been reported to range from 1 to 157 days [124]. The plasma concentration is strongly dependent on the administered dose and the duration of the treatment [124]. The therapeutic concentration in blood of hydroxychloroquine (HCQ), a less toxic analogue of CQ with similar pharmacokinetics, has been reported to be between 0.03 and 15 mg/L; the toxic and lethal concentration ranges were 3–26 mg/L and 20–104 mg/L, respectively [125]. CQ at high concentrations has been reported to have a number of effects in cell cultures, but clinically safe and achievable doses are in the low micromolar range [116]. Thus, it has been suggested that for clinical relevance, the CQ concentrations should not exceed 10 mg/L or 31 μM [116]. As shown in Figure 4, in long-term toxicity experiments (24 h with toxin), a 1 h preincubation with a concentration of 25 μM CQ gave approx. 20 fold protection against Stx2, indicating that CQ doses within the therapeutic window might be sufficient to protect against the toxin. HCQ also protected cells against Stx2 to a similar degree as CQ (Figure 4).

It should be noted that in animal models, CQ was shown to strongly accumulate in tissues, for instance in the uvea, liver, lungs, and kidneys [124]. A similar distribution has been reported in humans [124]. Thus, the CQ concentration in the target cells of Stx might be a lot higher than the plasma concentration.

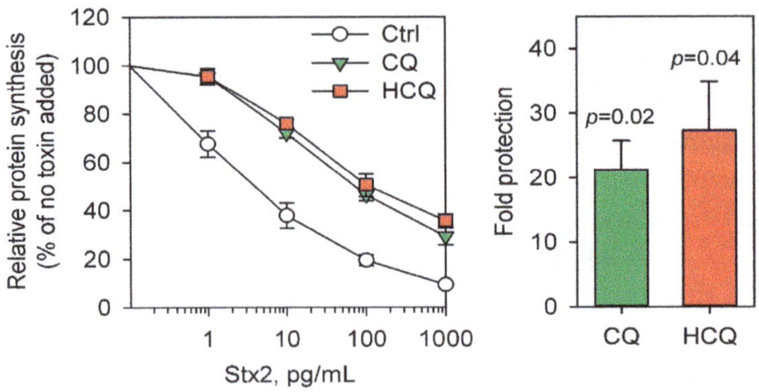

Figure 4. Cell protection against Stx2 by CQ and HCQ. HEp-2 cells were treated with or without 25 μM chloroquine (CQ) or 25 μM hydroxychloroquine (HCQ) in complete growth medium for 1 h prior to incubation with 10-fold serial dilutions of Stx2 for 24 h in the presence or absence of the drugs. The cells were then incubated in the presence of [^{3}H]leucine for 20 min, and protein synthesis was measured as described in [98]. The left panel shows relative protein synthesis as a percentage of the samples without Stx2 added. The right panel shows relative fold protection against Stx2. The protection was calculated as an increase in IC50 for treated samples compared to control. The error bars show SEM ($n = 4$). One sample *t*-test was used for statistical analysis of the protection data, and obtained *p* values are given in the figure.

2.2. The Glucose Analogues 2-Deoxy-D-glucose and 2-Fluoro-2-deoxy-D-glucose

2-Deoxy-D-glucose (2DG) is a structural analogue of glucose, and has been used in research as a glycolytic inhibitor since 1950s [126,127]. 2DG differs from glucose only by the absence of one oxygen atom at the second carbon. Similarly to glucose, 2DG is taken up through the glucose transporters and phosphorylated by hexokinase to form 2DG-6-phosphate (2DG-6-P). However, 2DG-6-P is not metabolized further and accumulates in the cells [127–129]. 2DG-6-P inhibits glycolysis by competing with glucose-6-P for phosphoglucose isomerase [127], and by acting as a non-competitive inhibitor of hexokinase [130]. However, although the inhibition of glycolysis has been a commonly exploited effect of 2DG, this compound has a much broader spectrum of activities. In addition to inhibiting glycolysis, 2DG inhibits *N*-linked protein glycosylation [131,132]. In turn, this leads to accumulation of misfolded proteins and triggers the unfolded protein response in the ER, leading to ER stress [133,134]. Interestingly, Okuda et al. recently discovered that 2DG inhibits the expression of the Gb3 synthase by yet unknown mechanisms, and thus reduces cellular Gb3 levels in the cells [135].

2-Fluoro-2-deoxy-D-glucose (FDG) is a structural analogue of glucose where the hydroxyl group at the second carbon is replaced by a fluorine atom. Like glucose and 2DG, FDG is transported into cells, where it is phosphorylated by hexokinase to yield FDG-6-P. However, FDG-6-P does not undergo isomerization to fructose and thus cannot be further catabolized, leading to accumulation of FDG-6-P in the cells [136]. Similarly to 2DG, FDG also inhibits glycolysis, but because the binding energy of FDG-6-P for the allosteric site of the hexokinase is lower than that of 2DG-6-P, and closely resembles the energy of glucose-6-P, FDG is a better inhibitor of glycolysis than 2DG [137]. FDG also interferes with *N*-linked protein glycosylation [131,138,139], but in contrast to 2DG, FDG does not become incorporated into dolichol-linked oligosaccharides [138]. FDG seems to slow down rather than prevent the assembly of the dolichol-linked oligosaccharide, and thus is a weaker inhibitor of *N*-glycosylation than 2DG. In addition, we have recently found that in contrast to 2DG, FDG does not become incorporated into newly synthesized glycolipids [48]. [^{18}F]FDG, which contains the ^{18}F radioisotope, is a commonly used imaging agent for positron emission tomography (PET). [^{18}F]FDG based PET is widely used for diagnosis and monitoring of oncological, neurological, and cardiological diseases (for review see [140,141]).

We have recently discovered that both 2DG and FDG reduce cell sensitivity to both Shiga toxin and Stx2. Four hours preincubation with either 10 mM 2DG or 1 mM FDG led to an increase in the IC50 for Shiga toxin by 13-fold in HEp-2 cells [48,99]. In addition, 24 h preincubation with 10 mM 2DG reduced cell sensitivity to Shiga toxin by 30-fold [99], while 24 h pretreatment with 1 mM FDG made HEp-2 cells fully resistant to both Shiga toxin and Stx2 (maximal concentration tested was 100 ng/mL with 3 h challenge) [48]. For FDG, we have also tested whether the protective effect is observed in non-cancer originated cells, and found that immortalized human brain microvascular cells (HBMEC) also became less sensitive to Shiga toxin after 4 h and 24 h pretreatment with 1 mM FDG [48]. However, all these toxicity assays have been performed with a relatively short 3 h challenge with the toxin. Thus, we have now also tested whether FDG protects HEp-2 cells against 24 h challenge with Stx2, and observed an essentially similar protection as shown for 3 h incubation with the toxin (Figure 5).

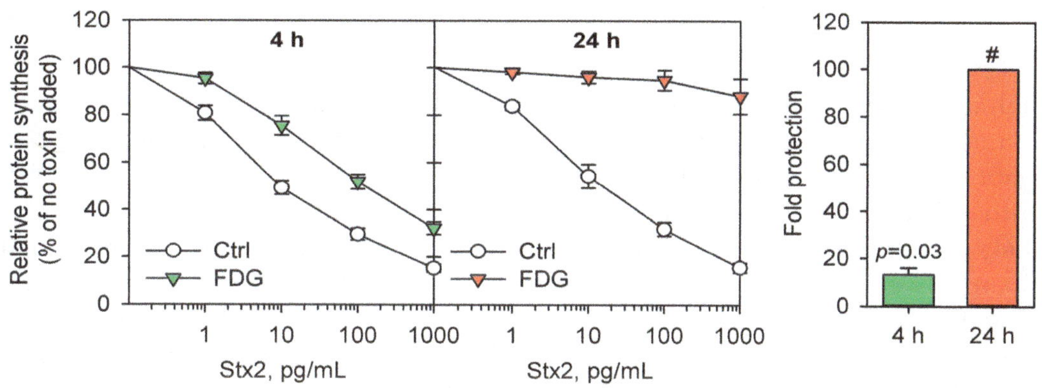

Figure 5. Cell protection against Stx2 by FDG. HEp-2 cells were treated with or without 1 mM 2-fluoro-2-deoxy-D-glucose (FDG) in complete growth medium for 4 h or 24 h prior to incubation with 10-fold serial dilutions of Stx2 for 24 h in the presence or absence of FDG. The cells were then incubated in the presence of [^{3}H]leucine for 20 min, and protein synthesis was measured as described in [98]. The left panel shows relative protein synthesis as a percentage of the samples without Stx2 added. The right panel shows relative fold protection against Stx2. The protection was calculated as an increase in IC50 for treated samples compared to control. In the samples treated with FDG for 24 h, the highest toxin concentration tested (1 ng/mL) did not reduce protein synthesis down to 50%, therefore the fold-protection could not be calculated and was estimated to be more than 100-fold (marked as #). The error bars show SEM for 4 h treatment (n = 4) and the deviation from the mean of two independent experiments for 24 h treatment. One sample t-test was used for statistical analysis of the protection data for 4 h treatment, and the obtained p value is given in the figure.

Interestingly, although both 2DG and FDG were found to reduce cellular Gb3 levels by 50% following 24 h treatment (Figure 6), it was only FDG that also led to reduction in Shiga toxin binding [48,99]. The mechanisms by which 2DG and FDG inhibit Gb3 synthesis are still not clear and seem to be different, as 2DG has been shown to inhibit the transcription of the Gb3 synthase gene [99,135], while FDG has no effect on the expression of Gb3 synthase [48]. In addition, we have found that cell treatment with FDG also reduces cellular levels of LacCer and glucosylceramide (GlcCer) (Figure 6) [48], indicating that FDG inhibits the synthesis of GlcCer and thus depletes cells for the precursors required for Gb3 synthesis, rather than inhibiting the synthesis of Gb3 directly.

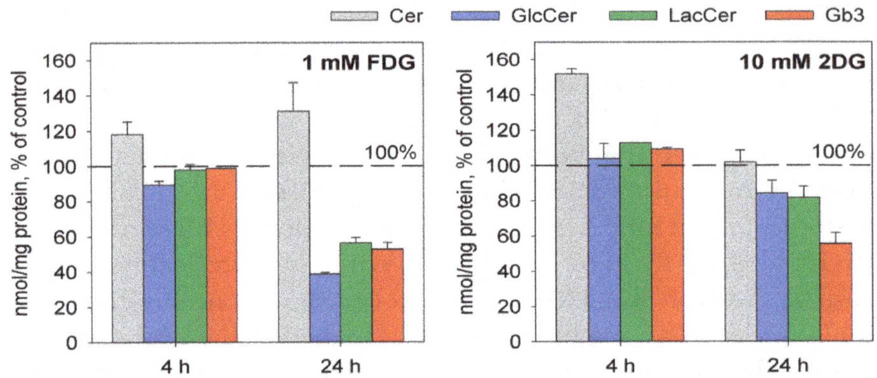

Figure 6. Effect of FDG and 2DG on total levels of Gb3 and its precursors. Cells were treated with or without 1 mM FDG or 10 mM 2DG for 4 h or 24 h, and lipids were analyzed by mass spectrometry in whole-cell lysates. The total amount of lipid was normalized to the total amount of protein in each sample (protein content was measured by BCA assay). The graph shows the levels of Cer, GlcCer, LacCer, and Gb3 in treated cells compared to control samples; the error bars show the deviation from the mean of two biological samples. For a detailed method description and the raw data see [48,99].

Although the inhibition of Gb3 synthesis might be an important factor for the protection against Stx by 2DG and FDG after long-term treatment, the protection observed after 4 h treatment with drugs does not seem to be mediated by changes in Gb3 [48,99]. We have shown that following 4 h treatment with either 10 mM 2DG or 1 mM FDG the intracellular transport of Shiga toxin is changed and most likely accounts for the protection observed at this time point. Four hours pretreatment with 10 mM 2DG almost completely blocked the release of Shiga toxin A_1-moeity in the ER in HEp-2 cells, which correlated well with the depletion of calcium from the ER [99]. Although it has been proposed that 2DG induces release of calcium from the ER via induction of ER stress [142], combined treatment with mannose, which rescues 2DG-mediated ER stress [143], does not prevent calcium leakage from the ER upon 2DG treatment and does not rescue cell sensitivity to Shiga toxin [99]. Three ER chaperones, HEDJ (also called ERdj3), BiP (also called GRP78) and GRP94 (glucose-regulated protein of 94 kDa), have been shown to bind the A-moiety of Stx1 [120], and thus are suggested to be involved in the release of $StxA_1$ in the ER. Substrate binding to GRP94 and BiP has been shown to be regulated by calcium [144,145], suggesting that 2DG might prevent the release of $StxA_1$ in the ER by inhibiting Shiga toxin interaction with chaperones.

Similar to 2DG, FDG was found to deplete calcium from the ER and to inhibit release of the A_1-moiety from the holotoxin [48]. However, since the concentrations for FDG and 2DG used were different, it is difficult to conclude whether the effect on ER calcium levels and the efficiency in blocking $StxA_1$ release are similar for these drugs. However, we have found that FDG also inhibits Stx1 transport from the Golgi to ER [48], which did not seem to be the case for 2DG [99]. This additional block in the Stx transport might explain why FDG is more efficient than 2DG in protecting cells against Shiga toxin and Stx2 following short-term incubation. Furthermore, since the long-term preincubation (24 h) with FDG, but not 2DG, also leads to reduced Stx1 binding, it makes FDG a promising candidate drug for STEC infections. However, in vivo studies are required to test whether FDG could potentially be used for treatment. The first challenge is to reach high enough concentrations of FDG in the Stx-targeted cells. When [^{18}F]FDG is used in the clinics for PET, the main limiting factor for the concentration of [^{18}F]FDG used in patients is the allowed maximal radiation dose. This is not the problem when using stable nonradioactive FDG. There are no clinical studies describing maximal FDG doses that could be safely achieved in human plasma/tissues, but there are several clinical studies that have analyzed the safety of 2DG when used in combination with chemotherapy or radiation. Based on the study by Raez et al. [146] the recommended daily dose of 2DG in combination with docetaxel was 63 mg/kg, which resulted in a median maximum plasma 2DG concentration of 0.7 mM, and caused tolerable adverse effects. The tolerable concentration and adverse effects of FDG might differ from those shown for 2DG and have to be assessed in the future, but if the achievable concentration is similar to that shown for 2DG (0.7 mM), it is then in the same range as the concentration (1 mM) shown to protect cells against Stx in vitro [48].

2.3. Retro-2 Substances

By performing a high-throughput screening of 16,480 drugs Stechmann et al. [100] identified two low molecular weight substances that reduced sensitivity to Stx1, Stx2, and the plant toxin ricin (which also follows a retrograde route to the ER), when added 30 min prior to challenge with toxins in A459 and HeLa cells. These substances were called Retro-1 and Retro-2. They were reported to inhibit retrograde transport of Stx1B from endosomes to the Golgi apparatus without affecting compartment integrity and endogenous retrograde cargo transport [100]. Other compounds, which may have similarities to the Retro compounds, have also been shown to block trafficking and toxicity of Stx1 and ricin [147]. Both Retro-1 and Retro-2 have been found to relocalize the SNARE proteins syntaxin 5 and, to a lesser extent, syntaxin 6 from their normal site of accumulation on perinuclear Golgi membranes [100]. Thus, it has been suggested that the inhibition of Stx1B transport by these compounds could be mediated by the relocalization of syntaxin 5 and 6, although additional studies are required to confirm this. Retro-2 was found to be the most effective of these drugs, and an intraperitoneal injection of 200 mg/kg Retro-2 given 1 h prior to toxin challenge completely protected mice that were given a lethal nasal instillation of ricin (animal experiments with Stx were not performed) [100]. Later, this group and others published data for several substances similar to Retro-2, and showed that it was a cyclic form (Retro-2^{cycl}) and not Retro-2 that was active [148–150]. Cyclization and modification of Retro-2 resulted in a compound with approx. 100-fold increased efficacy in inhibiting Stx1, and only one enantiomer was found to be active with an EC50 value (the concentration of the drug that gives 50% of its full inhibitory effect against the toxin) of approx. 300 nM in HeLa cells [149]. The most active of the Retro substances in counteracting the cytotoxicity of Stx1 (named (S)-Retro-2.1) has been reported to have an EC50 value of 54 nM in HeLa cells [151].

Secher et al. has investigated whether Retro-2^{cycl} could protect mice against the toxic effects of infection with *E. coli* O104:H4, the strain that was responsible for the deadly outbreak in Europe in 2011 [152,153]. The bacteria were given to mice by oral gavage, and on days 16 and 26 the mice received intraperitoneal injections of 100 mg/kg of Retro-2^{cycl} , which resulted in reduced mortality rate in the treated group (29 and 16 mice were dead in control and treated group, respectively, out of 40 mice per group) [101]. The same authors recently published a review article about the use of Retro-2 and similar compounds to protect against Stxs, ricin, multiple viruses, including different polyomaviruses, Ebola virus and poxviruses, and intracellular parasites, such as *Leishmania* [154], and we refer to this review for further discussions about these substances. The authors conclude that these lead compounds now need to be developed as drugs for human use. As these drugs, as far as we know, have not been given to humans or even been tested in formal preclinical drug safety studies, that task will be ongoing for many years. Furthermore, it will be interesting to see which compound will be selected for such a development as the most efficient compound may also be the most toxic. Finally, it should also be investigated whether all of these compounds have similar effects on the localization of syntaxin 5 as Retro-1 and Retro-2 [100], since targeting of a universal trafficking factor such as syntaxin 5 may prove challenging in a clinical setting.

2.4. Manganese

Mn^{2+}-ions have been described by several groups to protect cells against Shiga toxin [105] and Stx1 [103]. Mukhopadhyay and Lindstedt reported that the protection was due to redirection of the toxin to the degradative pathway; they also reported that mice injected with a lethal dose of Stx1 could be rescued by a nontoxic dose of Mn^{2+} that was injected five days before the toxin [103]. However, Mn^{2+} failed to protect cells against Stx2 under the same conditions [65]. These authors also showed that elevated levels of Mn^{2+} resulted in a down-regulation of GPP130, which in turn led to inhibition of retrograde transport of Stx1, but not Stx2 [65]. Furthermore, this group later showed that the increased Mn^{2+} levels induced GPP130 oligomerization and its sorting to lysosomes for degradation [102].

The protective effect of Mn^{2+} against Stx toxicity was also studied by another group, and there the authors failed to see any protective effect of Mn^{2+} against Stx1 and Stx2 in cultured Vero cells and CD-1 outbred mice at Mn^{2+}-doses that were not toxic to the cells and the animals [104]. They concluded that the ability of Mn^{2+} to protect against Stx toxicity might be dependent on the cell line and mouse strain, and that protection may be observed only at potentially toxic concentrations of Mn^{2+}. These authors and others have discussed that Mn^{2+}-ions are neurotoxic; for reviews see [155,156]. Due to this well-known toxic effect of Mn^{2+}, we believe that it is unlikely that Mn^{2+} can be developed as a drug against Stx toxicity.

2.5. Inhibitors of GlcCer Synthesis PDMP and C-9

The glycosphingolipid Gb3 is the sole functional receptor for Stx in humans, which makes it a potential target for preventing Stx toxicity to cell. The drawback of the approach directed towards the receptor is the time required to deplete Gb3 from cells, and this might limit the therapeutic potential. However, although the expression of Gb3 is a prerequisite for cell sensitivity to Stx, a specific Gb3 species composition [42–45] is required for efficient Stx binding and intracellular transport, meaning that a complete depletion of Gb3 might not be necessary to prevent Stx intoxication. To our knowledge, there are no compounds available that would specifically block Gb3 synthesis in the cells. However, multiple substances have been developed for inhibiting glucosylceramide synthesis, and thus prevent formation of more complex glycosphingolipids, including Gb3. Two such compounds, PDMP [43] and C-9 [106], have been shown to reduce cell sensitivity to Stx, and are discussed in this section.

PDMP (1-phenyl-2-decanoyl-amino-3-morpholino-1-propanol) is a ceramide analogue first developed in a search for drugs to treat individuals with Gaucher disease [157], which are deficient of the lysosomal enzyme glucosylceramidase and thus accumulate GlcCer in certain tissues [158]. PDMP has been shown to inhibit the synthesis of GlcCer and specifically affect the content and composition of glycosphingolipids, without perturbing other lipid profiles [43,159], and without significantly affecting the synthesis of glycoproteins in cells [159]. Raa et al. showed that treatment with 1 μM PDMP for 24 h had only a small (20%–30%) effect on Shiga toxin binding to HEp-2 cells, but led to an approx. 50% reduction in the toxin uptake and almost completely blocked (by 90%) the transport of StxB into the Golgi [43]. These effects were accompanied by a 6.5-fold increase in the IC50 value for Shiga toxin in HEp-2 cells. After 24 h incubation with 1 μM PDMP, the cellular levels of Gb3 and its precursors were reduced by approx. 50% in treated HEp-2 cells, and the Gb3 species with the fatty acyl group 16:0 were found to be degraded faster and to a larger extent than the Gb3 with the fatty acid 24:1 [43]. Results from studies on butyric acid-mediated cell sensitization to Shiga toxin, and comparison of cell lines with different sensitivities to Stx, have indicated that Gb3 with certain fatty acyl groups might be important for endosome-to-Golgi transport [49,91,160,161]. Thus, the changes in Gb3 species composition may play an important role in the PDMP-induced protection against Stx, at least at shorter treatment times, when the total Gb3 is not yet completely depleted from the cells. However, in addition to GlcCer synthase, PDMP has been found to target other lipid enzymes, such as ceramide glycanase and a lysosomal phospholipase A2 called 1-O-acylceramide synthase [162,163], which might limit the applicability of PDMP for specific depletion of Gb3 in vivo. On the other hand, the development of PDMP has boosted the synthesis of a variety of related compounds [162]. Some of these PDMP analogues exhibit a dramatic increase in potency and specificity for the ceramide-specific glucosyl transferase [162], and have been tested in β-galactosidase a-null mice (model of Fabry disease in which Gb3 accumulates in the vasculature and kidneys) [164]. The mice were injected intraperitoneally with 2 mg/kg of the PDMP analogue EtDO-P4 (D-threo-1-(3,4-ethylenedioxyphenyl)-2-(palmitoylamino)-3-(1-pyrrolidinyl)propanol) twice a day for three days, which led to approx. 50% reduction in GlcCer levels in the kidney, liver, and heart.

The reduction in Gb3 was little pronounced after three days of treatment, but following eight weeks of treatment with 10 mg/kg of EtDO-P4 twice a day, the total levels of Gb3 were reduced by approx. 50% in the kidney, liver and heart. Importantly, the treatment did not show apparent toxicity to the animals. However, a potential protection against Stx toxicity was not tested in this study.

The search for treatment of glycosphingolipid storage diseases has led to the development of another ceramide analogue named C-9 [(1R, 2R)-nonanoic acid [2-(2′,3′-dihydro-benzo [1–4]dioxin-6′-yl)-2-hydroxy-1-pyrrolidin-1-ylmethyl-ethyl]-amide-L-tartaric acid salt] (Genzyme Corp., Waltham, MA, USA). Silberstein et al. has demonstrated that a 48 h pretreatment with 5 μM C-9 reduced cellular Gb3 levels by approx. 80% in human renal tubular epithelial cells (HRTEC), which was accompanied by an almost complete cell protection against a 24 h challenge with 1 ng/mL Stx2 [106]. However, the drug had no effect on the cell sensitivity to Stx2 when added at the same time as the toxin and a 24 h preincubation was required to obtain the protective effect [106]. Essentially similar protection and reduction in Gb3 was also shown in human glomerular endothelial cells [108]. Silberstein et al. also investigated the potency of C-9 against Stx2 in an in vivo model [107]. Rats received C-9 orally two days prior and four days after the intraperitoneal injection of the supernatant from recombinant *E. coli* expressing Stx2. The treatment reduced rat mortality by 50% and prevented intestinal and renal tissue damage, which was observed in the group treated with Stx2 only. The failure of C-9 to completely prevent rat mortality after Stx2 challenge was attributed to the possibility that C-9 does not pass the blood- brain barrier, and thus the deaths in the C-9 group could be the outcome of neurological injuries. However, no histological analysis was performed to support this [107].

A related compound, which differs from C-9 only in the fatty acid part (contains octanoic acid instead of nonanoic acid), has been approved by FDA for treatment of Gaucher disease [165] and is now sold under the name Cerdelga® (Eliglustat tartrate; Genzyme Corp., Cambridge, MA, USA). Eliglustat tartrate (also called Genz-112638) is well tolerated and has a recommended dosing of 100 mg (contains 84 mg of eliglustat) twice daily in Gaucher patients [166,167]. When used at the recommended dosing, the average concentration of free eliglustat in the plasma is 12–25 ng/mL (but there is a great variation between individuals due to differences in the rate of metabolic degradation of eliglustat) (reviewed in [165]). Thus, it would be interesting to see whether eliglustat provides similar protection against Stx in cells and in animal models as C-9, and whether the tolerated dose leads to significant changes in Gb3 levels and/or species, which would indicate its potential use for the treatment of STEC infections. However, very little eliglustat is taken up into brain [164], which would limit its effect in the HUS cases with neuronal damage.

2.6. HG

The alkylglycerol 1-*O*-hexadecyl-*sn*-glycerol (HG, also called chimyl alcohol) is a precursor for ether-linked glycerophospholipids. HG enters the biosynthetic pathway of ether lipids after being phosphorylated by alkylglycerol kinase [168]. Addition of HG was shown to alter the lipidome of HEp-2 cells, with the most notable changes being an increase in the cellular level of ether-linked lipids with 16 carbon atoms at *sn*-1 position, an increase in lysophosphatidylinositol (LPI) and a decrease in all glycosphingolipid classes analyzed, amongst them the Stx receptor Gb3 [50].

We recently showed that a 24 h preincubation with 20 μM HG strongly protected HEp-2, HMEC-1 and HBMEC cells against Shiga toxin and Stx2, with an average 30-fold increase of IC50 [50]. There was a moderate reduction in Stx1 binding to HEp-2 cells after HG treatment, presumably due to reduced Gb3 levels, but this decrease was not sufficiently large to account for the strong protection against the toxin. Immunofluorescence confocal microscopy revealed that HG treatment led to an accumulation of Stx1 in the Golgi apparatus after 4 h of toxin challenge, while in control cells, the toxin clearly colocalized with both the ER-marker PDI and the Golgi-marker TGN46 at the same time point, suggesting that HG treatment inhibits Golgi-to-ER transport of Stx [50]. Thus, HG prevents Stx intoxication primarily by interfering with the Golgi-to-ER transport of Stx, but also to some degree by downregulating the Stx receptor Gb3.

The 1-*O*-alkylglycerols, including HG, are naturally occurring ether lipids that are present in human cells and body fluids, particularly in hematopoietic organs and in neutrophils and human milk [169]. Alkylglycerols can also be obtained from the diet, with marine oils, especially shark liver oil, being a good source [169,170]. Shark liver oil contains 9%–13% HG and has been used in traditional medicine in the Nordic countries for beneficial health effects and for wound healing [169,170]. Shark liver oil and alkylglycerol supplementation have been reported to mediate several biological effects, including the ability to boost the immune system and alleviate radiation therapy-induced side effects [169,170]. However, to evaluate the potential of HG as an inhibitor of Stx intoxication, further studies, first in animal models, are required.

2.7. Statins

Statins, also known as HMG-CoA (3-hydroxy-3-methylglutaryl coenzyme A) reductase inhibitors, are a class of compounds that inhibit cholesterol biosynthesis and prenylation of proteins (reviewed in [171]). Statins are widely prescribed to lower serum cholesterol levels for the prevention of cardiac diseases [172–174]. Recently, statins were found to inhibit Stx1B transport to the Golgi apparatus and to protect cells against Stx1 and Stx2 [110]. ACHN (epithelial carcinoma from renal tubular adenocarcinoma pleural metastasis) cells were pretreated with rosuvastatin (5–20 μM) for 24 h prior to addition of Stx1 or Stx2, and the number of surviving cells was measured after 72 h. Treatment with rosuvastatin increased cell survival by approx. 25% in the group treated with 5 μM rosuvastatin and then challenged with 1 pg/mL of Stx1 or Stx2, but there was essentially no improvement in cell survival in the group challenged with 100 pg/mL of the toxins. On the other hand, cell treatment with 10 μM of rosuvastatin increased cell survival by approx 70% in the groups challenged with 1 pg/mL of Stx1 or Stx2, and by approx. 25% in the groups challenged with100 pg/mL of the toxins. Although increasing the concentrations of rosuvastatin to 20 μM showed even better protection, there seemed to appear a difference in protection against Stx1 and Stx2, with the protection against Stx2 being lower than that against Stx1, especially at high toxin doses. As previously mentioned, Stx2 is suggested to exploit an additional transport pathway [65] which might not be blocked by rosuvastatin. In addition, a protective effect by lovastatin has previously been shown for several other protein toxins, including ricin, modeccin, *Pseudomonas* toxin, and diphtheria toxin, suggesting that statins might have a potential to be used against several toxins [175].

Cell treatment with statins was found to increase the protein level and activity of GlcCer synthase, which in turn led to increased cellular levels of GlcCer [110]. The fact that the cells were maintained in a medium containing cholesterol during the treatment with statins, and that the isoprenol precursor counteracted the upregulation of GlcCer synthesis induced by statins, indicated that the effects observed, at least for the upregulation of GlcCer synthesis, were mediated by the inhibition of isoprenylation rather than by depletion of cellular cholesterol. In addition, by the use of specific inhibitors of different prenylation enzymes, the authors were able to show that it was only the inhibition of geranylgeranyl transferase II, also known as Rab geranylgeranyl transferase, that led to similar upregulation of GlcCer as statin treatment. Rab prenylation facilitates their membrane association and activity [176,177]. Rab GTPases regulate intracellular vesicular trafficking events, and several Rabs have also been implicated in Stx retrograde transport (reviewed in [1]), suggesting that rosuvastatin-induced protection against Stx is mediated via aberrant Rab prenylation.

Importantly, although for the prevention of cholesterol-related cardiac diseases statins are prescribed in relatively low doses (recommended daily dose of lovastatin is 20–40 mg per day [174]) the achievable safe serum values might be as high as 0.2–12 μM (shown for lovastatin at doses of 133–412 mg/m^2, which corresponds to approx. 200–650 mg) [178], and are within the dose range found to protect cells against Stx1 and Stx2 [110].

2.8. Furin Inhibitors

Furin is a type I transmembrane serine protease that activates precursors of different physiologically important proteins. Although furin is mainly located in the Golgi and *trans*-Golgi network, it also circulates through the endosomal system to the cell surface and back to the Golgi [179], and may also be secreted as a soluble truncated active enzyme [180]. In addition to its physiological role, furin activates numerous toxic proteins, including Stxs [95]. Moreover, furin has been implicated in various human diseases, including cancer, osteoarthritis, atherosclerosis, diabetes and neurodegenerative disorders [181–184]. Although the complete knockout of the *fur* gene, which encodes furin, is embryonically lethal to mice, the specific inhibition of furin by polyarginine inhibitors has been shown to be well tolerated in adult animals [185,186], indicating that short term inhibition of furin might be exploited for treatment. Thus, various furin inhibitors have been developed and investigated for their potential therapeutic applications during the last years [111,112,183,184,187]. For instance, the furin inhibitor hexa-D-arginine amide has been demonstrated to improve the survival of mice challenged with *Pseudomonas aeruginosa* exotoxin A, which requires furin-mediated cleavage for toxicity [185]. Later, the same compound was shown to delay anthrax toxin-induced toxemia both in cells and in rats [186], supporting the therapeutic potential of furin inhibitors for the treatment of infectious diseases.

A number of peptidomimetic furin inhibitors have been developed and optimized to improve the activity and stability of the compounds [111], and some of these compounds have been found to protect against Shiga toxin without showing significant toxicity to cells [111]. The most effective analogue for the protection against Shiga toxin was 4-(guanidinomethyl)phenylacetyl-Arg-Val-Arg-4-amidinobenzylamide (called No. 24 in the original article), which reduced HEp-2 cells sensitivity to Shiga toxin by approx. 6-fold, when added to the cells 30 min before a 4 h incubation with unnicked Shiga toxin [111]. However, it has not yet been investigated whether cells are protected against longer challenges with the toxin. It has been shown that the Shiga toxin may be processed by other cellular proteases, but more slowly [39,95]. Thus, it is still not clear whether furin inhibitors help to prevent or treat HUS in humans infected with Stx-producing bacteria.

3. Concluding Remarks

A number of compounds that protect cells against Stxs are known, and in some cases they have also been found to protect animals against the challenge with purified toxin or toxic effects of infection with STEC. Most of these compounds have not been used in humans and thus need to be carefully evaluated for potential use in the clinic, and it is not clear to which extent one will be able to find compounds that are safe in humans. There are, however, some exceptions, for instance chloroquine and statins, which should be investigated for a possible protective effect in connection with Stx-induced disease. In addition, there is an ongoing search for efficient inhibitors of glycosphingolipid synthesis for the treatment of glycosphingolipid storage diseases [188] as well as cancer [189], meaning that new, more potentially useful compounds will be investigated in the near future. Such new compounds should not be overlooked for their potential in the treatment of STEC and *Shigella* infections.

Acknowledgments: The work has been supported by the Norwegian Cancer Society, the Research Council of Norway through its Centers of Excellence funding scheme, project number 179571, and the South-Eastern Norway Regional Health Authority.

Abbreviations

ACHN	epithelial carcinoma from renal tubular adenocarcinoma pleural metastasis
Baf	bafilomycin A1
ConA	concanamycin A
cPLA2	cytoplasmic phospholipase A2

CQ	chloroquine
C-9	(1R, 2R)-nonanoic acid [2-(2′,3′-dihydro-benzo [1–4]dioxin-6′-yl)-2-hydroxy-1-pyrrolidin-1-ylmethyl-ethyl]-amide-L-tartaric acid salt
EC50	the concentration of the drug that gives 50% of its full inhibitory effect against the toxin
FDG	2-fluoro-2-deoxy-D-glucose
Gal	galactose
Gb3	globotriaosylceramide
Glc	glucose
GlcCer	glucosylceramide
GPP130	Golgi phosphoprotein of 130 kDa
HBMEC	human brain microvascular endothelial cells
HCQ	hydroxychloroquine
HG	alkylglycerol 1-*O*-hexadecyl-*sn*-glycerol
HRTEC	human renal tubular epithelial cells
HUS	hemolytic-uremic syndrome
HuSAP	human serum amyloid component P
IC50	Stx concentration needed to inhibit protein synthesis by 50%
LacCer	lactosylceramide
LPI	lysophosphatidylinositol
Nig	nigericin
PDMP	1-phenyl-2-decanoyl-amino-3-morpholino-1- propanol
PET	positron emission tomography
Retro-2^{cycl}	cyclic form of Retro-2
STEC	Shiga toxin-producing *E. coli*
Stx	Shiga toxins (when referring to the whole family and common features)
StxB	B-moiety of Stx
Stx1	Shiga-like toxin 1
Stx2	Shiga-like toxin 2
TLC	thin layer chromatography
UDP	uridine diphosphate
[^{18}F]FDG	FDG which contains the ^{18}F radioisotope
2DG	2-deoxy-D-glucose

References

1. Bergan, J.; Dyve Lingelem, A.B.; Simm, R.; Skotland, T.; Sandvig, K. Shiga toxins. *Toxicon* **2012**, *60*, 1085–1107. [CrossRef] [PubMed]
2. Scheutz, F.; Teel, L.D.; Beutin, L.; Piérard, D.; Buvens, G.; Karch, H.; Mellmann, A.; Caprioli, A.; Tozzoli, R.; Morabito, S.; et al. Multicenter evaluation of a sequence-based protocol for subtyping Shiga toxins and standardizing Stx nomenclature. *J. Clin. Microbiol.* **2012**, *50*, 2951–2963. [CrossRef] [PubMed]
3. Tesh, V.L.; Burris, J.A.; Owens, J.W.; Gordon, V.M.; Wadolkowski, E.A.; O'Brien, A.D.; Samuel, J.E. Comparison of the relative toxicities of Shiga-like toxins type I and type II for mice. *Infect. Immun.* **1993**, *61*, 3392–3402. [PubMed]
4. Siegler, R.L.; Obrig, T.G.; Pysher, T.J.; Tesh, V.L.; Denkers, N.D.; Taylor, F.B. Response to Shiga toxin 1 and 2 in a baboon model of hemolytic uremic syndrome. *Pediatr. Nephrol.* **2003**, *18*, 92–96. [PubMed]
5. Karch, H.; Tarr, P.I.; Bielaszewska, M. Enterohaemorrhagic *Escherichia coli* in human medicine. *Int. J. Med. Microbiol.* **2005**, *295*, 405–418. [CrossRef] [PubMed]
6. Palermo, M.S.; Exeni, R.A.; Fernandez, G.C. Hemolytic uremic syndrome: pathogenesis and update of interventions. *Expert Rev. Anti Infect. Ther.* **2009**, *7*, 697–707. [CrossRef] [PubMed]
7. Agger, M.; Scheutz, F.; Villumsen, S.; Molbak, K.; Petersen, A.M. Antibiotic treatment of verocytotoxin-producing *Escherichia coli* (VTEC) infection: a systematic review and a proposal. *J. Antimicrob. Chemother.* **2015**, *70*, 2440–2446. [CrossRef] [PubMed]
8. Barton Behravesh, C.; Jones, T.F.; Vugia, D.J.; Long, C.; Marcus, R.; Smith, K.; Thomas, S.; Zansky, S.; Fullerton, K.E.; Henao, O.L.; et al. Deaths associated with bacterial pathogens transmitted commonly

through food: foodborne diseases active surveillance network (FoodNet), 1996–2005. *J. Infect. Dis.* **2011**, *204*, 263–267. [CrossRef] [PubMed]
9. Boisen, N.; Melton-Celsa, A.R.; Scheutz, F.; O'Brien, A.D.; Nataro, J.P. Shiga toxin 2a and Enteroaggregative *Escherichia coli*—a deadly combination. *Gut Microbes* **2015**, *6*, 272–278. [CrossRef] [PubMed]
10. Tarr, P.I.; Gordon, C.A.; Chandler, W.L. Shiga-toxin-producing *Escherichia coli* and haemolytic uraemic syndrome. *Lancet* **2005**, *365*, 1073–1086. [CrossRef]
11. Gould, L.H.; Demma, L.; Jones, T.F.; Hurd, S.; Vugia, D.J.; Smith, K.; Shiferaw, B.; Segler, S.; Palmer, A.; Zansky, S.; et al. Hemolytic uremic syndrome and death in persons with *Escherichia coli* O157:H7 infection, foodborne diseases active surveillance network sites, 2000–2006. *Clin. Infect. Dis.* **2009**, *49*, 1480–1485. [CrossRef] [PubMed]
12. Brandt, J.R.; Fouser, L.S.; Watkins, S.L.; Zelikovic, I.; Tarr, P.I.; Nazar-Stewart, V.; Avner, E.D. *Escherichia coli* O157:H7-associated hemolytic-uremic syndrome after ingestion of contaminated hamburgers. *J. Pediatr.* **1994**, *125*, 519–526. [CrossRef]
13. Hughes, D.A.; Beattie, T.J.; Murphy, A.V. Haemolytic uraemic syndrome: 17 years' experience in a Scottish paediatric renal unit. *Scott. Med. J.* **1991**, *36*, 9–12. [CrossRef] [PubMed]
14. Taylor, C.M.; White, R.H.; Winterborn, M.H.; Rowe, B. Haemolytic-uraemic syndrome: Clinical experience of an outbreak in the West Midlands. *Br. Med. J. (Clin. Res. Ed.)* **1986**, *292*, 1513–1516. [CrossRef]
15. Bale, J.F., Jr.; Brasher, C.; Siegler, R.L. CNS manifestations of the hemolytic-uremic syndrome. Relationship to metabolic alterations and prognosis. *Am. J. Dis. Child.* **1980**, *134*, 869–872. [CrossRef] [PubMed]
16. Obata, F.; Tohyama, K.; Bonev, A.D.; Kolling, G.L.; Keepers, T.R.; Gross, L.K.; Nelson, M.T.; Sato, S.; Obrig, T.G. Shiga toxin 2 affects the central nervous system through receptor globotriaosylceramide localized to neurons. *J. Infect. Dis.* **2008**, *198*, 1398–1406. [CrossRef] [PubMed]
17. Armstrong, G.D.; Fodor, E.; Vanmaele, R. Investigation of Shiga-like toxin binding to chemically synthesized oligosaccharide sequences. *J. Infect. Dis.* **1991**, *164*, 1160–1167. [CrossRef] [PubMed]
18. Trachtman, H.; Cnaan, A.; Christen, E.; Gibbs, K.; Zhao, S.; Acheson, D.W.; Weiss, R.; Kaskel, F.J.; Spitzer, A.; Hirschman, G.H. Effect of an oral Shiga toxin-binding agent on diarrhea-associated hemolytic uremic syndrome in children: a randomized controlled trial. *JAMA* **2003**, *290*, 1337–1344. [CrossRef] [PubMed]
19. Nishikawa, K.; Matsuoka, K.; Kita, E.; Okabe, N.; Mizuguchi, M.; Hino, K.; Miyazawa, S.; Yamasaki, C.; Aoki, J.; Takashima, S.; et al. A therapeutic agent with oriented carbohydrates for treatment of infections by Shiga toxin-producing *Escherichia coli* O157:H7. *Proc. Natl. Acad. Sci. USA* **2002**, *99*, 7669–7674. [CrossRef] [PubMed]
20. Mulvey, G.L.; Marcato, P.; Kitov, P.I.; Sadowska, J.; Bundle, D.R.; Armstrong, G.D. Assessment in mice of the therapeutic potential of tailored, multivalent Shiga toxin carbohydrate ligands. *J. Infect. Dis.* **2003**, *187*, 640–649. [CrossRef] [PubMed]
21. Melton-Celsa, A.R.; Carvalho, H.M.; Thuning-Roberson, C.; O'Brien, A.D. Protective efficacy and pharmacokinetics of human/mouse chimeric anti-Stx1 and anti-Stx2 antibodies in mice. *Clin. Vaccine Immunol.* **2015**, *22*, 448–455. [CrossRef] [PubMed]
22. Kimura, T.; Tani, S.; Matsumoto Yi, Y.; Takeda, T. Serum amyloid P component is the Shiga toxin 2-neutralizing factor in human blood. *J. Biol. Chem.* **2001**, *276*, 41576–41579. [CrossRef] [PubMed]
23. Armstrong, G.D.; Mulvey, G.L.; Marcato, P.; Griener, T.P.; Kahan, M.C.; Tennent, G.A.; Sabin, C.A.; Chart, H.; Pepys, M.B. Human serum amyloid P component protects against *Escherichia coli* O157:H7 Shiga toxin 2 in vivo: Therapeutic implications for hemolytic-uremic syndrome. *J. Infect. Dis.* **2006**, *193*, 1120–1124. [CrossRef] [PubMed]
24. Kielstein, J.T.; Beutel, G.; Fleig, S.; Steinhoff, J.; Meyer, T.N.; Hafer, C.; Kuhlmann, U.; Bramstedt, J.; Panzer, U.; Vischedyk, M.; et al. Best supportive care and therapeutic plasma exchange with or without eculizumab in Shiga-toxin-producing *E. coli* O104:H4 induced haemolytic-uraemic syndrome: an analysis of the German STEC-HUS registry. *Nephrol. Dial. Transplant.* **2012**, *27*, 3807–3815. [CrossRef] [PubMed]
25. Orth, D.; Khan, A.B.; Naim, A.; Grif, K.; Brockmeyer, J.; Karch, H.; Joannidis, M.; Clark, S.J.; Day, A.J.; Fidanzi, S.; et al. Shiga toxin activates complement and binds factor H: evidence for an active role of complement in hemolytic uremic syndrome. *J. Immunol.* **2009**, *182*, 6394–6400. [CrossRef] [PubMed]
26. Melton-Celsa, A.R.; O'Brien, A.D. New therapeutic developments against Shiga toxin-producing *Escherichia coli*. *Microbiol. Spectr.* **2014**, *2*. [CrossRef] [PubMed]

27. Stein, P.E.; Boodhoo, A.; Tyrrell, G.J.; Brunton, J.L.; Read, R.J. Crystal structure of the cell-binding B oligomer of verotoxin-1 from *E. coli*. *Nature* **1992**, *355*, 748–750. [CrossRef] [PubMed]
28. Fraser, M.E.; Chernaia, M.M.; Kozlov, Y.V.; James, M.N. Crystal structure of the holotoxin from *Shigella dysenteriae* at 2.5 Å resolution. *Nat. Struct. Biol.* **1994**, *1*, 59–64. [CrossRef] [PubMed]
29. Jacewicz, M.; Clausen, H.; Nudelman, E.; Donohue-Rolfe, A.; Keusch, G.T. Pathogenesis of shigella diarrhea. XI. Isolation of a shigella toxin-binding glycolipid from rabbit jejunum and HeLa cells and its identification as globotriaosylceramide. *J. Exp. Med.* **1986**, *163*, 1391–1404. [CrossRef] [PubMed]
30. Lindberg, A.A.; Brown, J.E.; Stromberg, N.; Westling-Ryd, M.; Schultz, J.E.; Karlsson, K.A. Identification of the carbohydrate receptor for Shiga toxin produced by *Shigella dysenteriae* type 1. *J. Biol. Chem.* **1987**, *262*, 1779–1785. [PubMed]
31. Lingwood, C.A.; Law, H.; Richardson, S.; Petric, M.; Brunton, J.L.; De, G.S.; Karmali, M. Glycolipid binding of purified and recombinant *Escherichia coli* produced verotoxin *in vitro*. *J. Biol. Chem.* **1987**, *262*, 8834–8839. [PubMed]
32. DeGrandis, S.; Law, H.; Brunton, J.; Gyles, C.; Lingwood, C.A. Globotetraosylceramide is recognized by the pig edema disease toxin. *J. Biol. Chem.* **1989**, *264*, 12520–12525. [PubMed]
33. Ling, H.; Boodhoo, A.; Hazes, B.; Cummings, M.D.; Armstrong, G.D.; Brunton, J.L.; Read, R.J. Structure of the Shiga-like toxin I B-pentamer complexed with an analogue of its receptor Gb3. *Biochemistry* **1998**, *37*, 1777–1788. [CrossRef] [PubMed]
34. Peter, M.G.; Lingwood, C.A. Apparent cooperativity in multivalent verotoxin-globotriaosyl ceramide binding: kinetic and saturation binding studies with [^{125}I]verotoxin. *Biochim. Biophys Acta* **2000**, *1501*, 116–124. [CrossRef]
35. Bast, D.J.; Banerjee, L.; Clark, C.; Read, R.J.; Brunton, J.L. The identification of three biologically relevant globotriaosyl ceramide receptor binding sites on the Verotoxin 1 B subunit. *Mol. Microbiol.* **1999**, *32*, 953–960. [CrossRef] [PubMed]
36. Marcato, P.; Mulvey, G.; Armstrong, G.D. Cloned Shiga toxin 2 B subunit induces apoptosis in Ramos Burkitt's lymphoma B cells. *Infect. Immun.* **2002**, *70*, 1279–1286. [CrossRef] [PubMed]
37. Kymre, L.; Simm, R.; Skotland, T.; Sandvig, K. Different roles of the C-terminal end of Stx1A and Stx2A for AB_5 complex integrity and retrograde transport of Stx in HeLa cells. *Pathog. Dis.* **2015**, *73*, ftv083. [CrossRef] [PubMed]
38. Fraser, M.E.; Fujinaga, M.; Cherney, M.M.; Melton-Celsa, A.R.; Twiddy, E.M.; O'Brien, A.D.; James, M.N. Structure of Shiga toxin type 2 (Stx2) from *Escherichia coli* O157:H7. *J. Biol. Chem.* **2004**, *279*, 27511–27517. [CrossRef] [PubMed]
39. Garred, O.; van Deurs, B.; Sandvig, K. Furin-induced cleavage and activation of Shiga toxin. *J. Biol. Chem.* **1995**, *270*, 10817–10821. [PubMed]
40. Tam, P.J.; Lingwood, C.A. Membrane cytosolic translocation of verotoxin A_1 subunit in target cells. *Microbiology* **2007**, *153*, 2700–2710. [CrossRef] [PubMed]
41. Zumbrun, S.D.; Hanson, L.; Sinclair, J.F.; Freedy, J.; Melton-Celsa, A.R.; Rodriguez-Canales, J.; Hanson, J.C.; O'Brien, A.D. Human intestinal tissue and cultured colonic cells contain globotriaosylceramide synthase mRNA and the alternate Shiga toxin receptor globotetraosylceramide. *Infect. Immun.* **2010**, *78*, 4488–4499. [CrossRef] [PubMed]
42. Lingwood, C.A.; Binnington, B.; Manis, A.; Branch, D.R. Globotriaosyl ceramide receptor function—where membrane structure and pathology intersect. *FEBS Lett.* **2010**, *584*, 1879–1886. [CrossRef] [PubMed]
43. Raa, H.; Grimmer, S.; Schwudke, D.; Bergan, J.; Walchli, S.; Skotland, T.; Shevchenko, A.; Sandvig, K. Glycosphingolipid requirements for endosome-to-Golgi transport of Shiga toxin. *Traffic* **2009**, *10*, 868–882. [CrossRef] [PubMed]
44. Kiarash, A.; Boyd, B.; Lingwood, C.A. Glycosphingolipid receptor function is modified by fatty acid content. Verotoxin 1 and verotoxin 2c preferentially recognize different globotriaosyl ceramide fatty acid homologues. *J. Biol. Chem.* **1994**, *269*, 11138–11146. [PubMed]
45. Pellizzari, A.; Pang, H.; Lingwood, C.A. Binding of verocytotoxin 1 to its receptor is influenced by differences in receptor fatty acid content. *Biochemistry* **1992**, *31*, 1363–1370. [CrossRef] [PubMed]
46. Sandvig, K.; Bergan, J.; Kavaliauskiene, S.; Skotland, T. Lipid requirements for entry of protein toxins into cells. *Prog. Lipid Res.* **2014**, *54*, 1–13. [CrossRef] [PubMed]

47. Kavaliauskiene, S.; Nymark, C.-M.; Bergan, J.; Simm, R.; Sylvanne, T.; Simolin, H.; Ekroos, K.; Skotland, T.; Sandvig, K. Cell density-induced changes in lipid composition and intracellular trafficking. *Cell. Mol. Life Sci.* **2013**, *71*, 1097–1116. [CrossRef] [PubMed]
48. Kavaliauskiene, S.; Torgersen, M.L.; Dyve Lingelem, A.B.; Klokk, T.I.; Lintonen, T.; Simolin, H.; Ekroos, K.; Skotland, T.; Sandvig, K. Cellular effects of fluorodeoxyglucose: Global changes in the lipidome and alteration in intracellular transport. *Oncotarget* **2016**, *7*, 79885–79900. [CrossRef] [PubMed]
49. Sandvig, K.; Ryd, M.; Garred, O.; Schweda, E.; Holm, P.K.; van Deurs, B. Retrograde transport from the Golgi complex to the ER of both Shiga toxin and the nontoxic Shiga B-fragment is regulated by butyric acid and cAMP. *J. Cell Biol.* **1994**, *126*, 53–64. [CrossRef] [PubMed]
50. Bergan, J.; Skotland, T.; Sylvänne, T.; Simolin, H.; Ekroos, K.; Sandvig, K. The ether lipid precursor hexadecylglycerol causes major changes in the lipidome of HEp-2 cells. *PLoS ONE* **2013**, *8*, e75904. [CrossRef] [PubMed]
51. Hughes, A.K.; Stricklett, P.K.; Schmid, D.; Kohan, D.E. Cytotoxic effect of Shiga toxin-1 on human glomerular epithelial cells. *Kidney Int.* **2000**, *57*, 2350–2359. [CrossRef] [PubMed]
52. Ramegowda, B.; Samuel, J.E.; Tesh, V.L. Interaction of Shiga toxins with human brain microvascular endothelial cells: Cytokines as sensitizing agents. *J. Infect. Dis.* **1999**, *180*, 1205–1213. [CrossRef] [PubMed]
53. Zumbrun, S.D.; Melton-Celsa, A.R.; O'Brien, A.D. When a healthy diet turns deadly. *Gut Microbes* **2014**, *5*, 40–43. [CrossRef] [PubMed]
54. Skotland, T.; Ekroos, K.; Kavaliauskiene, S.; Bergan, J.; Kauhanen, D.; Lintonen, T.; Sandvig, K. Determining the turnover of glycosphingolipid species by stable-isotope tracer lipidomics. *J. Mol. Biol.* **2016**, *428*, 4856–4866. [CrossRef] [PubMed]
55. Mahfoud, R.; Manis, A.; Lingwood, C.A. Fatty acid-dependent globotriaosyl ceramide receptor function in detergent resistant model membranes. *J. Lipid Res.* **2009**, *50*, 1744–1755. [CrossRef] [PubMed]
56. Lingwood, D.; Binnington, B.; Rog, T.; Vattulainen, I.; Grzybek, M.; Coskun, U.; Lingwood, C.A.; Simons, K. Cholesterol modulates glycolipid conformation and receptor activity. *Nat. Chem. Biol.* **2011**, *7*, 260–262. [CrossRef] [PubMed]
57. Mahfoud, R.; Manis, A.; Binnington, B.; Ackerley, C.; Lingwood, C.A. A major fraction of glycosphingolipids in model and cellular cholesterol-containing membranes is undetectable by their binding proteins. *J. Biol. Chem.* **2010**, *285*, 36049–36059. [CrossRef] [PubMed]
58. Boerlin, P.; McEwen, S.A.; Boerlin-Petzold, F.; Wilson, J.B.; Johnson, R.P.; Gyles, C.L. Associations between virulence factors of Shiga toxin-producing *Escherichia coli* and disease in humans. *J. Clin. Microbiol.* **1999**, *37*, 497–503. [PubMed]
59. Itoh, K.; Tezuka, T.; Inoue, K.; Tada, H.; Suzuki, T. Different binding property of verotoxin-1 and verotoxin-2 against their glycolipid receptor, globotriaosylceramide. *Tohoku J. Exp. Med.* **2001**, *195*, 237–243. [CrossRef] [PubMed]
60. Head, S.C.; Karmali, M.A.; Lingwood, C.A. Preparation of VT1 and VT2 hybrid toxins from their purified dissociated subunits. Evidence for B subunit modulation of a subunit function. *J. Biol. Chem.* **1991**, *266*, 3617–3621. [PubMed]
61. Rutjes, N.W.; Binnington, B.A.; Smith, C.R.; Maloney, M.D.; Lingwood, C.A. Differential tissue targeting and pathogenesis of verotoxins 1 and 2 in the mouse animal model. *Kidney Int.* **2002**, *62*, 832–845. [CrossRef] [PubMed]
62. Chark, D.; Nutikka, A.; Trusevych, N.; Kuzmina, J.; Lingwood, C. Differential carbohydrate epitope recognition of globotriaosyl ceramide by verotoxins and a monoclonal antibody. *Eur. J. Biochem.* **2004**, *271*, 405–417. [CrossRef] [PubMed]
63. Khan, F.; Proulx, F.; Lingwood, C.A. Detergent-resistant globotriaosyl ceramide may define verotoxin/glomeruli-restricted hemolytic uremic syndrome pathology. *Kidney Int.* **2009**, *75*, 1209–1216. [CrossRef] [PubMed]
64. Tam, P.; Mahfoud, R.; Nutikka, A.; Khine, A.A.; Binnington, B.; Paroutis, P.; Lingwood, C. Differential intracellular transport and binding of verotoxin 1 and verotoxin 2 to globotriaosylceramide-containing lipid assemblies. *J. Cell. Physiol.* **2008**, *216*, 750–763. [CrossRef] [PubMed]
65. Mukhopadhyay, S.; Redler, B.; Linstedt, A.D. Shiga toxin-binding site for host cell receptor GPP130 reveals unexpected divergence in toxin-trafficking mechanisms. *Mol. Biol. Cell* **2013**, *24*, 2311–2318. [CrossRef] [PubMed]

66. Selyunin, A.S.; Mukhopadhyay, S. A conserved structural motif mediates retrograde trafficking of Shiga toxin types 1 and 2. *Traffic* **2015**, *16*, 1270–1287. [CrossRef] [PubMed]
67. Ergonul, Z.; Clayton, F.; Fogo, A.B.; Kohan, D.E. Shigatoxin-1 binding and receptor expression in human kidneys do not change with age. *Pediatr. Nephrol.* **2003**, *18*, 246–253. [PubMed]
68. Lingwood, C.A. Verotoxin-binding in human renal sections. *Nephron* **1994**, *66*, 21–28. [CrossRef] [PubMed]
69. Boyd, B.; Lingwood, C. Verotoxin receptor glycolipid in human renal tissue. *Nephron* **1989**, *51*, 207–210. [CrossRef] [PubMed]
70. Ohmi, K.; Kiyokawa, N.; Takeda, T.; Fujimoto, J. Human microvascular endothelial cells are strongly sensitive to Shiga toxins. *Biochem. Biophys. Res. Commun.* **1998**, *251*, 137–141. [CrossRef] [PubMed]
71. Miyamoto, Y.; Iimura, M.; Kaper, J.B.; Torres, A.G.; Kagnoff, M.F. Role of Shiga toxin versus H7 flagellin in enterohaemorrhagic *Escherichia coli* signalling of human colon epithelium *in vivo*. *Cell. Microbiol.* **2006**, *8*, 869–879. [CrossRef] [PubMed]
72. Ren, J.; Utsunomiya, I.; Taguchi, K.; Ariga, T.; Tai, T.; Ihara, Y.; Miyatake, T. Localization of verotoxin receptors in nervous system. *Brain Res.* **1999**, *825*, 183–188. [CrossRef]
73. Cooling, L.L.; Walker, K.E.; Gille, T.; Koerner, T.A. Shiga toxin binds human platelets via globotriaosylceramide (P^k antigen) and a novel platelet glycosphingolipid. *Infect. Immun.* **1998**, *66*, 4355–4366. [PubMed]
74. Tao, R.V.; Sweeley, C.C.; Jamieson, G.A. Sphingolipid composition of human platelets. *J. Lipid Res.* **1973**, *14*, 16–25. [PubMed]
75. Steffensen, R.; Carlier, K.; Wiels, J.; Levery, S.B.; Stroud, M.; Cedergren, B.; Nilsson, S.B.; Bennett, E.P.; Jersild, C.; Clausen, H. Cloning and expression of the histo-blood group P^k UDP-galactose: Galβ1-4G1cβ1-Cer α1,4-galactosyltransferase. Molecular genetic basis of the p phenotype. *J. Biol. Chem.* **2000**, *275*, 16723–16729. [CrossRef] [PubMed]
76. Mangeney, M.; Richard, Y.; Coulaud, D.; Tursz, T.; Wiels, J. CD77: An antigen of germinal center B cells entering apoptosis. *Eur. J. Immunol.* **1991**, *21*, 1131–1140. [CrossRef] [PubMed]
77. Engedal, N.; Skotland, T.; Torgersen, M.L.; Sandvig, K. Shiga toxin and its use in targeted cancer therapy and imaging. *Microb. Biotechnol.* **2011**, *4*, 32–46. [CrossRef] [PubMed]
78. Okuda, T.; Tokuda, N.; Numata, S.; Ito, M.; Ohta, M.; Kawamura, K.; Wiels, J.; Urano, T.; Tajima, O.; Furukawa, K.; et al. Targeted disruption of Gb3/CD77 synthase gene resulted in the complete deletion of globo-series glycosphingolipids and loss of sensitivity to verotoxins. *J. Biol. Chem.* **2006**, *281*, 10230–10235. [CrossRef] [PubMed]
79. Lauvrak, S.U.; Walchli, S.; Iversen, T.G.; Slagsvold, H.H.; Torgersen, M.L.; Spilsberg, B.; Sandvig, K. Shiga toxin regulates its entry in a Syk-dependent manner. *Mol. Biol. Cell* **2006**, *17*, 1096–1109. [CrossRef] [PubMed]
80. Katagiri, Y.U.; Mori, T.; Nakajima, H.; Katagiri, C.; Taguchi, T.; Takeda, T.; Kiyokawa, N.; Fujimoto, J. Activation of Src family kinase yes induced by Shiga toxin binding to globotriaosyl ceramide (Gb3/CD77) in low density, detergent-insoluble microdomains. *J. Biol. Chem.* **1999**, *274*, 35278–35282. [CrossRef] [PubMed]
81. Mori, T.; Kiyokawa, N.; Katagiri, Y.U.; Taguchi, T.; Suzuki, T.; Sekino, T.; Sato, N.; Ohmi, K.; Nakajima, H.; Takeda, T.; et al. Globotriaosyl ceramide (CD77/Gb3) in the glycolipid-enriched membrane domain participates in B-cell receptor-mediated apoptosis by regulating lyn kinase activity in human B cells. *Exp. Hematol.* **2000**, *28*, 1260–1268. [CrossRef]
82. Torgersen, M.L.; Walchli, S.; Grimmer, S.; Skanland, S.S.; Sandvig, K. Protein kinase Cδ is activated by Shiga toxin and regulates its transport. *J. Biol. Chem.* **2007**, *282*, 16317–16328. [CrossRef] [PubMed]
83. Walchli, S.; Skanland, S.S.; Gregers, T.F.; Lauvrak, S.U.; Torgersen, M.L.; Ying, M.; Kuroda, S.; Maturana, A.; Sandvig, K. The mitogen-activated protein kinase p38 links Shiga toxin-dependent signaling and trafficking. *Mol. Biol. Cell* **2008**, *19*, 95–104. [CrossRef] [PubMed]
84. Klokk, T.I.; Kavaliauskiene, S.; Sandvig, K. Cross-linking of glycosphingolipids at the plasma membrane: consequences for intracellular signaling and traffic. *Cell. Mol. Life Sci.* **2016**, *73*, 1301–1316. [CrossRef] [PubMed]
85. Tcatchoff, L.; Andersson, S.; Utskarpen, A.; Klokk, T.I.; Skanland, S.S.; Pust, S.; Gerke, V.; Sandvig, K. Annexin A1 and A2: Roles in retrograde trafficking of Shiga toxin. *PLoS ONE* **2012**, *7*, e40429. [CrossRef] [PubMed]
86. Takenouchi, H.; Kiyokawa, N.; Taguchi, T.; Matsui, J.; Katagiri, Y.U.; Okita, H.; Okuda, K.; Fujimoto, J. Shiga toxin binding to globotriaosyl ceramide induces intracellular signals that mediate cytoskeleton remodeling in human renal carcinoma-derived cells. *J. Cell Sci.* **2004**, *117*, 3911–3922. [CrossRef] [PubMed]

87. Hehnly, H.; Longhini, K.M.; Chen, J.L.; Stamnes, M. Retrograde Shiga toxin trafficking is regulated by ARHGAP21 and Cdc42. *Mol. Biol. Cell* **2009**, *20*, 4303–4312. [CrossRef] [PubMed]
88. Hehnly, H.; Sheff, D.; Stamnes, M. Shiga toxin facilitates its retrograde transport by modifying microtubule dynamics. *Mol. Biol. Cell* **2006**, *17*, 4379–4389. [CrossRef] [PubMed]
89. Mallard, F.; Antony, C.; Tenza, D.; Salamero, J.; Goud, B.; Johannes, L. Direct pathway from early/recycling endosomes to the Golgi apparatus revealed through the study of Shiga toxin B-fragment transport. *J. Cell Biol.* **1998**, *143*, 973–990. [CrossRef] [PubMed]
90. Sandvig, K.; Olsnes, S.; Brown, J.E.; Petersen, O.W.; van Deurs, B. Endocytosis from coated pits of Shiga toxin: a glycolipid-binding protein from *Shigella dysenteriae* 1. *J. Cell Biol.* **1989**, *108*, 1331–1343. [CrossRef] [PubMed]
91. Sandvig, K.; Garred, O.; Prydz, K.; Kozlov, J.V.; Hansen, S.H.; van Deurs, B. Retrograde transport of endocytosed Shiga toxin to the endoplasmic reticulum. *Nature* **1992**, *358*, 510–512. [CrossRef] [PubMed]
92. Donta, S.T.; Tomicic, T.K.; Donohue-Rolfe, A. Inhibition of Shiga-like toxins by brefeldin A. *J. Infect. Dis.* **1995**, *171*, 721–724. [CrossRef] [PubMed]
93. Sandvig, K.; Prydz, K.; Ryd, M.; van Deurs, B. Endocytosis and intracellular transport of the glycolipid-binding ligand Shiga toxin in polarized MDCK cells. *J. Cell Biol.* **1991**, *113*, 553–562. [CrossRef] [PubMed]
94. Garred, O.; Dubinina, E.; Polesskaya, A.; Olsnes, S.; Kozlov, J.; Sandvig, K. Role of the disulfide bond in Shiga toxin A-chain for toxin entry into cells. *J. Biol. Chem.* **1997**, *272*, 11414–11419. [PubMed]
95. Garred, O.; Dubinina, E.; Holm, P.K.; Olsnes, S.; van Deurs, B.; Kozlov, J.V.; Sandvig, K. Role of processing and intracellular transport for optimal toxicity of Shiga toxin and toxin mutants. *Exp. Cell Res.* **1995**, *218*, 39–49. [CrossRef] [PubMed]
96. Johannes, L.; Romer, W. Shiga toxin—From cell biology to biomedical applications. *Nat. Rev. Microbiol.* **2010**, *8*, 105–116. [CrossRef] [PubMed]
97. Lee, M.S.; Koo, S.; Jeong, D.G.; Tesh, V.L. Shiga toxins as multi-functional proteins: Induction of host cellular stress responses, role in pathogenesis and therapeutic applications. *Toxins (Basel)* **2016**, *8*, 77. [CrossRef] [PubMed]
98. Dyve Lingelem, A.B.; Bergan, J.; Sandvig, K. Inhibitors of intravesicular acidification protect against Shiga toxin in a pH-independent manner. *Traffic* **2012**, *13*, 443–454. [CrossRef] [PubMed]
99. Kavaliauskiene, S.; Skotland, T.; Sylvanne, T.; Simolin, H.; Klokk, T.I.; Torgersen, M.L.; Lingelem, A.B.; Simm, R.; Ekroos, K.; Sandvig, K. Novel actions of 2-deoxy-d-glucose: protection against Shiga toxins and changes in cellular lipids. *Biochem. J.* **2015**, *470*, 23–37. [CrossRef] [PubMed]
100. Stechmann, B.; Bai, S.K.; Gobbo, E.; Lopez, R.; Merer, G.; Pinchard, S.; Panigai, L.; Tenza, D.; Raposo, G.; Beaumelle, B.; et al. Inhibition of retrograde transport protects mice from lethal ricin challenge. *Cell* **2010**, *141*, 231–242. [CrossRef] [PubMed]
101. Secher, T.; Shima, A.; Hinsinger, K.; Cintrat, J.C.; Johannes, L.; Barbier, J.; Gillet, D.; Oswald, E. Retrograde trafficking inhibitor of Shiga toxins reduces morbidity and mortality of mice infected with enterohemorrhagic *Esherichia coli*. *Antimicrob. Agents Chemother.* **2015**, *59*, 5010–5013. [CrossRef]
102. Tewari, R.; Jarvela, T.; Linstedt, A.D. Manganese induces oligomerization to promote down-regulation of the intracellular trafficking receptor used by Shiga toxin. *Mol. Biol. Cell* **2014**, *25*, 3049–3058. [CrossRef] [PubMed]
103. Mukhopadhyay, S.; Linstedt, A.D. Manganese blocks intracellular trafficking of Shiga toxin and protects against Shiga toxicosis. *Science* **2012**, *335*, 332–335. [CrossRef] [PubMed]
104. Gaston, M.A.; Pellino, C.A.; Weiss, A.A. Failure of manganese to protect from Shiga toxin. *PLoS ONE* **2013**, *8*, e69823. [CrossRef] [PubMed]
105. Sandvig, K.; Brown, J.E. Ionic requirements for entry of Shiga toxin from *Shigella dysenteriae* 1 into cells. *Infect. Immun.* **1987**, *55*, 298–303. [PubMed]
106. Silberstein, C.; Copeland, D.P.; Chiang, W.; Repetto, H.A.; Ibarra, C. A glucosylceramide synthase inhibitor prevents the cytotoxic effects of Shiga toxin-2 on human renal tubular epithelial cells. *J. Epithel. Biol. Pharmacol.* **2008**, *1*, 71–75. [CrossRef]
107. Silberstein, C.; Lucero, M.S.; Zotta, E.; Copeland, D.P.; Lingyun, L.; Repetto, H.A.; Ibarra, C. A glucosylceramide synthase inhibitor protects rats against the cytotoxic effects of Shiga toxin 2. *Pediatr. Res.* **2011**, *69*, 390–394. [CrossRef] [PubMed]

108. Amaral, M.M.; Sacerdoti, F.; Jancic, C.; Repetto, H.A.; Paton, A.W.; Paton, J.C.; Ibarra, C. Action of Shiga toxin type-2 and subtilase cytotoxin on human microvascular endothelial cells. *PLoS ONE* **2013**, *8*, e70431. [CrossRef] [PubMed]
109. Bergan, J.; Skotland, T.; Lingelem, A.B.; Simm, R.; Spilsberg, B.; Lindback, T.; Sylvanne, T.; Simolin, H.; Ekroos, K.; Sandvig, K. The ether lipid precursor hexadecylglycerol protects against Shiga toxins. *Cell. Mol. Life Sci.* **2014**, *71*, 4285–4300. [CrossRef] [PubMed]
110. Binnington, B.; Nguyen, L.; Kamani, M.; Hossain, D.; Marks, D.L.; Budani, M.; Lingwood, C.A. Inhibition of Rab prenylation by statins induces cellular glycosphingolipid remodeling. *Glycobiology* **2016**, *26*, 166–180. [CrossRef] [PubMed]
111. Becker, G.L.; Lu, Y.; Hardes, K.; Strehlow, B.; Levesque, C.; Lindberg, I.; Sandvig, K.; Bakowsky, U.; Day, R.; Garten, W.; et al. Highly potent inhibitors of proprotein convertase furin as potential drugs for treatment of infectious diseases. *J. Biol. Chem.* **2012**, *287*, 21992–22003. [CrossRef] [PubMed]
112. Hardes, K.; Becker, G.L.; Lu, Y.; Dahms, S.O.; Kohler, S.; Beyer, W.; Sandvig, K.; Yamamoto, H.; Lindberg, I.; Walz, L.; et al. Novel Furin inhibitors with potent anti-infectious activity. *ChemMedChem* **2015**, *10*, 1218–1231. [CrossRef] [PubMed]
113. Meshnick, S.R.; Dobson, M.J. The history of antimalarial drugs. In *Antimalarial Chemotherapy: Mechanisms of Action, Resistance, and New Directions in Drug Discovery*; Rosenthal, P.J., Ed.; Humana Press: Totowa, NJ, USA, 2001; pp. 15–25.
114. Kitchen, L.W.; Vaughn, D.W.; Skillman, D.R. Role of US military research programs in the development of US Food and Drug Administration—Approved antimalarial drugs. *Clin. Infect. Dis.* **2006**, *43*, 67–71. [CrossRef] [PubMed]
115. Al-Bari, M.A. Chloroquine analogues in drug discovery: New directions of uses, mechanisms of actions and toxic manifestations from malaria to multifarious diseases. *J. Antimicrob. Chemother.* **2015**, *70*, 1608–1621. [CrossRef] [PubMed]
116. Pascolo, S. Time to use a dose of chloroquine as an adjuvant to anti-cancer chemotherapies. *Eur. J. Pharmacol.* **2016**, *771*, 139–144. [CrossRef] [PubMed]
117. Keeling, D.J.; Herslof, M.; Ryberg, B.; Sjogren, S.; Solvell, L. Vacuolar H(+)-ATPases. Targets for drug discovery? *Ann. N. Y. Acad. Sci.* **1997**, *834*, 600–608. [CrossRef] [PubMed]
118. Parra, K.J. Vacuolar ATPase: A model proton pump for antifungal drug discovery. In *Antimicrobial Drug Discovery: Emerging Strategies*; Tegos, G., Mylonakis, E., Eds.; CAB International: Accra, Ghana, 2012; pp. 89–100.
119. Moreau, D.; Kumar, P.; Wang, S.C.; Chaumet, A.; Chew, S.Y.; Chevalley, H.; Bard, F. Genome-wide RNAi screens identify genes required for Ricin and PE intoxications. *Dev. Cell* **2011**, *21*, 231–244. [CrossRef] [PubMed]
120. Yu, M.; Haslam, D.B. Shiga toxin is transported from the endoplasmic reticulum following interaction with the luminal chaperone HEDJ/ERDj3. *Infect. Immun.* **2005**, *73*, 2524–2532. [CrossRef] [PubMed]
121. Rodriguez-Menchaca, A.A.; Navarro-Polanco, R.A.; Ferrer-Villada, T.; Rupp, J.; Sachse, F.B.; Tristani-Firouzi, M.; Sanchez-Chapula, J.A. The molecular basis of chloroquine block of the inward rectifier Kir2.1 channel. *Proc. Natl. Acad. Sci. USA* **2008**, *105*, 1364–1368. [CrossRef] [PubMed]
122. Orlik, F.; Schiffler, B.; Benz, R. Anthrax toxin protective antigen: inhibition of channel function by chloroquine and related compounds and study of binding kinetics using the current noise analysis. *Biophys J.* **2005**, *88*, 1715–1724. [CrossRef] [PubMed]
123. Bachmeyer, C.; Benz, R.; Barth, H.; Aktories, K.; Gilbert, M.; Popoff, M.R. Interaction of *Clostridium botulinum* C2 toxin with lipid bilayer membranes and Vero cells: inhibition of channel function by chloroquine and related compounds *in vitro* and intoxification *in vivo*. *FASEB J.* **2001**, *15*, 1658–1660. [CrossRef] [PubMed]
124. Browning, D.J. Pharmacology of chloroquine and hydroxychloroquine. In *Hydroxychloroquine and Chloroquine Retinopathy*; Springer: New York, NY, USA, 2014; pp. 35–63.
125. Molina, D.K. Postmortem hydroxychloroquine concentrations in nontoxic cases. *Am. J. Forensic Med. Pathol.* **2012**, *33*, 41–42. [CrossRef] [PubMed]
126. Brown, J. Effects of 2-deoxyglucose on carbohydrate metablism: review of the literature and studies in the rat. *Metabolism* **1962**, *11*, 1098–1112. [PubMed]
127. Wick, A.N.; Drury, D.R.; Nakada, H.I.; Wolfe, J.B. Localization of the primary metabolic block produced by 2-deoxyglucose. *J. Biol. Chem.* **1957**, *224*, 963–969. [PubMed]

128. Cramer, F.B.; Woodward, G.E. 2-Desoxy-D-glucose as an antagonist of glucose in yeast fermentation. *J. Frankl. Inst.* **1952**, *253*, 354–360. [CrossRef]
129. Sols, A.; Crane, R.K. Substrate specificity of brain hexokinase. *J. Biol. Chem.* **1954**, *210*, 581–595. [PubMed]
130. Chen, W.; Gueron, M. The inhibition of bovine heart hexokinase by 2-deoxy-D-glucose-6-phosphate: characterization by ^{31}P NMR and metabolic implications. *Biochimie* **1992**, *74*, 867–873. [CrossRef]
131. Datema, R.; Schwarz, R.T. Interference with glycosylation of glycoproteins. Inhibition of formation of lipid-linked oligosaccharides *in vivo*. *Biochem. J.* **1979**, *184*, 113–123. [CrossRef] [PubMed]
132. Desselle, A.; Chaumette, T.; Gaugler, M.H.; Cochonneau, D.; Fleurence, J.; Dubois, N.; Hulin, P.; Aubry, J.; Birkle, S.; Paris, F. Anti-Gb3 monoclonal antibody inhibits angiogenesis and tumor development. *PLoS ONE* **2012**, *7*, e45423. [CrossRef] [PubMed]
133. Watowich, S.S.; Morimoto, R.I. Complex regulation of heat shock- and glucose-responsive genes in human cells. *Mol. Cell. Biol.* **1988**, *8*, 393–405. [CrossRef] [PubMed]
134. Shinjo, S.; Mizotani, Y.; Tashiro, E.; Imoto, M. Comparative analysis of the expression patterns of UPR-target genes caused by UPR-inducing compounds. *Biosci. Biotechnol. Biochem.* **2013**, *77*, 729–735. [CrossRef] [PubMed]
135. Okuda, T.; Furukawa, K.; Nakayama, K.I. A novel, promoter-based, target-specific assay identifies 2-deoxy-D-glucose as an inhibitor of globotriaosylceramide biosynthesis. *FEBS J.* **2009**, *276*, 5191–5202. [CrossRef] [PubMed]
136. Kurtoglu, M.; Maher, J.C.; Lampidis, T.J. Differential toxic mechanisms of 2-deoxy-D-glucose versus 2-fluorodeoxy-D-glucose in hypoxic and normoxic tumor cells. *Antioxid. Redox Signal.* **2007**, *9*, 1383–1390. [CrossRef] [PubMed]
137. Lampidis, T.J.; Kurtoglu, M.; Maher, J.C.; Liu, H.; Krishan, A.; Sheft, V.; Szymanski, S.; Fokt, I.; Rudnicki, W.R.; Ginalski, K.; et al. Efficacy of 2-halogen substituted D-glucose analogs in blocking glycolysis and killing "hypoxic tumor cells". *Cancer Chemother. Pharmacol.* **2006**, *58*, 725–734. [CrossRef] [PubMed]
138. Datema, R.; Schwarz, R.T.; Jankowski, A.W. Fluoroglucose-inhibition of protein glycosylation *in vivo*. Inhibition of mannose and glucose incorporation into lipid-linked oligosaccharides. *Eur. J. Biochem.* **1980**, *109*, 331–341. [CrossRef] [PubMed]
139. Schmidt, M.F.; Biely, P.; Kratky, Z.; Schwarz, R.T. Metabolism of 2-deoxy-2-fluoro-D-[^{3}H]glucose and 2-deoxy-2-fluoro-D-[^{3}H]mannose in yeast and chick-embryo cells. *Eur. J. Biochem.* **1978**, *87*, 55–68. [CrossRef] [PubMed]
140. Hoh, C.K. Clinical use of FDG-PET. *Nucl. Med. Biol.* **2007**, *34*, 737–742. [CrossRef] [PubMed]
141. Kelloff, G.J.; Hoffman, J.M.; Johnson, B.; Scher, H.I.; Siegel, B.A.; Cheng, E.Y.; Cheson, B.D.; O'Shaughnessy, J.; Guyton, K.Z.; Mankoff, D.A.; et al. Progress and promise of FDG-PET imaging for cancer patient management and oncologic drug development. *Clin. Cancer. Res.* **2005**, *11*, 2785–2808. [CrossRef] [PubMed]
142. Xi, H.; Barredo, J.C.; Merchan, J.R.; Lampidis, T.J. Endoplasmic reticulum stress induced by 2-deoxyglucose but not glucose starvation activates AMPK through CAMKKβ leading to autophagy. *Biochem. Pharmacol.* **2013**, *85*, 1463–1477. [CrossRef] [PubMed]
143. Xi, H.; Kurtoglu, M.; Liu, H.; Wangpaichitr, M.; You, M.; Liu, X.; Savaraj, N.; Lampidis, T.J. 2-Deoxy-D-glucose activates autophagy via endoplasmic reticulum stress rather than ATP depletion. *Cancer Chemother. Pharmacol.* **2011**, *67*, 899–910. [CrossRef] [PubMed]
144. Biswas, C.; Ostrovsky, O.; Makarewich, C.A.; Wanderling, S.; Gidalevitz, T.; Argon, Y. The peptide-binding activity of GRP94 is regulated by calcium. *Biochem. J.* **2007**, *405*, 233–241. [CrossRef] [PubMed]
145. Hendershot, L.M. The ER function BiP is a master regulator of ER function. *Mt. Sinai J. Med.* **2004**, *71*, 289–297. [PubMed]
146. Raez, L.E.; Papadopoulos, K.; Ricart, A.D.; Chiorean, E.G.; Dipaola, R.S.; Stein, M.N.; Rocha Lima, C.M.; Schlesselman, J.J.; Tolba, K.; Langmuir, V.K.; et al. A phase I dose-escalation trial of 2-deoxy-D-glucose alone or combined with docetaxel in patients with advanced solid tumors. *Cancer Chemother. Pharmacol.* **2013**, *71*, 523–530. [CrossRef] [PubMed]
147. Saenz, J.B.; Doggett, T.A.; Haslam, D.B. Identification and characterization of small molecules that inhibit intracellular toxin transport. *Infect. Immun.* **2007**, *75*, 4552–4561. [CrossRef] [PubMed]
148. Park, J.G.; Kahn, J.N.; Tumer, N.E.; Pang, Y.P. Chemical structure of Retro-2, a compound that protects cells against ribosome-inactivating proteins. *Sci. Rep.* **2012**, *2*, 631. [CrossRef] [PubMed]

149. Noel, R.; Gupta, N.; Pons, V.; Goudet, A.; Garcia-Castillo, M.D.; Michau, A.; Martinez, J.; Buisson, D.A.; Johannes, L.; Gillet, D.; et al. *N*-methyldihydroquinazolinone derivatives of Retro-2 with enhanced efficacy against Shiga toxin. *J. Med. Chem.* **2013**, *56*, 3404–3413. [CrossRef] [PubMed]
150. Carney, D.W.; Nelson, C.D.; Ferris, B.D.; Stevens, J.P.; Lipovsky, A.; Kazakov, T.; DiMaio, D.; Atwood, W.J.; Sello, J.K. Structural optimization of a retrograde trafficking inhibitor that protects cells from infections by human polyoma- and papillomaviruses. *Bioorg. Med. Chem.* **2014**, *22*, 4836–4847. [CrossRef] [PubMed]
151. Gupta, N.; Pons, V.; Noel, R.; Buisson, D.A.; Michau, A.; Johannes, L.; Gillet, D.; Barbier, J.; Cintrat, J.C. (*S*)-*N*-Methyldihydroquinazolinones are the active enantiomers of Retro-2 derived compounds against toxins. *ACS Med. Chem. Lett.* **2014**, *5*, 94–97. [CrossRef] [PubMed]
152. Frank, C.; Werber, D.; Cramer, J.P.; Askar, M.; Faber, M.; an der Heiden, M.; Bernard, H.; Fruth, A.; Prager, R.; Spode, A.; et al. Epidemic profile of Shiga-toxin-producing *Escherichia coli* O104:H4 outbreak in Germany. *N. Engl. J. Med.* **2011**, *365*, 1771–1780. [CrossRef] [PubMed]
153. Bielaszewska, M.; Mellmann, A.; Zhang, W.; Kock, R.; Fruth, A.; Bauwens, A.; Peters, G.; Karch, H. Characterisation of the *Escherichia coli* strain associated with an outbreak of haemolytic uraemic syndrome in Germany, 2011: a microbiological study. *Lancet Infect. Dis.* **2011**, *11*, 671–676. [CrossRef]
154. Gupta, N.; Noel, R.; Goudet, A.; Hinsinger, K.; Michau, A.; Pons, V.; Abdelkafi, H.; Secher, T.; Shima, A.; Shtanko, O.; et al. Inhibitors of retrograde trafficking active against ricin and Shiga toxins also protect cells from several viruses, *Leishmania* and Chlamydiales. *Chem. Biol. Interact.* **2016**. [CrossRef] [PubMed]
155. Aschner, M.; Erikson, K.M.; Herrero Hernandez, E.; Tjalkens, R. Manganese and its role in Parkinson's disease: from transport to neuropathology. *Neuromol. Med.* **2009**, *11*, 252–266. [CrossRef] [PubMed]
156. Racette, B.A.; Aschner, M.; Guilarte, T.R.; Dydak, U.; Criswell, S.R.; Zheng, W. Pathophysiology of manganese-associated neurotoxicity. *Neurotoxicology* **2012**, *33*, 881–886. [CrossRef] [PubMed]
157. Vunnam, R.R.; Radin, N.S. Analogs of ceramide that inhibit glucocerebroside synthetase in mouse brain. *Chem. Phys. Lipids* **1980**, *26*, 265–278. [CrossRef]
158. Jmoudiak, M.; Futerman, A.H. Gaucher disease: pathological mechanisms and modern management. *Br. J. Haematol.* **2005**, *129*, 178–188. [CrossRef] [PubMed]
159. Barbour, S.; Edidin, M.; Felding-Habermann, B.; Taylor-Norton, J.; Radin, N.S.; Fenderson, B.A. Glycolipid depletion using a ceramide analogue (PDMP) alters growth, adhesion, and membrane lipid organization in human A431 cells. *J. Cell Physiol.* **1992**, *150*, 610–619. [CrossRef] [PubMed]
160. Arab, S.; Lingwood, C.A. Intracellular targeting of the endoplasmic reticulum/nuclear envelope by retrograde transport may determine cell hypersensitivity to verotoxin via globotriaosyl ceramide fatty acid isoform traffic. *J. Cell. Physiol.* **1998**, *177*, 646–660. [CrossRef]
161. Sandvig, K.; Garred, O.; van Helvoort, A.; van Meer, G.; van Deurs, B. Importance of glycolipid synthesis for butyric acid-induced sensitization to Shiga toxin and intracellular sorting of toxin in A431 cells. *Mol. Biol. Cell* **1996**, *7*, 1391–1404. [CrossRef] [PubMed]
162. Lee, L.; Abe, A.; Shayman, J.A. Improved inhibitors of glucosylceramide synthase. *J. Biol. Chem.* **1999**, *274*, 14662–14669. [CrossRef] [PubMed]
163. Basu, M.; Dastgheib, S.; Girzadas, M.A.; O'Donnell, P.H.; Westervelt, C.W.; Li, Z.; Inokuchi, J.; Basu, S. Hydrophobic nature of mammalian ceramide glycanases: purified from rabbit and rat mammary tissues. *Acta Biochim. Pol.* **1998**, *45*, 327–342. [PubMed]
164. Abe, A.; Gregory, S.; Lee, L.; Killen, P.D.; Brady, R.O.; Kulkarni, A.; Shayman, J.A. Reduction of globotriaosylceramide in Fabry disease mice by substrate deprivation. *J. Clin. Invest.* **2000**, *105*, 1563–1571. [CrossRef] [PubMed]
165. Poole, R.M. Eliglustat: first global approval. *Drugs* **2014**, *74*, 1829–1836. [CrossRef] [PubMed]
166. Belmatoug, N.; Di Rocco, M.; Fraga, C.; Giraldo, P.; Hughes, D.; Lukina, E.; Maison-Blanche, P.; Merkel, M.; Niederau, C.; Plckinger, U.; et al. Management and monitoring recommendations for the use of eliglustat in adults with type 1 Gaucher disease in Europe. *Eur. J. Intern. Med.* **2016**. [CrossRef] [PubMed]
167. Lukina, E.; Watman, N.; Arreguin, E.A.; Banikazemi, M.; Dragosky, M.; Iastrebner, M.; Rosenbaum, H.; Phillips, M.; Pastores, G.M.; Rosenthal, D.I.; et al. A phase 2 study of eliglustat tartrate (Genz-112638), an oral substrate reduction therapy for Gaucher disease type 1. *Blood* **2010**, *116*, 893–899. [CrossRef] [PubMed]
168. Das, A.K.; Holmes, R.D.; Wilson, G.N.; Hajra, A.K. Dietary ether lipid incorporation into tissue plasmalogens of humans and rodents. *Lipids* **1992**, *27*, 401–405. [CrossRef] [PubMed]

169. Iannitti, T.; Palmieri, B. An update on the therapeutic role of alkylglycerols. *Mar. Drugs* **2010**, *8*, 2267–2300. [CrossRef] [PubMed]
170. Deniau, A.L.; Mosset, P.; Pedrono, F.; Mitre, R.; Le Bot, D.; Legrand, A.B. Multiple beneficial health effects of natural alkylglycerols from shark liver oil. *Mar. Drugs* **2010**, *8*, 2175–2184. [CrossRef] [PubMed]
171. Matusewicz, L.; Meissner, J.; Toporkiewicz, M.; Sikorski, A.F. The effect of statins on cancer cells—Review. *Tumour Biol.* **2015**, *36*, 4889–4904. [CrossRef] [PubMed]
172. Correale, M.; Abruzzese, S.; Greco, C.A.; Concilio, M.; Di Biase, M.; Brunetti, N.D. Statins in heart failure. *Curr. Vasc. Pharmacol.* **2012**, *69*, 232. [CrossRef]
173. Paliani, U.; Ricci, S. The role of statins in stroke. *Intern. Emerg. Med.* **2012**, *7*, 305–311. [CrossRef] [PubMed]
174. US Preventive Services Task Force; Bibbins-Domingo, K.; Grossman, D.C.; Curry, S.J.; Davidson, K.W.; Epling, J.W.; Garcia, F.A.; Gillman, M.W.; Kemper, A.R.; Krist, A.H.; et al. Statin use for the primary prevention of cardiovascular disease in adults: US preventive services task force recommendation statement. *JAMA* **2016**, *316*, 1997–2007. [PubMed]
175. Oda, T.; Wu, H.C. Effect of lovastatin on the cytotoxicity of ricin, modeccin, *Pseudomonas* toxin, and diphtheria toxin in brefeldin A-sensitive and -resistant cell lines. *Exp. Cell Res.* **1994**, *212*, 329–337. [CrossRef] [PubMed]
176. Gomes, A.Q.; Ali, B.R.; Ramalho, J.S.; Godfrey, R.F.; Barral, D.C.; Hume, A.N.; Seabra, M.C. Membrane targeting of Rab GTPases is influenced by the prenylation motif. *Mol. Biol. Cell* **2003**, *14*, 1882–1899. [CrossRef] [PubMed]
177. Pfeffer, S.; Aivazian, D. Targeting Rab GTPases to distinct membrane compartments. *Nat. Rev. Mol. Cell Biol.* **2004**, *5*, 886–896. [CrossRef] [PubMed]
178. Holstein, S.A.; Knapp, H.R.; Clamon, G.H.; Murry, D.J.; Hohl, R.J. Pharmacodynamic effects of high dose lovastatin in subjects with advanced malignancies. *Cancer Chemother. Pharmacol.* **2006**, *57*, 155–164. [CrossRef] [PubMed]
179. Schafer, W.; Stroh, A.; Berghofer, S.; Seiler, J.; Vey, M.; Kruse, M.L.; Kern, H.F.; Klenk, H.D.; Garten, W. Two independent targeting signals in the cytoplasmic domain determine trans-Golgi network localization and endosomal trafficking of the proprotein convertase furin. *EMBO J.* **1995**, *14*, 2424–2435. [PubMed]
180. Plaimauer, B.; Mohr, G.; Wernhart, W.; Himmelspach, M.; Dorner, F.; Schlokat, U. 'Shed' furin: Mapping of the cleavage determinants and identification of its C-terminus. *Biochem. J.* **2001**, *354*, 689–695. [CrossRef] [PubMed]
181. Thomas, G. Furin at the cutting edge: from protein traffic to embryogenesis and disease. *Nat. Rev. Mol. Cell Biol.* **2002**, *3*, 753–766. [CrossRef] [PubMed]
182. Seidah, N.G.; Prat, A. The biology and therapeutic targeting of the proprotein convertases. *Nat. Rev. Drug Discov.* **2012**, *11*, 367–383. [CrossRef] [PubMed]
183. Couture, F.; D'Anjou, F.; Day, R. On the cutting edge of proprotein convertase pharmacology: from molecular concepts to clinical applications. *Biomol. Concepts* **2011**, *2*, 421–438. [CrossRef] [PubMed]
184. De Cicco, R.L.; Bassi, D.E.; Benavides, F.; Conti, C.J.; Klein-Szanto, A.J. Inhibition of proprotein convertases: approaches to block squamous carcinoma development and progression. *Mol. Carcinog.* **2007**, *46*, 654–659. [CrossRef] [PubMed]
185. Sarac, M.S.; Cameron, A.; Lindberg, I. The furin inhibitor hexa-D-arginine blocks the activation of *Pseudomonas aeruginosa* exotoxin A in vivo. *Infect. Immun.* **2002**, *70*, 7136–7139. [CrossRef] [PubMed]
186. Sarac, M.S.; Peinado, J.R.; Leppla, S.H.; Lindberg, I. Protection against anthrax toxemia by hexa-D-arginine in vitro and in vivo. *Infect. Immun.* **2004**, *72*, 602–605. [CrossRef] [PubMed]
187. Basak, A. Inhibitors of proprotein convertases. *J. Mol. Med. (Berl.)* **2005**, *83*, 844–855. [CrossRef] [PubMed]
188. Shayman, J.A. The design and clinical development of inhibitors of glycosphingolipid synthesis: Will invention be the mother of necessity? *Trans. Am. Clin. Climatol. Assoc.* **2013**, *124*, 46–60. [PubMed]
189. Liu, Y.Y.; Hill, R.A.; Li, Y.T. Ceramide glycosylation catalyzed by glucosylceramide synthase and cancer drug resistance. *Adv. Cancer Res.* **2013**, *117*, 59–89. [PubMed]

12

Evaluating the Potential for Cross-Interactions of Antitoxins in Type II TA Systems

Chih-Han Tu †, Michelle Holt †, Shengfeng Ruan † and Christina Bourne *

Department of Chemistry and Biochemistry, University of Oklahoma, Norman, OK 73019, USA; chihhantu@ou.edu (C.-H.T.); mnholt1217@ou.edu (M.H.); shengfengruan@ou.edu (S.R.)

* Correspondence: cbourne@ou.edu.

† All authors contributed equally to this article.

Abstract: The diversity of Type-II toxin–antitoxin (TA) systems in bacterial genomes requires tightly controlled interaction specificity to ensure protection of the cell, and potentially to limit cross-talk between toxin–antitoxin pairs of the same family of TA systems. Further, there is a redundant use of toxin folds for different cellular targets and complexation with different classes of antitoxins, increasing the apparent requirement for the insulation of interactions. The presence of Type II TA systems has remained enigmatic with respect to potential benefits imparted to the host cells. In some cases, they play clear roles in survival associated with unfavorable growth conditions. More generally, they can also serve as a "cure" against acquisition of highly similar TA systems such as those found on plasmids or invading genetic elements that frequently carry virulence and resistance genes. The latter model is predicated on the ability of these highly specific cognate antitoxin–toxin interactions to form cross-reactions between chromosomal antitoxins and invading toxins. This review summarizes advances in the Type II TA system models with an emphasis on antitoxin cross-reactivity, including with invading genetic elements and cases where toxin proteins share a common fold yet interact with different families of antitoxins.

Keywords: cognate interactions; cross-interactions; molecular insulation; toxin; antitoxin; TA systems; addiction; anti-addiction

Key Contribution: Current models posit TA systems can be used for both cellular addiction to mobile extra-chromosomal DNA, as well as for anti-addiction functions. This review examines known cases of antitoxin cross-interactions with non-cognate toxins, and explores the potential for additional cross-interactions based on conserved structures of toxin families.

1. Introduction

Toxin–antitoxin (TA) systems are abundant in bacterial and archaeal chromosomes as well as extra-chromosomal genetic elements including plasmids, phages, and transposons [1–7]. TA systems have been the subject of numerous reviews that describe their typically bicistronic operon encoding a toxin that targets essential cellular process and the cognate neutralizing antitoxin [8–16]. Toxin proteins are able to manipulate their bacterial host cells in powerful ways. This drives great interest in understanding their functions and the potential to utilize them for biotechnology and bacterial control strategies [17–26].

Depending on the molecular identity of the antitoxin, as well as the mechanism by which it neutralizes the toxin, the known TA systems have been classified into six different types. While the toxin is typically a protein, the antitoxin is either a noncoding RNA (in Type I and III) or a protein (in Type II, IV, V and VI) [15,18,25,27–30]. Limited numbers of Type III TA pairs [31–34] and IV TA pairs have been identified [35–37]. Type V systems identified to date are the GhoST system in *Escherichia coli* K strains and the orphan OrtT toxin from *Salmonella* [38–40]. Similarly, the Type VI system is currently comprised of the Soc system in *Caulobacter crescentus* [41]. However, continuing surprises are challenging the canonical TA system paradigms. These include the integration of some TA systems into alarmone signaling pathways [42,43] and recognition of alternatives to protease-dependent regulation such as acetylation of antitoxins and/or chaperone protection [44–46], in addition to widespread variations on the bicistronic antitoxin–toxin arrangement [44,47]. In comparison, Type I and II TA systems are well-studied, and Type II systems appear in thousands of bacterial loci that are accessible and searchable through tailored web portals; to-date, the TADB webserver lists more than 6000 Type II systems with approximately 10% of these located on plasmids [1,2,7,48–51]. In Type II systems the antitoxin partner plays a dual role by neutralizing the cognate toxin and by mediating transcriptional regulation through binding to its promoter, making it the lynchpin of Type II TA system functions [52] (Figure 1). As such, many of the proposed functions of these systems are predicated on the ability of antitoxins to potentially neutralize multiple homologous toxins, as explored in the current review.

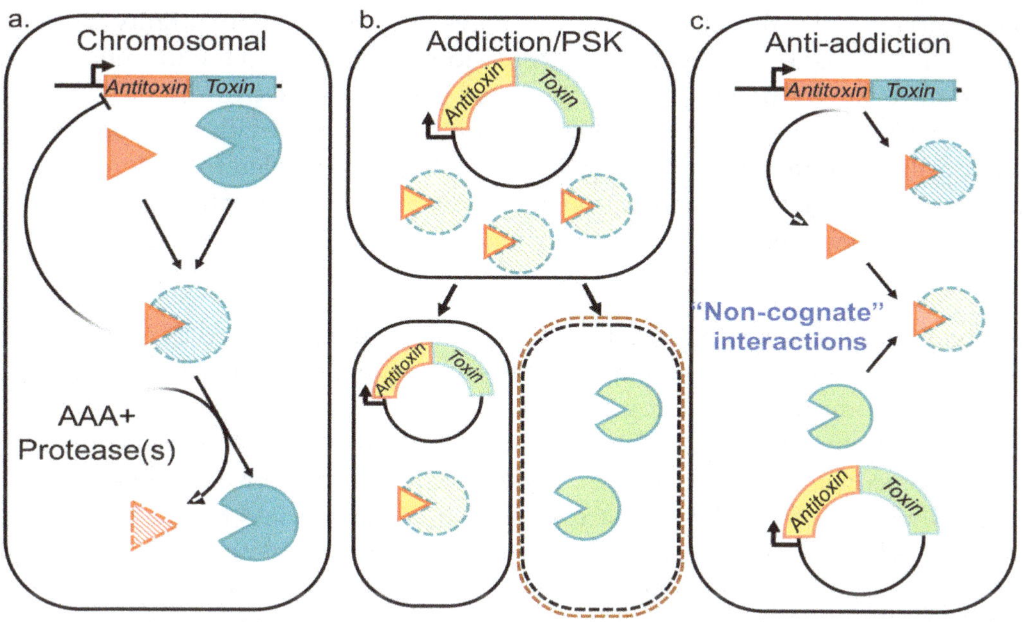

Figure 1. Paradigms for Type II toxin–antitoxin (TA) system functions span roles in (**a**) physiology mediated by chromosomally-encoded systems, in addition to potential roles (**b**) as addiction modules on mobile genetic elements. An alternative may arise (**c**) when multiple TA systems are present within the same cell, giving rise to an anti-addiction role. Experimental evidence has demonstrated that chromosomally-encoded systems can protect individual cells from external stressors (**a**). When TA systems are encoded on mobile genetic elements (pictured on a plasmid) they can function as "Addiction" elements that enforce retention (also referred to as post-segregation killing, or PSK) (**b**). A hybrid paradigm combines these two functionalities, wherein a chromosomally-encoded antitoxin (or toxin) can neutralize an invading toxin, thus providing an anti-addiction function (**c**).

2. Paradigms for Type II TA Systems

Type II TA systems were originally recognized as mediators of plasmid addiction, also termed "post-segregational killing" (PSK) (Figure 1b) [53–57]. Subsequent genome sequencing efforts identified them throughout bacteria and archaea [1,2,7], with those presenting some levels of similarity to the plasmidic versions but not other related genomic versions referred to as "xenologs" to highlight their likelihood of arising through horizontal gene transfer [58]. Analysis of xenolog distribution led to insight that these TA systems are a form of bacterial immunity with implications for protection from invading genetic material [28,59–61]. Therefore, the roles of given Type II TA systems described in Figure 1 remain unsettled and are likely overlapping.

2.1. Functions Attributed to Chromosomal Type II TA Systems

The prevailing evidence favors the stochastic accumulation of TA systems through horizontal gene transfer, with a predominance of phage-derived systems at "hotspots" of genetic diversity within a given bacterial species [2,7,58,62–65]. As such, their potential role has been a subject of intense debate. For many, it has been demonstrated that they are located within integrated (at least previously) mobile genetic elements, and further, some of these function as addiction modules for those integrated elements (see Section 2.2, below). Some have proposed that integration of a TA system into a bacterial chromosome has allowed it to be co-opted to maintain normal physiological homeostasis in response to environmental or other changes in growth conditions, sometimes termed "domestication" or linked to "accessory" genomic content [8,66]. Further, for some TA systems there likely exists a functional overlap of these ideas, encompassing both genetic addiction and the potential for usefulness to the cells (Figure 1) [27,67].

Some investigations have documented an attenuation of chromosomally-encoded TA system toxicity. This has arisen either through sequence changes resulting in lower affinity for interaction with the cellular target [68–72] or by attenuation of expression via changes in the promoter regions [73–75]. A loss of function, termed "degeneration", has also been observed for some systems, such as a chromosomal CcdAB system in *E. coli* [62]. Attenuation of toxicity provides an opportunity for toxins to regulate their cellular target without killing the host cell. While this led to the long-touted idea of TA system involvement in persister cell generation, it can also be more generalized simply as the ability of cells to withstand external stressors by reducing metabolism or protecting critical cellular targets from damage [16,76,77].

Gyrase-targeting systems CcdAB and ParDE, each originally identified as mediators of PSK, have been documented to provide useful advantages to their host cells. The CcdB toxin, carried on the F plasmid and the chromosomal xenolog in *E. coli*, both were observed to increase cellular survival to heat, anti-gyrase compounds, and other antibiotics [68,78]. An analogous protective effect from anti-gyrase antibiotics was observed for the chromosomal ParE from *Pseudomonas aeruginosa* [71]. The protection of host cells from thermal stress has also been observed for the ParE toxin carried on the RK2 plasmid [79]. *Caulobacter crescentus* carries three functional chromosomal ParDE systems, and these were observed to offer protection to stressors, as did one of a ParDE system from *Mycobacteria tuberculosis* [70,80]. We note that many studies rely on "over-expression" of toxin proteins to analyze the cellular impacts, resulting in non-native concentrations. However, effects can be observed at very low levels of toxin protein expression barely detectable by Western blotting, leading to the suggestion these could be reasonable estimates for the free toxin of an "activated" system [69]. Overall, it seems that gyrase-targeting TA systems may occupy a specialized niche that bridges protection at low-levels of expression and/or with attenuated toxicity, to a higher-level toxicity at prolonged exposure or concentration levels. Given that FicT toxins have also been identified as modulating both DNA gyrase and topoisomerase IV via adenylation [81], it remains to be determined if they will have similar impacts to heat tolerance or target protection.

Protection from thermal stress is noted for multiple members of RNA-degrading ("RNase") type toxins, particularly those of the RelE-type fold. The YoeB toxin from *E. coli* protects thermal stress, and further, this is dependent on the proteases needed for antitoxin degradation, implying this protection arises from some direct action of the toxin [82]. Studies in the gram-positive bacteria *Streptococcus pneumonia* demonstrated genetic deletions of two *yoeB* loci and related RNase toxins were less hardy when exposed to oxidative conditions, while complemented mutants recapitulated wild type survival levels [83]. The structurally related GraT toxin from *Pseudomonads* similarly protects from temperature stresses but appears to function in response to lower rather than elevated temperatures [72].

Subsequent work with GraT highlighted the overall depression in metabolism in response to this toxin [84]. YafQ, again within the same structural class as the RelE/ParE family, was also found to be protective for bacterial growth at sub-optimal temperatures [85].

Other classes of toxin activities, particularly those that impact RNA lifetimes, have also proven to impart beneficial functions to host physiology. Three RelE toxins in the genome of *Mycobacterium tuberculosis* were identified as upregulated at the transcript level in response to altered oxygen levels or limiting nitrogen [86]. This study linked RNase toxin activation to proteome alteration by two-dimensional gel electrophoresis and mass spectrometry analysis of differential products. The RelBE TA system has been documented to be activated in response to nutritional stress [87], likely triggered in natural settings by high cell density [88] and similar to findings with *Acinetobacter baumannii* and *P. aeruginosa* HigBA systems [89,90]. These systems in *Mycobacterium tuberculosis* also provide protection to antibiotics but, importantly, did not induce persister cell formation [76]. It remains unclear if this is the "native" use for these; or, in other words, if the bacterial cells experience these conditions as part of their normal growth cycle and in turn utilize these toxins to slow growth and thus hedge survival. Further, considerable controversy still surrounds many earlier findings for specific details of the RNase toxin–starvation models [12,13,91].

Many of these studies rely on a reductive approach in defined growth conditions, whereas the use of TA systems within a natural ecological setting has been harder to access. The TA systems YefM-YoeB, Hha-YbaJ (Hha-TomB), and PasTI (also named RatA [92] to reflect its Ribosome Association toxin, or inhibition) harbored on the chromosome of extraintestinal pathogenic *E. coli* (ExPEC) strains were noted to promote increases in recoverable viable bacteria from the bladder and kidneys of a mouse model of infection [93]. The deletion of MazEF systems in *M. tuberculosis* reduced virulence in an animal model, as did Vap systems in *Haemophilus influenza* [94,95]. The Hha-TomB Type II system, which impacts translation through ribosomal interactions, has also been associated to increased *Salmonella typhimurium* survival in infection models [96]. Other studies have highlighted a role for TA systems in the survival of *Salmonella* within macrophages [97]. Subsequent studies have complicated this model and suggested that the effect resulted from slowed bacterial metabolism, which is a known impact from TA system activation [16,77]. Similar complexities that produce different outcomes have been highlighted for the MqsRA and MazEF systems in *E. coli* [98–103]. The impacts of chromosomal TA systems on their host cells are clearly affected by very specific interactions and by interwoven pathways. These have been well reviewed elsewhere, and are brought up here to illustrate the unsettled nature of these on-going studies and evolving conclusions [104–107].

The continued discovery of new TA systems highlights their broad capacity for integration into bacterial cells. The *Streptococcus pneumoniae* chromosomal PezAT system, a xenolog of the Epsilon-Zeta system [108], is located within an integrated pathogenicity island, and when deleted induced phenotypic changes that produced both beneficial (harder to lyse) and negative (more sensitive to cell wall antibiotics) effects for the cell [109]. The ParST, an mART-type that transfers an ADP-ribosyl group onto an enzyme involved in phosphoribosyl pyrophosphate synthetase, induces bacteriostasis when transplanted from its native *Sphingobium* host to *E. coli* [110]. This study notes the widespread distribution of the ParST system with enrichment in Proteobacterial classes. The diverse and highly integrated nature of TA systems and bacterial physiology is exemplified by the recently published work demonstrating some alarmone synthases are housed as TA loci [43]. These systems, named toxSAS, deviate from the traditional canon in encoding multiple antitoxins per synthase, or toxin, with both cognate and universal neutralization interactions [43]. It seems, then, that the diversity and functions of TA systems are expansive and may yet reveal new secrets of bacterial growth in the coming years.

2.2. *Plasmid Selection and/or Addiction via Type II TA Systems*

As mentioned above, chromosomal Type II TA systems were almost certainly acquired by invading genetic material. For example, a ParDE system in *P. aeruginosa* is located within the Pf1 prophage [66], and the widely-studied RelBE system in *E. coli* is located within a Qin prophage [63]. The *Vibrio cholera* superintegron on chromosome II, including an integrated and conjugative element (ICE) called SXT, is enriched with both antibiotic resistance cassettes as well as Type II TA systems [111,112]. *Neisseriaceae* and *Klebsiella* species have similar integrated genetic elements, including Type-IV Secretion System components connected to TA systems [47,113], as well as other polymorphic toxin systems resembling Type II TA systems [104,114–116]. The chromosomal localization of these TA systems coupled with demonstrated toxicity in the absence of the antitoxin strongly implies these likely functioned as an addiction system for loss of the genetic material, and as are mentioned here in the "addiction" category rather than above with other chromosomal TA systems (see Table 1).

The idea of plasmid addiction, also referred to as PSK, is generally found with low copy number plasmids and is predicated on the shorter half-life of the neutralizing antitoxin, allowing daughter cells to inherit portions of the parental cytoplasmic material including more stable toxins (Figure 1b) [57,117–120]. While this is also feasible for phage to utilize and thus mediate infections, as recently described for the Pf1 prophage in *P. aeruginosa* [121,122], this is more commonly used by Type III TA systems and therefore Table 1 focuses on well characterized examples of plasmid-based Type II systems. When the host bacterial cell lacks a corresponding mechanism to neutralize the inherited toxin proteins, a negative impact on growth is realized. However, some bacteria encode chromosomal antitoxins that can neutralize the plasmidic counterparts, referred to as "anti-addiction" (see Section 2.3, below).

Type II TA systems are widespread on plasmids where they mediate either addiction or plasmidic competition [8,12,123–125]. Further, because TA systems select for plasmid maintenance, they also then contribute to spread of AMR [8,11]. For example, the pUM505 plasmid (an IncI-type) from *P. aeruginosa* contains a pathogenicity island with an encoded *pumAB* TA system (a RelBE homolog) and numerous resistance genes, including a ciprofloxacin-modifying enzyme CrpC [126]. Strains containing this plasmid were more virulent in both *Caenorhabditis elegans* and mouse model infections. A recent analysis of plasmids carried in *Klebsiella* strains found a strong association of Type II TA systems and resistance genes, with enrichment for ParE, ParE-like, CcdB, and Vap-type PIN domain toxins within IncA/C- and IncH-type plasmids [47].

A recent preprint report nicely summarizes TA systems found on different Inc plasmids in *Enterobacteriaceae* [5]. For *Klebsiella pneumonia* strains, they can contain up to 11 different Inc types of plasmids, with approximately half of these being IncF-types [5]. These included 27 different Type-II families with *ccdAB* and *pemIK* the most common, consistent with previous reports on the pOXA-48 IncL/M-type plasmid [5,127] as well as previous work on *E. coli*-derived plasmids [128]. Interestingly, when compared to their counterparts in *E. coli*, the *ccdAB* loci show greater sequence divergence, while the *pemIK* systems are relatively well conserved [5]. Previous reports on IncX-type plasmids noted enrichment for RelE/ParE-type TA systems [123].

The literature presents at least one well-documented case of a CcdAB xenolog in *E. coli* O157:H7 that is neutralized by its plasmidic counterpart from the F plasmid, whereas the plasmid-borne toxin is only neutralized by the same (cognate) plasmidic antitoxin [73]. These homologous systems co-exist stably in the population, perhaps mediating a reverse addiction wherein the plasmid is retained specifically to neutralize the chromosomally-integrated copy of the toxin. What is clear is that this type of addiction requires an antitoxin protein to interact with multiple different toxins, and that this standard PSK "addiction" model would not typically work if chromosomal antitoxin xenologs could cross-interact with plasmid-derived toxins, thus requiring a type of directional insulation for cross-interactions.

Table 1. Plasmidic TA systems characterized in the literature.

TA System	Plasmid	Host Bacteria	Chromosomal Homolog/Xenolog	Citations
ParDE	RK2 */RP4 *	*(Enterobacteriaceae) Escherichia coli*	ParDE	[55,79,129]
RelBE	P307, IncB/C	*E. coli*	RelBE	[130,131]
VapBC	IncF	*E. coli*		[131]
StbDE	R485	*Morganella morganii*		[132]
CcdAB, Kid/Kis (PemKI)	F, R1/R100, pCHP91	*E. coli, Erwinia chrysanthemi, Staphylococcus*	MazEF	[118,133–136]
PhD-Doc	Bacteriophage P1	*E. coli*		[137]
HigBA	pRTS1	*Proteus vulgaris*		[138]
Tad-ata	pAMI2	*Paracoccus aminophilus*		[139]
Axe-txe	pRUM	*Enterococcus faecalis*	YoeB-YefM	[140,141]
TasA-TasB	pGI1	*Bacillus thuringiensis*		[125]
VapBC	pMYSH6000	*Shigella flexneri*		[142]
AtaRT	pB171-like plasmid	*E.coli*	AtaRT	[143]
Omega-epsilon-zeta	pSM19035 *, pVER1/2	*Streptococcus pyogenes. Enterococcus faecium*	PezAT	[141,144]
RelBE, HigBA, HTH/GNAT, SplTA	p3ABAYE	*Acinetobacter baumannii*		[145]
YacAB	PWR100	*E. coli*		[146]

* Indicates a broad-host range plasmid.

2.3. *Type II TA Systems Mediating Anti-Addition through Antitoxin Cross-Interactions*

An alternative outcome to the Addiction model discussed in Section 2.2 is Anti-addiction, wherein chromosomally encoded TA systems protect the host bacteria against PSK mediated by their plasmid-encoded counterparts through cross-interactions (Figure 1c) [7,59,63]. Recent work highlights an analogous model for Tn3 transposons [4] and prophage sequences [64]. Other experiments have highlighted that PSK is not necessarily the driving force for retention of TA systems, and instead it is due to plasmid competition, such that a TA system with homologs carried on two plasmids would "compete" for cross-toxin neutralization, leading to "survival" or retention of the "winning" plasmid [147]. Additional support comes from observations that PSK does not actually result in complete sterilization of the culture, but rather a decreased viability that recovers over time [148].

Anti-addiction was clearly demonstrated for the *Erwinia chrysanthemi* chromosomal antitoxin CcdA, which interacts with and neutralizes the incoming F1 plasmidic CcdB toxin [59]. A similar modality is mediated by the phage-derived protein Dmd, which serves as an antitoxin for the RnlA and LsoA toxins in *E. coli* [149]. Anti-addiction is closely linked to "abortive infection" of phages, which center more on the Type IV type of TA system including ToxIN and AbiEI [31,35,150–152]. The presence of orphan antitoxins that appear to encode a protein, such as used in Type II TA systems, readily supports the anti-addiction model, but this remains to be demonstrated as a common usage of these orphans [13].

This model of TA system functions is dependent on the ability of antitoxin proteins to cross-react, such that they can neutralize the toxin on invading genetic material. Given the dogma of cognate toxin–antitoxin interactions, the current review will revisit examples of known non-cognate interactions (with some previously reviewed in [28]) as well as the feasibility of this occurring in selected systems.

3. Conservation of Type II TA System Folds and Cognate Antitoxin Interactions

Numerous experimentally determined structures are available for Type II TA systems, and these confirm that although sequence conservation is low, their structures are highly conserved and can be used to group them into superfamilies [28,51,153]. These classifications highlight the modular nature of toxin family interactions with different antitoxin families, consistent with propagation by horizontal gene transfer [28,154]. Antitoxins are the critical regulatory component for Type II systems, wherein they (typically) contain an N-terminal DNA binding motif used for autoregulation of the TA operon, and a less structured C-terminal domain to neutralize the toxin's activity. Selected systems that have the antitoxin domains reversed, such that the N-terminal region mediates toxin neutralization [89,155–159], while some antitoxins are limited to only the toxin-binding domain [160]. Antitoxins neutralize toxins by either by blocking the active site or by causing conformation changes that prevent toxin interaction with the cellular target [51,154,161].

3.1. Toxin Families Share Conserved Folds but Interact with Different Families of Antitoxins

The Pfam database utilizes Hidden Markov-models to associate similar protein families exhibiting a conserved fold or annotated function into "clans" [162]. When viewed for Type II TA systems, there are three large mostly toxin-containing clans, in addition to three common antitoxin-containing clans (Figure 2). A key feature of these is the mix-and-match nature of different antitoxin families with toxins from the same family, and vice versa. The toxin proteins are categorized into a PIN-type clan, encompassing Vap and Fit family toxins, a RelE/ParE-type clan, and a CcdB/PemK-type clan. Among these, the PIN-type VapBC systems and the RelBE-like systems appear to be the two largest families [49]. Some toxins are self-contained families that do not correlate into a larger clan (at least to date), including those involved in post-translation modifications such as the HEPN, FIC, and GNAT types (for these, see other recent reviews [7,12,27]). Previous work demonstrated the strong conservation of DNA binding motifs, generally a helix-turn-helix (HTH, see Figure 2) or ribbon-helix-helix (RHH, see "CL0057", Figure 2), in addition to the Abr-like DNA binding domain for different antitoxin families (Figure 2) [28,154,163].

The CL0280 group of toxins contains families that interact with either CL0132 or CL0057, depending on the toxin family. The toxins in this family all share a PIN domain that forms a compact RNA-binding with three highly conserved acidic residues required for metal-dependent endonuclease activity [153,164–166]. Within this fold, the VapC-type toxins found in *M. tuberculosis*, *Shigella flexneri* and *Rickettsia felis* have minor variations in numbers of specific secondary structure elements [167,168]. This fold also encompasses the FitB toxin family [169]. While the VapC and FitB toxins share structural homology, their cognate antitoxins, VapB and FitA, are located in different Pfam clans. These mediate similar yet distinct interactions their cognate toxins, with distinct structures at the more N-terminal part of FitA as compared to the C-terminus of VapB (red versus tan ribbons, Figure 3a). The VapB antitoxin is part of the RHH antitoxin family (CL0057), while the FitA antitoxins more closely correlate with the Mnt-like repressors (CL0132). Within this family, the conserved toxin fold thus interacts with antitoxins from two different structural Pfam "clans" [168–170].

Overall, toxins segregated in the CL0136 group are paired with distinct antitoxin families in the CL0057 group with a few notable exceptions. This toxin group is comprised of the RelE/ParE family of toxins containing a shared microbial RNase fold but variability in the specific active site amino acids as well as extensions at the C-terminus [13,171,172]. While most of the toxin members mediate RNA cleavage, as recently reviewed [13], the ParE toxins are unique in inhibiting DNA gyrase through an as yet unknown mechanism [71,80,173]. Despite the shared fold and mechanism, subfamilies appear to exist with both ParE-types and RelE-types, including a wide range of RelE-like RNases including HigB, YoeB, YafQ, BrnT, and MqsR (Figure 2) [27,171,172,174]. This family also includes an integrated phage-derived tripartite ParE system that appears to not mediate RNA cleavage or DNA gyrase inhibition [160,175]. Interactions with antitoxins span five different specific families in CL0057, as well as antitoxin members of CL0136 and CL0123 (Figure 2) [1,51,174]. These interactions are mediated by analogous surfaces among these distinct families, yet with distinct sequence differences and, in particular, the ParD antitoxins have longer C-terminal regions versus the RelB-type antitoxins (Figure 3b).

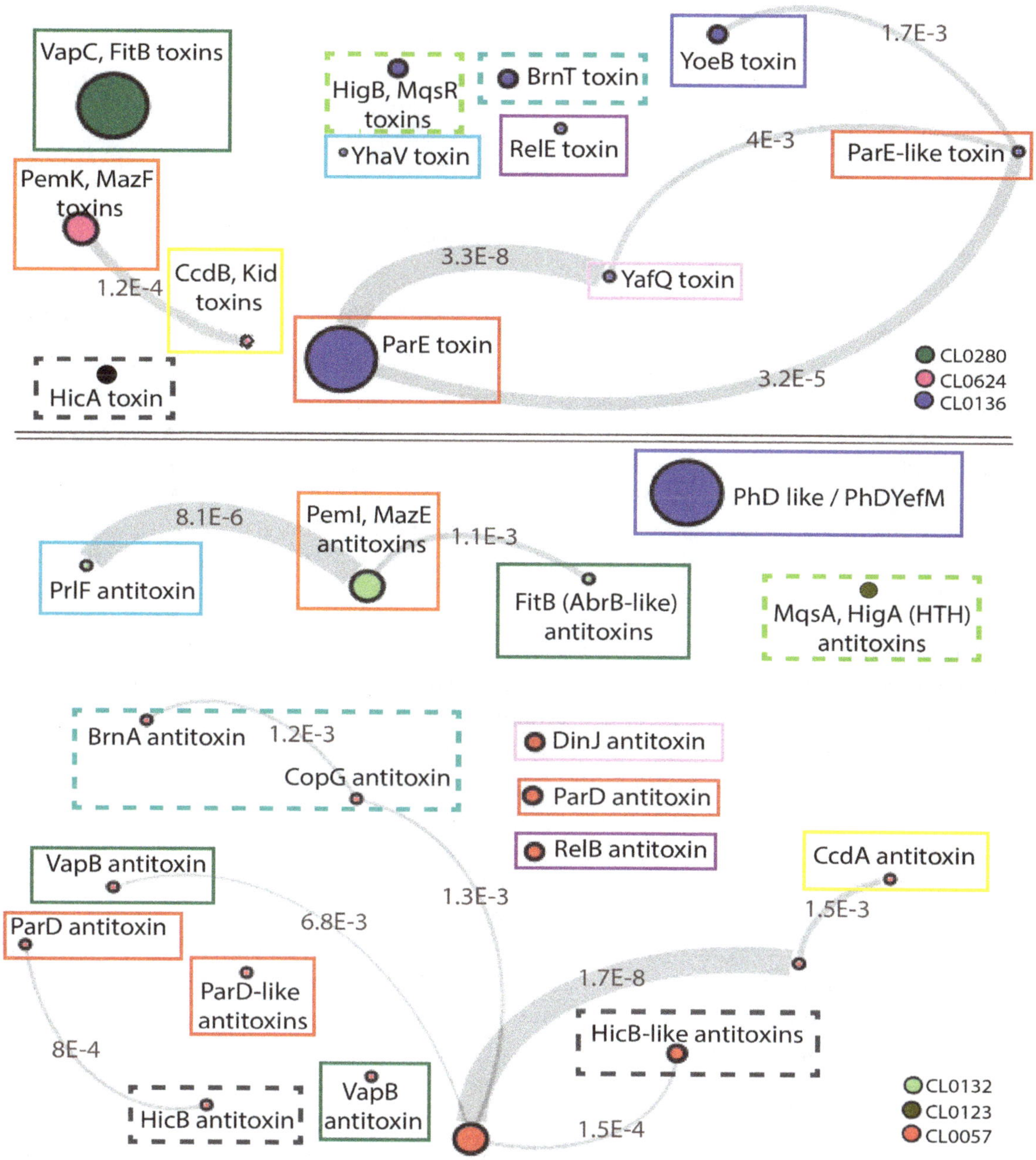

Figure 2. TA families can be grouped into Pfam "clans" (here connected by grey lines) based on Hidden Markov models, although toxin families can interact with antitoxin families from disparate clans. The force-directed diagram indicates the families that toxins and antitoxins belong to providing a predictive effect for noncognate toxin-antitoxin interaction. Patterns are adapted from the Pfam database, where the size of the circle corresponds to the number of sequences within a given family while the color corresponds to the clan. Colored boxes (dashed or solid lines) indicate cognate toxin (top) and antitoxin (bottom) pairs. The linkage between each circle indicates the E value between related families, with a lower limit of 10^{-3} to be considered significant.

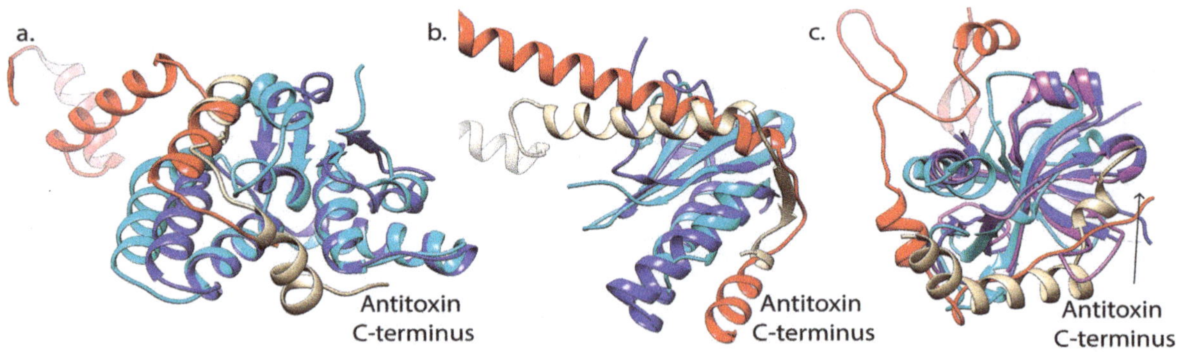

Figure 3. Shared toxin structures are neutralized by distinct antitoxin family structures, with variation evident in the C-terminal ends of the antitoxins. (**a**) Superposition of VapC (cyan) and FitB (blue) toxins reveals an overlap in interaction surfaces with antitoxin VapB (tan) and FitA (red). (PDB IDs 3DBO [167] and 2H1O [169]) (**b**) Superposition of RelE (cyan) and ParE (blue) toxins reveals an overlap in interaction surfaces with antitoxin RelB (tan) and ParD (red). (PDB IDs 4FXE [176] and 3KXE [177]) (**c**) Superposition of Kis toxin (light blue) with MazF (cyan) and CcdB (blue) toxins reveals an overlap in interaction surfaces with antitoxin MazE (tan) and CcdA (red). (PDB IDs 1M1F [178], 1UB4 [179], and 3HPW [180]).

The CL0624 toxin group similarly interacts with antitoxins in either CL0132 or CL0057. This Pfam "clan" of toxins includes the CcdAB, Kis-Kid, PemIK, and MazEF Type II TA systems. CcdB toxins act by inhibiting DNA gyrase, although using distinct mechanisms as the ParE-type toxins [181–183]. CcdB toxins have a striking structural similarity with the toxins Kid and PemK, which are endoribonuclease encoded by the Kis-Kid (*parD*) TA system found on the R1 plasmid and the PemIK system found on the R100 plasmid [133–135,184,185]. This structural family is further expanded by PemIK xenologs in *Bacillus anthracis* [186] and *E. coli* (named ChpAB) [187]. The MazF toxin is a structurally homologous RNase, although its interaction with the antitoxin is distinct among this clan (Figure 2) [179]. The MazE antitoxin consisting of a looped–hinge–helix (LHH) fold, the N-terminal region of Kis antitoxin has a unique LHH fold [188], while the CcdA antitoxin has an RHH fold [189]. It is clear that cross-interactions with non-cognate toxins in theory could occur, as the antitoxins are classified based on the DNA binding domain rather than the toxin-binding domain, and the toxin surfaces complexed with antitoxin are largely overlapping (Figure 3c).

3.2. Interactions with Cognate Antitoxins

For type II TA systems, the neutralization between the toxin and the antitoxin mediate direct protein-protein interaction. Under normal condition, the antitoxin neutralizes its cognate toxin as well as its expression to prevent its toxicity, whereas under environmental stressors, the antitoxins are believed to follow a common proteolytic degradation to release toxins, allowing it to kill cells or return cells to a dormant state [12,154,163,190]. The insulation within homologous systems has been well documented [190–192], such that the direct protein-protein interactions of toxins and antitoxins, the central mechanism of toxin control for Type II TA systems, appear to be highly specific [175,193,194].

This is exemplified within *Caulobacter crescentus*, which contains four chromosomal RelBE systems in addition to three functional ParDE systems; while these toxin families have structural similarity, they are neutralized by different classes of antitoxins [70,154]. Using a deletion approach, each cognate pair was demonstrated to have no cross-reactivity by virtue of a lack of survival when cognate antitoxins were deleted [70]. Similarly, seventeen TA systems in *V. cholera* were demonstrated to have no cross-reactivity of antitoxins [195]. Further, the relatively unique tripartite systems paaA-ParE found in integrated prophage regions in *E. coli* O157:H7 also do not cross-react [175].

M. tuberculosis encodes up to 55 different cognate VapBC systems, which were the subject of a recent study that, guided by available crystal structures, made predictions on the amino acids in the interface of each cognate pair [196]. They were able to identify sub-clusters within both toxins and antitoxins that were more likely to contain cross-interacting pairs. These predictions corroborate previous work that experimentally demonstrated high insulation thus limited crosstalk between these systems in different Vap sub-clusters [197]. Within a sub-cluster, however, possibility for cross-talk increases [196], and has previously been predicated on the identity of the C-terminal (30 amino acids) of at least a few of these VapB antitoxins [198]

4. Feasibility for Cross-Interactions of Type II TA Systems

A recent review provided a concise view of the models for TA system functions [12], and we will attempt to not repeat those here. The Anti-Addiction/Plasmid Competition model presents an attractive explanation for the long-sought functional significance of chromosomal TA systems. Based on this, we can draw the following suppositions that should be fulfilled for this model: That the chromosomal TA system will have a "match" on some invading genetic material (transposons, phage, integrons, and plasmids), and, that the chromosomal antitoxin will match the invading toxin well enough to neutralize it.

Instead, our intent is to more closely examine some of these ideas from the structural and molecular standpoint, and in particular, is predicated on the feasibility of antitoxin cross-reactions required to fulfill prevailing models of TA system addiction and anti-addiction functions.

4.1. Examples of Antitoxin Cross-Reactivity

Many TA systems, despite having low sequence similarity, exhibit a similar folding structure [25, 153,154,199]. Typically these structures enforce interactions limited to cognate partners, providing insulation from cross-reactivity. However, some examples of cross-interactions have been noted to occur both between chromosomal and plasmid-borne TA systems.

Such complex cross-regulations were observed between three different *M. tuberculosis* RelBE-like modules, RelBE (Rv1246c-Rv1247c), RelFG (Rv2865-Rv2866) and RelJK (Rv3357-Rv3358). Using in vitro and cell survival assays, it was demonstrated that the RelB antitoxin can cross-neutralize the non-cognate RelG toxin, but RelB can also enhance the toxicity of the RelK toxin in cell survival assays, although the molecular basis for this remains unclear. On the other hand, RelF, the antitoxin of RelG, is able to enhance the toxicity of RelE which causes severe inhibition on bacterial growth compared to the set only expressing RelE [191]. A similar form of cross-interaction was noted for two RelB proteins encoded in the *Y. pestis* CO92 genome, although these only differ by three amino acids [200]. Cross-reactivity has been documented for the CcdAB system carried on the F plasmid [59,73]. The *Erwinia chrysanthemi* chromosomal CcdA antitoxin interacts and neutralizes the incoming F1 CcdB toxin that would otherwise kill the cell [59]. Subsequent studies noted that amino acid changes Asn 69 to Tyr in the chromosomal CcdA antitoxin and the plasmid-derived Tyr 8 Arg in the CcdB toxin affects cross-interactions, resulting in weak plasmid-derived toxin binding to chromosomal antitoxin [68,184].

Interestingly, the Kid and MazE antitoxins are able to mutually interact and partially neutralize the toxicity mediated by the non-cognate family [188]. Similarly, the CcdA and Kis antitoxins are able to cross-interact with non-cognate toxins from non-cognate family members (Kid and CcdB, respectively) [176]. CcdA binding enhances the endoribonuclease activity of Kid by triggering a conformational change that promotes interaction with its target RNA, while the Kis antitoxin effectively neutralizes the toxicity of the CcdB toxin [184]. The different effect of antitoxins on the non-cognate toxins results from their overlapping yet distinct binding sites along the toxin (Figure 3c), and as well as from potential differences in their DNA binding regions [184,188,189].

The VapBC systems in *M. tuberculosis* are highly specific for their cognate pairs; a given VapB antitoxin is not able to neutralize a non-cognate VapC toxin [190]. This was also demonstrated for different ParDE families that interact in a highly specific manner [161]. However, both studies identify the determinants for antitoxin recognition as well as the specificity and insulation of crosstalk between different TA systems. Mutating a single tryptophan amino acid in the VapB1 antitoxin from non-typeable *Haemophilus influenza* renders it to antagonize both its cognate VapC1 toxin and its non-cognate VapC2 toxin [190]. In the ParDE family, switching antitoxin residues 60, 61 and 64 in ParD3 is sufficient to alter the specificity from the cognate ParE3 to non-cognate ParE2 [161]. Those results suggested the possibility of breakage to the insulation of crosstalk and specificity between different TA systems.

Other experiments were able to generate lab-derived cross-reaction of antitoxins. Chromosomal antitoxins MazF and ChpB were mutated and constructs were selected for their ability to then neutralize plasmid-derived PemK toxin located on R100/R1 plasmids [201]. Similarly, using chemical mutagenesis of the ChpBI system that then selected for cross-neutralization against the Kis toxin and noted this mutated version could also still neutralize the ChpK toxin [201]. Chromosomal MazE is a homolog of Kis on the R1 plasmid, and can neutralize the plasmid-derived Kid although the interaction is weaker than with the cognate pair. [188,202] *M. tuberculosis* encodes seven annotated MazEF systems and two additional MazE antitoxin homologs of the *E. coli* systems, although these are not paired with the normal cognate MazF toxin [203]. They identified a "network" of non-cognate interactions, such that one of the tested MazE antitoxins interacts with two different MazF toxins. Another set of non-cognate interactions is completely reciprocated, such that a MazE antitoxin can interact with non-cognate VapC type toxins, and their cognate VapB antitoxins interact with the cognate MazF toxin as well as the non-cognate VapC toxin [203].

4.2. Orphan Antitoxins

Given the widespread nature of TA systems, it is not surprising that many partial systems are annotated. However, antitoxins that lack a cognate toxin may be particularly important for Anti-Addiction functions, as they can provide a source of toxin neutralization independent of any inherent addiction properties themselves. Orphan antitoxins have been noted in bioinformatics studies of bacterial genomes, including pathogens associated with high incidences of antibacterial resistance. The *Bartonella schoenbuchensis* type-IV secretion system is encoded on a conjugative plasmid, pVbh, where it encodes fourteen canonical TA systems as well as four orphan antitoxins [104]. While the TA systems likely mediate PSK, which is an addiction function, the orphan antitoxins are inferred to likely function as anti-addiction modules.

A recent study analyzed the genome sequences of 259 species of *Klebsiella pneumonia* complex strains and was able to predict up to 2253 orphan antitoxins [47]. These were proposed to encode remnants of degraded TA systems, or to be regulators of other (unidentified) TA pairs, or as anti-addiction modules to prevent foreign genetic material from being retained in the cell [47]. The sequences surrounding these orphan antitoxins were screened for similarities to known TA system arrangements, which revealed a high percentage were likely to be degraded from previously intact canonical systems. However, around 20% of those identified appeared to be genuine orphan genes, and further, around half of these open reading frames encoded a protein with canonical features of antitoxins [47].

A similar study mined *Mycobacterium tuberculosis* (Mt) genome sequences and identified both VapB orphan antitoxins and VapC orphan toxins [196]. The orphan toxins were more closely related to other paralogous toxins, whereas the orphan antitoxins were not similar to other known antitoxins. *Acinetobacter baumannii* also encodes numerous orphan antitoxins, as well as pairings of canonical antitoxin or toxins with non-TA system proteins [145]. A subset of the Mt orphan toxins is closely related to TA systems in *Mycobacterium marinum,* implying horizontal gene transfer between the two organisms [196]. However, variations in the active site residue raised questions about the potential for these orphan toxins to remain active [196]; however, it is not clear if these amino acid changes would impact interactions with antitoxins. Similarly, a *parDE* loci in *Caulobacter crescentus* encodes a *parE* pseudogene, resulting in a ParD antitoxin with no apparent mate, technically an "orphan" [70]. A TA system found in *Shigella flexneri* encodes a non-functional toxin, YacB, due to a frameshift mutation causing a premature stop codon [146]. The cognate antitoxin, Orf176, has previously been identified as the YacA antitoxin paired with a YacB toxin on *E. coli* plasmid pWR100 [146]. While it can be considered an "orphan" antitoxin, interestingly it is one of three such orphans that were validated substrates for type-III secretion in *Shigella* [204].

4.3. Predictions of Antitoxin Cross-Reactivity

The idea of cross-interactions in anti-addiction (Figure 1c) is predicated on the ability of one partner to interact with multiple others; the likelihood of this increases as structures or sequences are more conserved. We undertook an examination of 14 available structures, identified using tools at the Protein Databank interface [205,206], for chromosomal RelE-type toxin interactions with their corresponding antitoxins to identify sequence conservation that may indicate the ability for cross-reactions [154,159,174,176,207–216]. However, we note that within this set, the RelBE complex from *E. coli* is represented by two structures (PDB IDs 2KC8 and 4FXE, [176,207], respectively), as is the Doc toxin from the P1 bacteriophage (PDB IDs 3DD7 and 3KH2, [154,210], respectively), limiting the structure set to twelve unique complexes. The overall secondary structure and interaction of cognate RelB-like antitoxins are conserved, with two α-helices separated by a β strand and wherein this β strand typically pairs with a strand from the toxin to form an extended cross-molecule β-sheet.

We utilized the PISA webserver [217] to list the interacting amino acids, visualized these with UCSF Chimera [218], and then inspected the different interactions by eye. Within these structures we identified two sets of antitoxins that have very similar sequences at the toxin binding sites (Figure 4a). In particular, the RelJ antitoxin from *M. tuberculosis* and the *E. coli* YefM antitoxin have high sequence similarity at the region of interaction with toxins (residues 39–77 of RelJ and residues 51–89 of *E. coli* YefM) at 56.4%(or 22 out of 39 amino acids). Further, the distribution of polar versus hydrophobic amino acids at the antitoxin–toxin interfaces are also highly correlated (Figure 4b,c). We also identified that the sequences of *E. coli* HigA and *S. flexneri* HigA are identical throughout the toxin interactions regions. Overall, the complexes adopt the same structure with some minor differences in toxin loops visualized in the crystal structures (PDB IDs 4FXE and 2KC8, [159,216]). Not surprisingly, the HigB toxins from these two organisms are identical [159,216]. These examples lend some support to the idea that a structure-based approach should be able to predict cross-interacting pairs.

With a similar objective in mind, we undertook a search of phage-derived sequences to assess any potential cross-reactive TA systems as compared to *M. tuberculosis*, such as would be expected for an anti-addiction function of chromosomal systems (Figure 1). The toxin and antitoxin sequences for *M. tuberculosis* (Mtb) strain H37Rv were obtained from the toxin-antitoxin database and included the RelBE, HigBA, and VapBC families [49]. The sequences were used as a query to search the Actinobacteriophage (previously "Mycobacteriophage") Database that contains approximately 3400 sequenced phage, of which 1900 are known to originate from Mycobacterial hosts (https://phagesdb.org; [219]). The resulting BlastP resulting sequences were surprisingly poorly matched; as such, any with an E-value of 0.10 or less were aligned with the H37Rv toxin or antitoxin to which a similarity was indicated [176,213,220,221].

Among the antitoxin sequences, Mtb VapB8, VapB13, and VapB40 all had a potential match with the phage database; however, each had only one conserved amino acid likely to be in the interface. Two HigA antitoxins (Rv2021c and Rv3183) presented sequence matches in the E 10^{-3} to E 10^{-4} range, and similarities were limited to approximately 30 amino acids at the toxin-binding interface. For Rv3183, also known as HigA3, 18 of 31 amino acids present an exact match while an addition four are conserved; further, this antitoxin sequence is derived from a *Mycobacterium*-derived phage. However, this appears to be an orphan antitoxin, as the annotated open reading frames on either side of this gene do not contain similarity to the HigB (or any other) toxin. *M. tuberculosis* HigA2, annotated as Rc2021c, matches a sequence derived from a phage originating in Propionibacterium. Similar to the HigA3 antitoxin, this HigA2 antitoxin contains 22 identical and 12 similar amino acids out of 53 total, although these matches are more central to the protein so would be expected to span the toxin binding domain and the DNA binding domain. While it contains a reasonably-sized open reading frame in the toxin position (150 amino acids), the encoded protein failed to match any known sequences in the TADB or any named protein in a Blast search at the NCBI (all were "hypothetical"), indicating perhaps this is a novel toxin or an orphan HigA antitoxin embedded within a different genetic context.

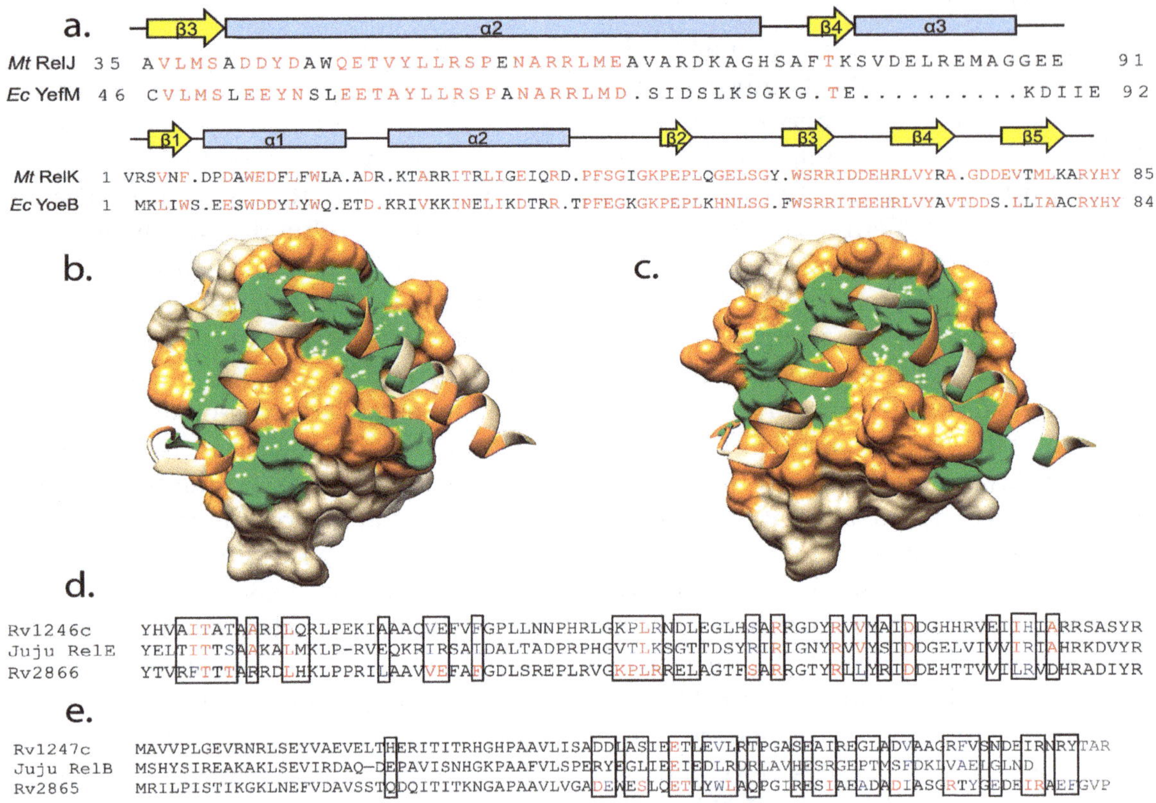

Figure 4. Potential cross-interactions of RelB/YefM antitoxins identified from analysis of sequence and structural conservation among available experimentally determined structures. (**a**) Sequence alignments (top, antitoxins; bottom, toxins) depict conserved amino acids (red), with secondary structure indicated above. (**b**) Surface of *Mycobacteria tuberculosis* toxin RelK with ribbon backbone of *M. tuberculosis* antitoxin RelJ (C-terminus is to the right); coloring is orange for hydrophobic interaction points, and green for polar interaction points. (PDB ID 3OEI, [212]). (**c**) Surface of *E. coli* toxin YoeB with ribbon backbone of *E. coli* antitoxin YefM; coloring as in b. (PDB ID 2A6Q, [211]). (**d**) Sequence alignment for Rv1246c, a RelE toxin found in the chromosome of *M. tuberculosis* and exhibiting similarity to a phage-derived sequences. This is compared to Rv2866, another *M. tuberculosis* RelE toxin with an available crystal structure (PDB ID 3G50, [212]), which was used to delineate likely protein–protein contacts for the RelE toxin and RelB antitoxin (black boxes). Red text indicates conservation, while blue indicates similarity. Note, Rv1246c is colored relative to the phage derived sequence. (**e**) The same type of sequence analysis was carried out for the respective RelB antitoxin sequences (colored and labeled as in d).

Three VapC toxins (VapC5, VapC28, and VapC42) had an E-value of 0.10 or less with the matching phage sequences. The most successful homologs were to an experimentally verified RelE toxin (Rv1246c) with an E-value to the best-matched sequence of 2 E 10^{-5}. This sequence originated from the "Juju" phage, with 33% identity and 50% conservation, as well as strong matches with 33 other phage entries [191,222]. Among these phage, the top twelve matches originated from a Gordonia host, while the remainder are from *Mycobacterium smegmatis*. Each of these phage (where genomes were annotated) contains an open reading frame for a RelB antitoxin just preceding the RelE-type toxin.

These sequences were analyzed for conservation of interaction sites based on the closest matching RelBE complex, the RelBE2 system (Rv2865-2866, PDB ID 3G5O, [212]). Assuming a conservation of interaction sites between RelE toxin and RelB antitoxin, at these sites of interaction the Rv1246c RelE toxin and the Juju phage RelE toxin have the same amino acids at 11 positions, conserved amino acids at 7 positions, and different amino acids at the remaining 9 positions (Figure 4d). When comparing the same chromosomal RelE toxin to its closest match in the PDB (Rv2866) there are 15 identical amino acids, 3 conserved, and 9 different amino acids at the interface (PDB ID 3G5O, [212]).

Given that the match was based on toxin sequences, we speculated that the corresponding RelB antitoxin sequences would be similar. The same type of analysis revealed limited conservation of antitoxins, with identical amino acids at one position, conserved amino acids at only 3 positions, and different amino acids at the remaining 22 positions (Figure 4e). As a marker of comparison, when the chromosomal Rv1246c antitoxin was compared to the crystal structure of Rv2865 there are 10 identical amino acids, 8 conserved, and 8 different amino acids at the toxin-contacting interface (PDB ID 3G5O, [212]).

This leaves an open question of—do toxins with similar antitoxin-interaction amino acids cross-react with disparate antitoxin sequences? Or, another way to phrase this might be, could the *M. tuberculosis* chromosomal RelB antitoxin (Rv1247c) neutralize an invading phage RelE toxin, and in so doing mediate anti-addiction? Further, the recovery of the phage from related Gordonia bacterial species could indicate an anti-addiction function when this chromosomal matching RelE toxin is present, or, simply could represent a sampling bias for environmental phage collection comprising the sequence database [219].

5. Discussion

TA systems are abundant on bacterial chromosomes with a seemingly high insulation from non-homologous interactions. To date, such cross-interactions have been detected when changes to either partner accumulate. Given that the premise of anti-addition offered by chromosomal systems towards mobile genetic elements requires cross-reactivity, it seems likely that it would be present in some integrated chromosomal systems as well. However, only limited cross-reactions have been observed between chromosomal and mobile genetic elements systems, perhaps due to bias in the type of experiments and functions of TA systems pursued.

It is not possible to reliably predict cross-interactions based on the currently available molecular characterizations, particularly if only a few amino acid changes are required in either partner. It may be that the natural variation that leads to such cross-reactivity does not follow such a minimalistic approach, with instead many more mutations providing a cumulative basis for an anti-addiction paradigm. It is clear that a more comprehensive understanding and screening for cross-interactions of non-cognate TA system partners would provide welcome insights into the functional possibilities for these intriguing systems.

Acknowledgments: We thank Kevin Snead for helpful input during the early stages of this manuscript.

References

1. Anantharaman, V.; Aravind, L. New connections in the prokaryotic toxin-antitoxin network: Relationship with the eukaryotic nonsense-mediated RNA decay system. *Genome Biol.* **2003**, *4*, R81. [CrossRef] [PubMed]
2. Pandey, P.D.; Gerde, K.S. Toxin-antitoxin loci are highly abundant in free-living but lost from host-associated prokaryotes. *Nucleic Acids Res.* **2005**, *33*, 966–976. [CrossRef]
3. Melderen, V.L.; Jurenas, D.; Garcia-Pino, A. Messing up translation from the start: How AtaT inhibits translation initiation in *E. coli*. *RNA Biol.* **2018**, *15*, 303–307. [CrossRef]

4. Lima-Mendez, G.; Alvarenga, D.O.; Ross, K.; Hallet, B.; Van Melderen, L.; Varani, A.M.; Chandler, M. Toxin-Antitoxin Gene Pairs Found in Tn3 Family Transposons Appear To Be an Integral Part of the Transposition Module. *mBio* **2020**, *11*, e00452-20. [CrossRef] [PubMed]
5. Wu, Y.A.; Kamruzzaman, M.; Iredell, J.R. Specialised functions of two common plasmid mediated toxin-antitoxin systems, ccdAB and pemIK, in Enterobacteriaceae. *bioRxiv* **2020**. [CrossRef]
6. Huguet, K.T.; Gonnet, M.; Doublet, B.; Cloeckaert, A. A toxin antitoxin system promotes the maintenance of the IncA/C-mobilizable Salmonella Genomic Island 1. *Sci. Rep.* **2016**, *6*, 32285. [CrossRef] [PubMed]
7. Makarova, S.K.; Wolf, Y.I.; Koonin, E.V. Comprehensive comparative-genomic analysis of type 2 toxin-antitoxin systems and related mobile stress response systems in prokaryotes. *Biol. Direct* **2009**, *4*, 19–56. [CrossRef] [PubMed]
8. Fernández-García, L.; Blasco, L.; Lopez, M.; Bou, G.; García-Contreras, R.; Wood, T.; Tomas, M. Toxin-antitoxin systems in clinical pathogens. *Toxins (Basel)* **2016**, *8*, 227–249. [CrossRef]
9. Kedzierska, B.; Hayes, F. Emerging Roles of Toxin-Antitoxin Modules in Bacterial Pathogenesis. *Molecules* **2016**, *21*, 790. [CrossRef]
10. Ramisetty, M.B.C.; Santhosh, R.S. Endoribonuclease type II toxin-antitoxin systems: Functional or selfish? *Microbiology* **2017**, *163*, 931–939. [CrossRef]
11. Yang, E.Q.; Walsh, T.R. Toxin-antitoxin systems and their role in disseminating and maintaining antimicrobial resistance. *FEMS Microbiol. Rev.* **2017**, *41*, 343–353. [CrossRef] [PubMed]
12. Fraikin, N.; Goormaghtigh, F.; van Melderen, L. Type II Toxin-Antitoxin Systems: Evolution and Revolutions. *J. Bacteriol.* **2020**, *202*, e00763-19. [CrossRef] [PubMed]
13. Jurėnas, D.; van Melderen, L. The Variety in the Common Theme of Translation Inhibition by Type II Toxin–Antitoxin Systems. *Front. Genet.* **2020**, *11*, 262. [CrossRef]
14. Hall, M.A.; Gollan, B.; Helaine, S. Toxin-antitoxin systems: Reversible toxicity. *Curr. Opin. Microbiol.* **2017**, *36*, 102–110. [CrossRef]
15. Page, R.; Peti, W. Toxin-antitoxin systems in bacterial growth arrest and persistence. *Nat. Chem. Biol.* **2016**, *12*, 208–214. [CrossRef]
16. Ronneau, S.; Helaine, S. Clarifying the Link between Toxin-Antitoxin Modules and Bacterial Persistence. *J. Mol. Biol* **2019**, *431*, 3462–3471. [CrossRef]
17. Unterholzner, J.S.; Poppenberger, B.; Rozhon, W. Toxin-antitoxin systems: Biology, identification, and application. *Mob. Genet. Elem.* **2013**, *3*, e26219. [CrossRef]
18. Schuster, F.C.; Bertram, R. Toxin-antitoxin systems are ubiquitous and versatile modulators of prokaryotic cell fate. *FEMS Microbiol. Lett.* **2013**, *340*, 73–85. [CrossRef]
19. Beyer, M.H.; Iwai, H. Off-Pathway-Sensitive Protein-Splicing Screening Based on a Toxin/Antitoxin System. *ChemBioChem* **2019**, *20*, 1933–1938. [CrossRef] [PubMed]
20. Klimina, K.M.; Kasianov, A.S.; Poluektova, E.U.; Emelyanov, K.V.; Voroshilova, V.N.; Zakharevich, N.V.; Kudryavtseva, A.V.; Makeev, V.J.; Danilenko, V.N. Employing toxin-antitoxin genome markers for identification of Bifidobacterium and Lactobacillus strains in human metagenomes. *PeerJ* **2019**, *7*, e6554. [CrossRef] [PubMed]
21. Kang, S.M.; Kim, D.H.; Jin, C.; Lee, B.J. A Systematic Overview of Type II and III Toxin-Antitoxin Systems with a Focus on Druggability. *Toxins (Basel)* **2018**, *10*, 515. [CrossRef] [PubMed]
22. Równicki, M.; Pieńko, T.; Czarnecki, J.; Kolanowska, M.; Bartosik, D.; Trylska, J. Artificial Activation of Escherichia coli mazEF and hipBA Toxin-Antitoxin Systems by Antisense Peptide Nucleic Acids as an Antibacterial Strategy. *Front. Microbiol.* **2018**, *9*, 2870. [CrossRef]
23. Lioy, V.S.; Rey, O.; Balsa, D.; Pellicer, T.; Alonso, J.C. A toxin-antitoxin module as a target for antimicrobial development. *Plasmid* **2010**, *63*, 31–39. [CrossRef]
24. Fedorec, A.J.; Ozdemir, T.; Doshi, A.; Ho, Y.K.; Rosa, L.; Rutter, J.; Velazquez, O.; Pinheiro, V.B.; Danino, T.; Barnes, C.P. Two New Plasmid Post-segregational Killing Mechanisms for the Implementation of Synthetic Gene Networks in Escherichia coli. *iScience* **2019**, *14*, 323–334. [CrossRef] [PubMed]
25. Lee, Y.K.; Lee, B.J. Structure, Biology, and Therapeutic Application of Toxin-Antitoxin Systems in Pathogenic Bacteria. *Toxins (Basel)* **2016**, *8*, 305. [CrossRef] [PubMed]
26. Magnuson, R.D. Hypothetical functions of toxin-antitoxin systems. *J. Bacteriol.* **2007**, *189*, 6089–6092. [CrossRef]

27. Harms, A.; Brodersen, D.E.; Mitarai, N.; Gerdes, K. Toxins, Targets, and Triggers: An Overview of Toxin-Antitoxin Biology. *Mol. Cell* **2018**, *70*, 768–784. [CrossRef]
28. Goeders, N.; van Melderen, L. Toxin-antitoxin systems as multilevel interaction systems. *Toxins (Basel)* **2014**, *6*, 304–324. [CrossRef]
29. Fozo, M.E.; Hemm, M.R.; Storz, G. Small Toxic Proteins and the Antisense RNAs That Repress Them. *Microbiol. Mol. Biol. Rev.* **2008**, *72*, 579–589. [CrossRef]
30. Blower, R.T.; Salmond, G.P.; Luisi, B.F. Balancing at survival's edge: The structure and adaptive benefits of prokaryotic toxin-antitoxin partners. *Curr. Opin. Struct. Biol.* **2011**, *21*, 109–118. [CrossRef]
31. Fineran, P.C.; Blower, T.R.; Foulds, I.J.; Humphreys, D.P.; Lilley, K.S.; Salmond, G.P. The phage abortive infection system, ToxIN, functions as a protein–RNA toxin–antitoxin pair. *Proc. Natl. Acad. Sci. USA* **2009**, *106*, 894. [CrossRef] [PubMed]
32. Samson, J.E.; Spinelli, S.; Cambillau, C.; Moineau, S. Structure and activity of AbiQ, a lactococcal endoribonuclease belonging to the type III toxin-antitoxin system. *Mol. Microbiol.* **2013**, *87*, 756–768. [CrossRef] [PubMed]
33. Blower, T.R.; Short, F.L.; Rao, F.; Mizuguchi, K.; Pei, X.Y.; Fineran, P.C.; Luisi, B.F.; Salmond, G.P. Identification and classification of bacterial Type III toxin-antitoxin systems encoded in chromosomal and plasmid genomes. *Nucleic Acids Res.* **2012**, *40*, 6158–6173. [CrossRef] [PubMed]
34. Goeders, N.; Chai, R.; Chen, B.; Day, A.; Salmond, G.P. Structure, Evolution, and Functions of Bacterial Type III Toxin-Antitoxin Systems. *Toxins (Basel)* **2016**, *8*, 282. [CrossRef] [PubMed]
35. Dy, R.L.; Przybilski, R.; Semeijn, K.; Salmond, G.P.; Fineran, P.C. A widespread bacteriophage abortive infection system functions through a Type IV toxin-antitoxin mechanism. *Nucleic Acids Res.* **2014**, *42*, 4590–4605. [CrossRef]
36. Masuda, H.; Tan, Q.; Awano, N.; Wu, K.P.; Inouye, M. YeeU enhances the bundling of cytoskeletal polymers of MreB and FtsZ, antagonizing the CbtA (YeeV) toxicity in Escherichia coli. *Mol. Microbiol.* **2012**, *84*, 979–989. [CrossRef]
37. Masuda, H.; Tan, Q.; Awano, N.; Yamaguchi, Y.; Inouye, M. A novel membrane-bound toxin for cell division, CptA (YgfX), inhibits polymerization of cytoskeleton proteins, FtsZ and MreB, in Escherichia coli. *FEMS Microbiol. Lett.* **2012**, *328*, 174–181. [CrossRef]
38. Wang, X.; Lord, D.M.; Cheng, H.Y.; Osbourne, D.O.; Hong, S.H.; Sanchez-Torres, V.; Quiroga, C.; Zheng, K.; Herrmann, T.; Peti, W.; et al. A new type V toxin-antitoxin system where mRNA for toxin GhoT is cleaved by antitoxin GhoS. *Nat. Chem. Biol.* **2012**, *8*, 855–861. [CrossRef]
39. Barbosa, L.C.B.; dos Santos Carrijo, R.; da Conceição, M.B.; Campanella, J.E.M.; Júnior, E.C.; Secches, T.O.; Bertolini, M.C.; Marchetto, R. Characterization of an OrtT-like toxin of Salmonella enterica serovar Houten. *Braz. J. Microbiol.* **2019**, *50*, 839–848. [CrossRef]
40. Islam, S.; Benedik, M.J.; Wood, T.K. Orphan toxin OrtT (YdcX) of Escherichia coli reduces growth during the stringent response. *Toxins (Basel)* **2015**, *7*, 299–321. [CrossRef]
41. Aakre, C.D.; Phung, T.N.; Huang, D.; Laub, M.T. A bacterial toxin inhibits DNA replication elongation through a direct interaction with the beta sliding clamp. *Mol. Cell* **2013**, *52*, 617–628. [CrossRef]
42. Gabrisko, M.; Barak, I. Evolution of the SpoIISABC Toxin-Antitoxin-Antitoxin System in Bacilli. *Toxins (Basel)* **2016**, *8*, 180. [CrossRef]
43. Jimmy, S.; Saha, C.K.; Kurata, T.; Stavropoulos, C.; Oliveira, S.R.A.; Koh, A.; Cepauskas, A.; Takada, H.; Rejman, D.; Tenson, T.; et al. A widespread toxin-antitoxin system exploiting growth control via alarmone signaling. *Proc. Natl. Acad. Sci. USA* **2020**, *117*, 10500–10510. [CrossRef] [PubMed]
44. Sala, A.; Calderon, V.; Bordes, P.; Genevaux, P. TAC from Mycobacterium tuberculosis: A paradigm for stress-responsive toxin-antitoxin systems controlled by SecB-like chaperones. *Cell Stress Chaperones* **2013**, *18*, 129–135. [CrossRef] [PubMed]
45. Wilcox, B.; Osterman, I.; Serebryakova, M.; Lukyanov, D.; Komarova, E.; Gollan, B.; Morozova, N.; Wolf, Y.I.; Makarova, K.S.; Helaine, S.; et al. Escherichia coli ItaT is a type II toxin that inhibits translation by acetylating isoleucyl-tRNAIle. *Nucleic Acids Res.* **2018**, *46*, 7873–7885. [CrossRef] [PubMed]
46. VanDrisse, M.C.; Parks, A.R.; Escalante-Semerena, J.C. A Toxin Involved in Salmonella Persistence Regulates Its Activity by Acetylating Its Cognate Antitoxin, a Modification Reversed by CobB Sirtuin Deacetylase. *mBio* **2017**, *8*, e00708-17. [CrossRef] [PubMed]

47. Horesh, G.; Fino, C.; Harms, A.; Dorman, M.J.; Parts, L.; Gerdes, K.; Heinz, E.; Thomson, N.R. Type II and type IV toxin-antitoxin systems show different evolutionary patterns in the global Klebsiella pneumoniae population. *Nucleic Acids Res.* **2020**, *48*, 4357–4370. [CrossRef]
48. Yamaguchi, Y.; Park, J.H.; Inouye, M. Toxin-antitoxin systems in bacteria and archaea. *Ann. Rev. Genet.* **2011**, *45*, 61–79. [CrossRef]
49. Xie, Y.; Wei, Y.; Shen, Y.; Li, X.; Zhou, H.; Tai, C.; Deng, Z.; Ou, H.Y. TADB 2.0: An updated database of bacterial type II toxin-antitoxin loci. *Nucleic Acids Res.* **2018**, *46*, D749–D753. [CrossRef]
50. Akarsu, H.; Bordes, P.; Mansour, M.; Bigot, D.J.; Genevaux, P.; Falquet, L. TASmania: A bacterial Toxin-Antitoxin Systems database. *PLoS Comput. Biol.* **2019**, *15*, e1006946. [CrossRef]
51. Leplae, R.; Geeraerts, D.; Hallez, R.; Guglielmini, J.; Dreze, P.; Van Melderen, L. Diversity of bacterial type II toxin-antitoxin systems: A comprehensive search and functional analysis of novel families. *Nucleic Acids Res.* **2011**, *39*, 5513–5525. [CrossRef] [PubMed]
52. Chan, T.W.; Balsa, D.; Espinosa, M. One cannot rule them all: Are bacterial toxins-antitoxins druggable? *FEMS Microbiol. Rev.* **2015**, *39*, 522–540. [CrossRef] [PubMed]
53. Bernard, P.; Couturier, M. Cell killing by the F plasmid CcdB protein involves poisoning of DNA-topoisomerase II complexes. *J. Mol. Biol.* **1992**, *226*, 735–745. [CrossRef]
54. Melderen, V.L.; Bernard, P.; Couturier, M. Lon-dependent proteolysis of CcdA is the key control for activation of CcdB in plasmid-free segregant bacteria. *Mol. Microbiol.* **1994**, *11*, 1151–1157. [CrossRef] [PubMed]
55. Roberts, C.R.; Strom, A.R.; Helinski, D.R. The parDE operon of the broad-host-range plasmid RK2 specifies growth inhibition associated with plasmid loss. *J. Mol. Biol.* **1994**, *237*, 35–51. [CrossRef]
56. Sobecky, P.A.; Easter, C.L.; Bear, P.D.; Helinski, D.R. Characterization of the stable maintenance properties of the par region of broad-host-range plasmid RK2. *J. Bacteriol.* **1996**, *178*, 2086–2093. [CrossRef]
57. Yarmolinsky, M.B. Programmed cell death in bacterial populations. *Science* **1995**, *267*, 836. [CrossRef]
58. Koonin, E.V.; Makarova, K.S.; Aravind, L. Horizontal Gene Transfer in Prokaryotes: Quantification and Classification. *Annu. Rev. Microbiol.* **2001**, *55*, 709–742. [CrossRef]
59. de Bast, S.M.; Mine, N.; van Melderen, L. Chromosomal toxin-antitoxin systems may act as antiaddiction modules. *J. Bacteriol.* **2008**, *190*, 4603–4609. [CrossRef]
60. Makarova, S.K.; Wolf, Y.I.; Koonin, E.V. Comparative genomics of defense systems in archaea and bacteria. *Nucleic Acids Res.* **2013**, *41*, 4360–4377. [CrossRef]
61. Hernández-Arriaga, A.M.; Chan, W.T.; Espinosa, M.; Díaz-Orejas, R. Conditional Activation of Toxin-Antitoxin Systems: Postsegregational Killing and Beyond. *Microbiol. Spectr.* **2014**, *2*, 175–192. [CrossRef] [PubMed]
62. Mine, N.; Guglielmini, J.; Wilbaux, M.; Van Melderen, L. The decay of the chromosomally encoded ccdO157 toxin-antitoxin system in the Escherichia coli species. *Genetics* **2009**, *181*, 1557–1566. [CrossRef]
63. Ramisetty, M.B.C.; Santhosh, R.S. Horizontal gene transfer of chromosomal Type II toxin–antitoxin systems of Escherichia coli. *FEMS Microbiol. Lett.* **2016**, *363*. [CrossRef] [PubMed]
64. Ramisetty, B.C.M.; Sudhakari, P.A. Bacterial 'Grounded' Prophages: Hotspots for Genetic Renovation and Innovation. *Front. Genet.* **2019**, *10*, 65. [CrossRef] [PubMed]
65. Koonin, V.E.; Wolf, Y.I. Genomics of bacteria and archaea: The emerging dynamic view of the prokaryotic world. *Nucleic Acids Res.* **2008**, *36*, 6688–6719. [CrossRef]
66. Andersen, S.B.; Ghoul, M.; Griffin, A.S.; Petersen, B.; Johansen, H.K.; Molin, S. Diversity, Prevalence, and Longitudinal Occurrence of Type II Toxin-Antitoxin Systems of Pseudomonas aeruginosa Infecting Cystic Fibrosis Lungs. *Front. Microbiol.* **2017**, *8*, 1180. [CrossRef] [PubMed]
67. Diaz-Orejas, R.; Espinosa, M.; Yeo, C.C. The Importance of the Expendable: Toxin-Antitoxin Genes in Plasmids and Chromosomes. *Front. Microbiol.* **2017**, *8*, 1479. [CrossRef]
68. Gupta, K.; Tripathi, A.; Sahu, A.; Varadarajan, R. Contribution of the chromosomal ccdAB operon to bacterial drug tolerance. *J. Bacteriol.* **2017**, *199*, e00397-17. [CrossRef]
69. Ames, J.R.; Muthuramalingam, M.; Murphy, T.; Najar, F.Z.; Bourne, C.R. Expression of different ParE toxins results in conserved phenotypes with distinguishable classes of toxicity. *MicrobiologyOpen* **2019**, *8*, e902. [CrossRef]
70. Fiebig, A.; Castro Rojas, C.M.; Siegal-Gaskins, D.; Crosson, S. Interaction specificity, toxicity and regulation of a paralogous set of ParE/RelE-family toxin-antitoxin systems. *Mol. Microbiol.* **2010**, *77*, 236–251. [CrossRef]

71. Muthuramalingam, M.; White, J.C.; Murphy, T.; Ames, J.R.; Bourne, C.R. The toxin from a ParDE toxin-antitoxin system found in Pseudomonas aeruginosa offers protection to cells challenged with anti-gyrase antibiotics. *Mol. Microbiol.* **2018**, *111*, 441–454. [CrossRef] [PubMed]
72. Tamman, H.; Ainelo, A.; Ainsaar, K.; Hõrak, R. A moderate toxin, GraT, modulates growth rate and stress tolerance of Pseudomonas putida. *J. Bacteriol.* **2014**, *196*, 157–169. [CrossRef] [PubMed]
73. Wilbaux, M.; Mine, N.; Guérout, A.M.; Mazel, D.; Van Melderen, L. Functional interactions between coexisting toxin-antitoxin systems of the ccd family in Escherichia coli O157:H7. *J. Bacteriol.* **2007**, *189*, 2712–2719. [CrossRef] [PubMed]
74. Fernandez-Garcia, L.; Kim, J.S.; Tomas, M.; Wood, T.K. Toxins of toxin/antitoxin systems are inactivated primarily through promoter mutations. *J. Appl. Microbiol.* **2019**, *127*, 1859–1868. [CrossRef] [PubMed]
75. Deter, H.S.; Jensen, R.V.; Mather, W.H.; Butzin, N.C. Mechanisms for Differential Protein Production in Toxin-Antitoxin Systems. *Toxins (Basel)* **2017**, *9*, 211. [CrossRef]
76. Singh, R.; Barry, C.E., 3rd; Boshoff, H.I. The three RelE homologs of Mycobacterium tuberculosis have individual, drug-specific effects on bacterial antibiotic tolerance. *J. Bacteriol.* **2010**, *192*, 1279–1291. [CrossRef]
77. Pontes, H.M.; Groisman, E.A. Slow growth determines nonheritable antibiotic resistance in Salmonella enterica. *Sci. Signal.* **2019**, *12*, eaax3938. [CrossRef] [PubMed]
78. Tripathi, A.; Dewan, P.C.; Barua, B.; Varadarajan, R. Additional role for the ccd operon of F-plasmid as a transmissible persistence factor. *Proc. Natl Acad. Sci. USA* **2012**, *109*, 12497–12502. [CrossRef] [PubMed]
79. Kamruzzaman, M.; Iredell, J. A ParDE-family toxin antitoxin system in major resistance plasmids of Enterobacteriaceae confers antibiotic and heat tolerance. *Sci. Rep.* **2019**, *9*, 9872. [CrossRef]
80. Gupta, M.; Nayyar, N.; Chawla, M.; Sitaraman, R.; Bhatnagar, R.; Banerjee, N. The chromosomal parDE2 toxin-antitoxin system of Mycobacterium tuberculosis H37Rv: Genetic and functional characterization. *Front. Microbiol.* **2016**, *7*, 886. [CrossRef]
81. Harms, A.; Stanger, F.V.; Scheu, P.D.; de Jong, I.G.; Goepfert, A.; Glatter, T.; Gerdes, K.; Schirmer, T.; Dehio, C. Adenylylation of Gyrase and Topo IV by FicT Toxins Disrupts Bacterial DNA Topology. *Cell Rep.* **2015**, *12*, 1497–1507. [CrossRef] [PubMed]
82. Janssen, D.B.; Garza-Sanchez, F.; Hayes, C.S. YoeB toxin is activated during thermal stress. *MicrobiologyOpen* **2015**, *4*, 682–697. [CrossRef]
83. Chan, W.T.; Domenech, M.; Moreno-Córdoba, I.; Navarro-Martínez, V.; Nieto, C.; Moscoso, M.; García, E.; Espinosa, M. The Streptococcus pneumoniae yefM-yoeB and relBE Toxin-Antitoxin Operons Participate in Oxidative Stress and Biofilm Formation. *Toxins (Basel)* **2018**, *10*, 378. [CrossRef] [PubMed]
84. Ainelo, A.; Porosk, R.; Kilk, K.; Rosendahl, S.; Remme, J.; Hõrak, R. Pseudomonas putida Responds to the Toxin GraT by Inducing Ribosome Biogenesis Factors and Repressing TCA Cycle Enzymes. *Toxins (Basel)* **2019**, *11*, 103. [CrossRef] [PubMed]
85. Zhao, Y.; McAnulty, M.J.; Wood, T.K. Toxin YafQ Reduces Escherichia coli Growth at Low Temperatures. *PLoS ONE* **2016**, *11*, e0161577. [CrossRef] [PubMed]
86. Korch, S.B.; Malhotra, V.; Contreras, H.; Clark-Curtiss, J.E. The Mycobacterium tuberculosis relBE toxin:antitoxin genes are stress-responsive modules that regulate growth through translation inhibition. *J. Microbiol.* **2015**, *53*, 783–795. [CrossRef]
87. Christensen, K.S.; Gerdes, K. Delayed-relaxed response explained by hyperactivation of RelE. *Mol. Microbiol.* **2004**, *53*, 587–597. [CrossRef]
88. Tashiro, Y.; Kawata, K.; Taniuchi, A.; Kakinuma, K.; May, T.; Okabe, S. RelE-mediated dormancy is enhanced at high cell density in Escherichia coli. *J. Bacteriol.* **2012**, *194*, 1169–1176. [CrossRef]
89. Armalytė, J.; Jurėnas, D.; Krasauskas, R.; Čepauskas, A.; Sužiedėlienė, E. The higBA Toxin-Antitoxin Module From the Opportunistic Pathogen Acinetobacter baumannii—Regulation, Activity, and Evolution. *Front. Microbiol.* **2018**, *9*, 732. [CrossRef]
90. Wood, L.T.; Wood, T.K. The HigB/HigA toxin/antitoxin system of Pseudomonas aeruginosa influences the virulence factors pyochelin, pyocyanin, and biofilm formation. *MicrobiolOpen* **2016**, *5*, 499–511. [CrossRef]
91. Ramisetty, B.; Ghosh, D.; Roy Chowdhury, M.; Santhosh, R.S. What Is the Link between Stringent Response, Endoribonuclease Encoding Type II Toxin-Antitoxin Systems and Persistence? *Front. Microbiol.* **2016**, *7*, 1882. [CrossRef] [PubMed]
92. Zhang, Y.; Inouye, M. RatA (YfjG), an Escherichia coli toxin, inhibits 70S ribosome association to block translation initiation. *Mol. Microbiol.* **2011**, *79*, 1418–1429. [CrossRef] [PubMed]

93. Norton, P.J.; Mulvey, M.A. Toxin-antitoxin systems are important for niche-specific colonization and stress resistance of uropathogenic Escherichia coli. *PLoS Pathog.* **2012**, *8*, e1002954. [CrossRef] [PubMed]
94. Tiwari, P.; Arora, G.; Singh, M.; Kidwai, S.; Narayan, O.P.; Singh, R. MazF ribonucleases promote Mycobacterium tuberculosis drug tolerance and virulence in guinea pigs. *Nat. Commun.* **2015**, *6*, 6059. [CrossRef]
95. Ren, D.; Walker, A.N.; Daines, D.A. Toxin-antitoxin loci vapBC-1 and vapXD contribute to survival and virulence in nontypeable Haemophilus influenzae. *BMC Microbiol.* **2012**, *12*, 263. [CrossRef]
96. Jaiswal, S.; Paul, P.; Padhi, C.; Ray, S.; Ryan, D.; Dash, S.; Suar, M. The Hha-TomB Toxin-Antitoxin System Shows Conditional Toxicity and Promotes Persister Cell Formation by Inhibiting Apoptosis-Like Death in S. *Typhimurium. Sci. Rep.* **2016**, *6*, 38204. [CrossRef]
97. Helaine, S.; Cheverton, A.M.; Watson, K.G.; Faure, L.M.; Matthews, S.A.; Holden, D.W. Internalization of Salmonella by macrophages induces formation of nonreplicating persisters. *Science* **2014**, *343*, 204–208. [CrossRef]
98. Fraikin, N.; Rousseau, C.J.; Goeders, N.; Van Melderen, L. Reassessing the Role of the Type II MqsRA Toxin-Antitoxin System in Stress Response and Biofilm Formation: mqsA Is Transcriptionally Uncoupled from mqsR. *mBio* **2019**, *10*, e02678-19. [CrossRef]
99. Gerdes, K.; Christensen, S.K.; Lobner-Olesen, A. Prokaryotic toxin-antitoxin stress response loci. *Nat. Rev. Microbiol.* **2005**, *3*, 371–382. [CrossRef]
100. Kim, Y.; Wang, X.; Zhang, X.S.; Grigoriu, S.; Page, R.; Peti, W.; Wood, T.K. Escherichia coli toxin/antitoxin pair MqsR/MqsA regulate toxin CspD. *Environ. Microbiol.* **2010**, *12*, 1105–1121. [CrossRef]
101. Wade, T.J.; Laub, M.T. Concerns about "Stress-Induced MazF-Mediated Proteins in *Escherichia coli*". *mBio* **2019**, *10*, e00825-19. [CrossRef] [PubMed]
102. Culviner, H.P.; Laub, M.T. Global Analysis of the *E. coli* Toxin MazF Reveals Widespread Cleavage of mRNA and the Inhibition of rRNA Maturation and Ribosome Biogenesis. *Mol. Cell* **2018**, *70*, 868–880. [CrossRef] [PubMed]
103. Nigam, A.; Ziv, T.; Oron-Gottesman, A. Engelberg-Kulka Stress-Induced MazF-Mediated Proteins in *Escherichia coli. mBio* **2019**, *10*, e00340-19. [PubMed]
104. Harms, A.; Liesch, M.; Körner, J.; Québatte, M.; Engel, P.; Dehio, C. A bacterial toxin-antitoxin module is the origin of inter-bacterial and inter-kingdom effectors of Bartonella. *PLoS Genet.* **2017**, *13*, e1007077. [CrossRef] [PubMed]
105. Harms, A.; Maisonneuve, E.; Gerdes, K. Mechanisms of bacterial persistence during stress and antibiotic exposure. *Science* **2016**, *354*, aaf4268. [CrossRef] [PubMed]
106. Goormaghtigh, F.; Fraikin, N.; Putrinš, M.; Hallaert, T.; Hauryliuk, V.; Garcia-Pino, A.; Sjödin, A.; Kasvandik, S.; Udekwu, K.; Tenson, T.; et al. Reassessing the Role of Type II Toxin-Antitoxin Systems in Formation of *Escherichia coli* Type II Persister Cells. *mBio* **2018**, *9*, e00640-18. [CrossRef] [PubMed]
107. Goormaghtigh, F.; Fraikin, N.; Putrinš, M.; Hauryliuk, V.; Garcia-Pino, A.; Udekwu, K.; Tenson, T.; Kaldalu, N.; Van Melderen, L. Reply to Holden and Errington, "Type II Toxin-Antitoxin Systems and Persister Cells". *mBio* **2018**, *9*, e01838-18. [CrossRef] [PubMed]
108. Zielenkiewicz, U.; Ceglowski, P. The toxin-antitoxin system of the streptococcal plasmid pSM19035. *J. Bacteriol.* **2005**, *187*, 6094–6105. [CrossRef]
109. Chan, T.W.; Espinosa, M. The Streptococcus pneumoniae pezAT Toxin-Antitoxin System Reduces beta-Lactam Resistance and Genetic Competence. *Front. Microbiol.* **2016**, *7*, 1322. [CrossRef]
110. Piscotta, J.F.; Jeffrey, P.D.; Link, A.J. ParST is a widespread toxin-antitoxin module that targets nucleotide metabolism. *Proc. Natl. Acad. Sci. USA* **2019**, *116*, 826–834. [CrossRef]
111. Yuan, J.; Yamaichi, Y.; Waldor, M.K. The three Vibrio cholerae chromosome II-encoded ParE toxins degrade chromosome I following loss of chromosome II. *J. Bacteriol.* **2011**, *193*, 611–619. [CrossRef] [PubMed]
112. Wozniak, A.R.; Waldor, M.K. A toxin-antitoxin system promotes the maintenance of an integrative conjugative element. *PLoS Genet.* **2009**, *5*, e1000439. [CrossRef] [PubMed]
113. Hughes-Games, A.; Roberts, A.P.; Davis, S.A.; Hill, D.J. Identification of integrative and conjugative elements in pathogenic and commensal Neisseriaceae species via genomic distributions of DNA uptake sequence dialects. *Microb. Genom.* **2020**, *6*, e000372.

114. Berni, B.; Soscia, C.; Djermoun, S.; Ize, B.; Bleves, S. A Type VI Secretion System Trans-Kingdom Effector Is Required for the Delivery of a Novel Antibacterial Toxin in Pseudomonas aeruginosa. *Front. Microbiol.* **2019**, *10*, 1218. [CrossRef]
115. Cao, Z.; Casabona, M.G.; Kneuper, H.; Chalmers, J.D.; Palmer, T. The type VII secretion system of Staphylococcus aureus secretes a nuclease toxin that targets competitor bacteria. *Nat. Microbiol.* **2016**, *2*, 16183. [CrossRef]
116. Whitney, J.C.; Peterson, S.B.; Kim, J.; Pazos, M.; Verster, A.J.; Radey, M.C.; Kulasekara, H.D.; Ching, M.Q.; Bullen, N.P.; Bryant, D.; et al. A broadly distributed toxin family mediates contact-dependent antagonism between gram-positive bacteria. *Elife* **2017**, *6*, e26938. [CrossRef]
117. Gerdes, K.; Bech, F.W.; Jørgensen, S.T.; Løbner-Olesen, A.; Rasmussen, P.B.; Atlung, T.; Boe, L.; Karlstrom, O.; Molin, S.; Von Meyenburg, K. Mechanism of postsegregational killing by the hok gene product of the parB system of plasmid R1 and its homology with the relF gene product of the *E. coli* relB operon. *EMBO J.* **1986**, *5*, 2023–2029. [CrossRef]
118. Ogura, T.; Hiraga, S. Mini-F plasmid genes that couple host cell division to plasmid proliferation. *Proc. Natl. Acad. Sci. USA* **1983**, *80*, 4784–4788. [CrossRef]
119. Aizenman, E.; Engelberg-Kulka, H.; Glaser, G. An *Escherichia coli* chromosomal "addiction module" regulated by guanosine 3′,5′-bispyrophosphate: A model for programmed bacterial cell death. *Proc. Natl. Acad. Sci. USA* **1996**, *93*, 6059–6063. [CrossRef]
120. Jensen, B.R.; Gerdes, K. Programmed cell death in bacteria: Proteic plasmid stabilization systems. *Mol. Microbiol.* **1995**, *17*, 205–210. [CrossRef]
121. Zander, I.; Shmidov, E.; Roth, S.; Ben-David, Y.; Shoval, I.; Shoshani, S.; Danielli, A.; Banin, E. Characterization of PfiT/PfiA toxin-antitoxin system of Pseudomonas aeruginosa that affects cell elongation and prophage induction. *Environ. Microbiol.* **2020**. [CrossRef] [PubMed]
122. Li, Y.; Liu, X.; Tang, K.; Wang, W.; Guo, Y.; Wang, X. Prophage encoding toxin/antitoxin system PfiT/PfiA inhibits Pf4 production in Pseudomonas aeruginosa. *Microb. Biotechnol.* **2020**, *13*, 1132–1144. [CrossRef] [PubMed]
123. Bustamante, P.; Iredell, J.R. Carriage of type II toxin-antitoxin systems by the growing group of IncX plasmids. *Plasmid* **2017**, *91*, 19–27. [CrossRef] [PubMed]
124. Cooper, F.T.; Paixao, T.; Heinemann, J.A. Within-host competition selects for plasmid-encoded toxin-antitoxin systems. *Proc. Biol. Sci.* **2010**, *277*, 3149–3155. [CrossRef] [PubMed]
125. Fico, S.; Mahillon, J. TasA-tasB, a new putative toxin-antitoxin (TA) system from Bacillus thuringiensis pGI1 plasmid is a widely distributed composite mazE-doc TA system. *BMC Genom.* **2006**, *7*, 259. [CrossRef]
126. Hernández-Ramírez, K.C.; Chávez-Jacobo, V.M.; Valle-Maldonado, M.I.; Patiño-Medina, J.A.; Díaz-Pérez, S.P.; Jácome-Galarza, I.E.; Ortiz-Alvarado, R.; Meza-Carmen, V.; Ramírez-Díaz, M.I. Plasmid pUM505 encodes a Toxin-Antitoxin system conferring plasmid stability and increased Pseudomonas aeruginosa virulence. *Microb. Pathog.* **2017**, *112*, 259–268. [CrossRef] [PubMed]
127. Poirel, L.; Bonnin, R.A.; Nordmann, P. Genetic features of the widespread plasmid coding for the carbapenemase OXA-48. *Antimicrob. Agents Chemother.* **2012**, *56*, 559–562. [CrossRef]
128. Mnif, B.; Vimont, S.; Boyd, A.; Bourit, E.; Picard, B.; Branger, C.; Denamur, E.; Arlet, G. Molecular characterization of addiction systems of plasmids encoding extended-spectrum beta-lactamases in *Escherichia coli. J. Antimicrob. Chemother.* **2010**, *65*, 1599–1603. [CrossRef]
129. Davis, L.T.; Helinski, D.R.; Roberts, R.C. Transcription and autoregulation of the stabilizing functions of broad-host-range plasmid RK2 in *Escherichia coli*, Agrobacterium tumefaciens and Pseudomonas aeruginosa. *Mol. Microbiol.* **1992**, *6*, 1981–1994. [CrossRef]
130. Gronlund, H.; Gerdes, K. Toxin-antitoxin systems homologous with relBE of *Escherichia coli* plasmid P307 are ubiquitous in prokaryotes. *J. Mol. Biol.* **1999**, *285*, 1401–1415. [CrossRef]
131. Wang, J.; Stephan, R.; Zurfluh, K.; Hächler, H.; Fanning, S. Characterization of the genetic environment of bla ESBL genes, integrons and toxin-antitoxin systems identified on large transferrable plasmids in multi-drug resistant *Escherichia coli. Front. Microbiol.* **2014**, *5*, 716. [PubMed]
132. Hayes, F. A family of stability determinants in pathogenic bacteria. *J. Bacteriol.* **1998**, *180*, 6415–6418. [CrossRef]

133. Bravo, A.; de Torrontegui, G.; Diaz, R. Identification of components of a new stability system of plasmid R1, ParD, that is close to the origin of replication of this plasmid. *Mol. Gen. Genet.* **1987**, *210*, 101–110. [CrossRef] [PubMed]
134. Diago-Navarro, E.; Hernandez-Arriaga, A.M.; López-Villarejo, J.; Muñoz-Gómez, A.J.; Kamphuis, M.B.; Boelens, R.; Lemonnier, M.; Díaz-Orejas, R. parD toxin-antitoxin system of plasmid R1–basic contributions, biotechnological applications and relationships with closely-related toxin-antitoxin systems. *FEBS J.* **2010**, *277*, 3097–3117. [CrossRef] [PubMed]
135. Tsuchimoto, S.; Ohtsubo, H.; Ohtsubo, E. Two genes, pemK and pemI, responsible for stable maintenance of resistance plasmid R100. *J. Bacteriol.* **1988**, *170*, 1461–1466. [CrossRef]
136. Bukowski, M.; Hyz, K.; Janczak, M.; Hydzik, M.; Dubin, G.; Wladyka, B. Identification of novel mazEF/pemIK family toxin-antitoxin loci and their distribution in the Staphylococcus genus. *Sci. Rep.* **2017**, *7*, 13462. [CrossRef]
137. Lehnherr, H.; Maguin, E.; Jafri, S.; Yarmolinsky, M.B. Plasmid addiction genes of bacteriophage P1: Doc, which causes cell death on curing of prophage, and phd, which prevents host death when prophage is retained. *J. Mol. Biol.* **1993**, *233*, 414–428. [CrossRef]
138. Tian, Q.B.; Hayashi, T.; Murata, T.; Terawaki, Y. Gene product identification and promoter analysis of hig locus of plasmid Rts1. *Biochem. Biophys Res. Commun.* **1996**, *225*, 679–684. [CrossRef]
139. Dziewit, L.; Jazurek, M.; Drewniak, L.; Baj, J.; Bartosik, D. The SXT conjugative element and linear prophage N15 encode toxin-antitoxin-stabilizing systems homologous to the tad-ata module of the Paracoccus aminophilus plasmid pAMI2. *J. Bacteriol.* **2007**, *189*, 1983–1997. [CrossRef]
140. Grady, R.; Hayes, F. Axe-Txe, a broad-spectrum proteic toxin-antitoxin system specified by a multidrug-resistant, clinical isolate of Enterococcus faecium. *Mol. Microbiol.* **2003**, *47*, 1419–1432. [CrossRef]
141. Rosvoll, T.C.; Pedersen, T.; Sletvold, H.; Johnsen, P.J.; Sollid, J.E.; Simonsen, G.S.; Jensen, L.B.; Nielsen, K.M.; Sundsfjord, A. PCR-based plasmid typing in Enterococcus faecium strains reveals widely distributed pRE25-, pRUM-, pIP501- and pHTbeta-related replicons associated with glycopeptide resistance and stabilizing toxin-antitoxin systems. *FEMS Immunol. Med. Microbiol.* **2010**, *58*, 254–268. [CrossRef] [PubMed]
142. Zhang, Y.X.; Guo, X.K.; Chuan, W.U.; Bo, B.I.; Ren, S.X.; Wu, C.F.; Zhao, G.P. Characterization of a novel toxin-antitoxin module, VapBC, encoded by Leptospira interrogans chromosome. *Cell Res.* **2004**, *14*, 208–216. [CrossRef] [PubMed]
143. Jurėnas, D.; Chatterjee, S.; Konijnenberg, A.; Sobott, F.; Droogmans, L.; Garcia-Pino, A.; Van Melderen, L. AtaT blocks translation initiation by N-acetylation of the initiator tRNAfMet. *Nat. Chem. Biol.* **2017**, *13*, 640. [CrossRef] [PubMed]
144. Wen, Y.; Behiels, E.; Devreese, B. Toxin-Antitoxin systems: Their role in persistence, biofilm formation, and pathogenicity. *Pathog. Dis.* **2014**, *70*, 240–249. [CrossRef] [PubMed]
145. Jurenaite, M.; Markuckas, A.; Suziedeliene, E. Identification and characterization of type II toxin-antitoxin systems in the opportunistic pathogen Acinetobacter baumannii. *J. Bacteriol.* **2013**, *195*, 3165–3172. [CrossRef] [PubMed]
146. McVicker, G.; Tang, C.M. Deletion of toxin-antitoxin systems in the evolution of Shigella sonnei as a host-adapted pathogen. *Nat. Microbiol.* **2016**, *2*, 16204. [CrossRef]
147. Cooper, F.T.; Heinemann, J.A. Postsegregational killing does not increase plasmid stability but acts to mediate the exclusion of competing plasmids. *Proc. Natl. Acad. Sci. USA* **2000**, *97*, 12643–12648. [CrossRef]
148. Song, S.; Wood, T.K. Post-segregational Killing and Phage Inhibition Are Not Mediated by Cell Death Through Toxin/Antitoxin Systems. *Front. Microbiol.* **2018**, *9*, 814. [CrossRef]
149. Otsuka, Y.; Yonesaki, T. Dmd of bacteriophage T4 functions as an antitoxin against *Escherichia coli* LsoA and RnlA toxins. *Mol. Microbiol.* **2012**, *83*, 669–681. [CrossRef]
150. Blower, T.R.; Fineran, P.C.; Johnson, M.J.; Toth, I.K.; Humphreys, D.P.; Salmond, G.P.C. Mutagenesis and functional characterization of the RNA and protein components of the toxIN abortive infection and toxin-antitoxin locus of Erwinia. *J. Bacteriol.* **2009**, *191*, 6029–6039. [CrossRef]
151. Dy, R.L.; Richter, C.; Salmond, G.P.; Fineran, P.C. Remarkable Mechanisms in Microbes to Resist Phage Infections. *Annu. Rev. Virol.* **2014**, *1*, 307–331. [CrossRef] [PubMed]
152. Chen, B.; Akusobi, C.; Fang, X.; Salmond, G.P. Environmental T4-Family Bacteriophages Evolve to Escape Abortive Infection via Multiple Routes in a Bacterial Host Employing "Altruistic Suicide" through Type III Toxin-Antitoxin Systems. *Front. Microbiol.* **2017**, *8*, 1006. [CrossRef] [PubMed]

153. Park, J.S.; Son, W.S.; Lee, B.J. Structural overview of toxin-antitoxin systems in infectious bacteria: A target for developing antimicrobial agents. *Biochim. Biophys. Acta* **2013**, *1843*, 1155–1167. [CrossRef]
154. Arbing, M.A.; Handelman, S.K.; Kuzin, A.P.; Verdon, G.; Wang, C.; Su, M.; Rothenbacher, F.P.; Abashidze, M.; Liu, M.; Hurley, J.M.; et al. Crystal structures of Phd-Doc, HigA, and YeeU establish multiple evolutionary links between microbial growth-regulating toxin-antitoxin systems. *Structure (London)* **2010**, *18*, 996–1010. [CrossRef] [PubMed]
155. Meinhart, A.; Alonso, J.C.; Sträter, N.; Saenger, W. Crystal structure of the plasmid maintenance system epsilon/zeta: Functional mechanism of toxin zeta and inactivation by epsilon 2 zeta 2 complex formation. *Proc. Natl. Acad. Sci. USA* **2003**, *100*, 1661–1666. [CrossRef]
156. Jaen-Luchoro, D.; Aliaga-Lozano, F.; Gomila, R.M.; Gomila, M.; Salva-Serra, F.; Lalucat, J.; Bennasar-Figueras, A. First insights into a type II toxin-antitoxin system from the clinical isolate Mycobacterium sp. MHSD3, similar to epsilon/zeta systems. *PLoS ONE* **2017**, *12*, e0189459. [CrossRef]
157. Fernández-Bachiller, M.I.; Brzozowska, I.; Odolczyk, N.; Zielenkiewicz, U.; Zielenkiewicz, P.; Rademann, J. Mapping Protein-Protein Interactions of the Resistance-Related Bacterial Zeta Toxin-Epsilon Antitoxin Complex (epsilon(2)zeta(2)) with High Affinity Peptide Ligands Using Fluorescence Polarization. *Toxins (Basel)* **2016**, *8*, 222. [CrossRef]
158. Budde, P.P.; Davis, B.M.; Yuan, J.; Waldor, M.K. Characterization of a higBA toxin-antitoxin locus in Vibrio cholerae. *J. Bacteriol.* **2007**, *189*, 491–500. [CrossRef]
159. Yang, J.; Zhou, K.; Liu, P.; Dong, Y.; Gao, Z.; Zhang, J.; Liu, Q. Structural insight into the *E. coli* HigBA complex. *Biochem. Biophys. Res. Commun.* **2016**, *478*, 1521–1527. [CrossRef]
160. Sterckx, Y.G.J.; Jové, T.; Shkumatov, A.V.; Garcia-Pino, A.; Geerts, L.; De Kerpel, M.; Lah, J.; De Greve, H.; Van Melderen, L.; Loris, R. A unique hetero-hexadecameric architecture displayed by the *Escherichia coli* O157 PaaA2-ParE2 antitoxin-toxin complex. *J. Mol. Biol.* **2016**, *428*, 1589–1603. [CrossRef]
161. Aakre, C.D.; Herrou, J.; Phung, T.N.; Perchuk, B.S.; Crosson, S.; Laub, M.T. Evolving new protein-protein interaction specificity through promiscuous intermediates. *Cell* **2015**, *163*, 594–606. [CrossRef] [PubMed]
162. El-Gebali, S.; Mistry, J.; Bateman, A.; Eddy, S.R.; Luciani, A.; Potter, S.C.; Qureshi, M.; Richardson, L.J.; Salazar, G.A.; Smart, A.; et al. The Pfam protein families database in 2019. *Nucleic Acids Res.* **2019**, *47*, D427–D432. [CrossRef]
163. Chan, T.W.; Espinosa, M.; Yeo, C.C. Keeping the Wolves at Bay: Antitoxins of Prokaryotic Type II Toxin-Antitoxin Systems. *Front. Mol. Biosci.* **2016**, *3*, 9. [CrossRef] [PubMed]
164. Senissar, M.; Manav, M.C.; Brodersen, D.E. Structural conservation of the PIN domain active site across all domains of life. *Protein Sci.* **2017**, *26*, 1474–1492. [CrossRef] [PubMed]
165. Matelska, D.; Steczkiewicz, K.; Ginalski, K. Comprehensive classification of the PIN domain-like superfamily. *Nucleic Acids Res.* **2017**, *45*, 6995–7020. [CrossRef]
166. Arcus, V.L.; McKenzie, J.L.; Robson, J.; Cook, G.M. The PIN-domain ribonucleases and the prokaryotic VapBC toxin-antitoxin array. *Protein Eng. Des. Sel.* **2011**, *24*, 33–40. [CrossRef]
167. Miallau, L.; Faller, M.; Chiang, J.; Arbing, M.; Guo, F.; Cascio, D.; Eisenberg, D. Structure and proposed activity of a member of the VapBC family of toxin-antitoxin systems. VapBC-5 from Mycobacterium tuberculosis. *J. Biol. Chem.* **2009**, *284*, 276–283. [CrossRef]
168. Maté, M.J.; Vincentelli, R.; Foos, N.; Raoult, D.; Cambillau, C.; Ortiz-Lombardía, M. Crystal structure of the DNA-bound VapBC2 antitoxin/toxin pair from Rickettsia felis. *Nucleic Acids Res.* **2012**, *40*, 3245–3258.
169. Mattison, K.; Wilbur, J.S.; So, M.; Brennan, R.G. Structure of FitAB from Neisseria gonorrhoeae bound to DNA reveals a tetramer of toxin-antitoxin heterodimers containing pin domains and ribbon-helix-helix motifs. *J. Biol. Chem.* **2006**, *281*, 37942–37951. [CrossRef]
170. Min, A.B.; Miallau, L.; Sawaya, M.R.; Habel, J.; Cascio, D.; Eisenberg, D. The crystal structure of the Rv0301-Rv0300 VapBC-3 toxin-antitoxin complex from M. *tuberculosis reveals a Mg(2)(+) ion in the active site and a putative RNA-binding site. Protein Sci.* **2012**, *21*, 1754–1767.
171. Schureck, M.A.; Dunkle, J.A.; Maehigashi, T.; Miles, S.J.; Dunham, C.M. Defining the mRNA recognition signature of a bacterial toxin protein. *Proc. Natl. Acad. Sci. USA* **2015**, *112*, 13862–13867. [CrossRef] [PubMed]
172. Schureck, M.A.; Repack, A.; Miles, S.J.; Marquez, J.; Dunham, C.M. Mechanism of endonuclease cleavage by the HigB toxin. *Nucleic Acids Res.* **2016**, *44*, 7944–7953. [CrossRef] [PubMed]

173. Yuan, J.; Sterckx, Y.; Mitchenall, L.A.; Maxwell, A.; Loris, R.; Waldor, M.K. Vibrio cholerae ParE2 poisons DNA gyrase via a mechanism distinct from other gyrase inhibitors. *J. Biol. Chem.* **2010**, *285*, 40397–40408. [CrossRef] [PubMed]
174. Brown, B.L.; Grigoriu, S.; Kim, Y.; Arruda, J.M.; Davenport, A.; Wood, T.K.; Peti, W.; Page, R. Three dimensional structure of the MqsR:MqsA complex: A novel TA pair comprised of a toxin homologous to RelE and an antitoxin with unique properties. *PLoS Pathog.* **2009**, *5*, e1000706. [CrossRef] [PubMed]
175. Hallez, R.; Geeraerts, D.; Sterckx, Y.; Mine, N.; Loris, R.; Van Melderen, L. New toxins homologous to ParE belonging to three-component toxin-antitoxin systems in *Escherichia coli* O157:H7. *Mol. Microbiol.* **2010**, *76*, 719–732. [CrossRef]
176. Bøggild, A.; Sofos, N.; Andersen, K.R.; Feddersen, A.; Easter, A.D.; Passmore, L.A.; Brodersen, D.E. The crystal structure of the intact *E. coli* RelBE toxin-antitoxin complex provides the structural basis for conditional cooperativity. *Structure (London)* **2012**, *20*, 1641–1658.
177. Dalton, M.K.; Crosson, S. A conserved mode of protein recognition and binding in a ParD-ParE toxin-antitoxin complex. *Biochemistry* **2010**, *49*, 2205–2215. [CrossRef]
178. Hargreaves, D.; Santos-Sierra, S.; Giraldo, R.; Sabariegos-Jareño, R.; de la Cueva-Méndez, G.; Boelens, R.; Díaz-Orejas, R.; Rafferty, J.B. Structural and functional analysis of the kid toxin protein from *E. coli* plasmid R1. *Structure* **2002**, *10*, 1425–1433.
179. Kamada, K.; Hanaoka, F.; Burley, S.K. Crystal structure of the MazE/MazF complex: Molecular bases of antidote-toxin recognition. *Mol. Cell* **2003**, *11*, 875–884. [CrossRef]
180. De Jonge, N.; Garcia-Pino, A.; Buts, L.; Haesaerts, S.; Charlier, D.; Zangger, K.; Wyns, L.; De Greve, H.; Loris, R. Rejuvenation of CcdB-poisoned gyrase by an intrinsically disordered protein domain. *Mol. Cell* **2009**, *35*, 154–163. [CrossRef]
181. Bahassi, E.M.; O'Dea, M.H.; Allali, N.; Messens, J.; Gellert, M.; Couturier, M. Interactions of CcdB with DNA gyrase. Inactivation of Gyra, poisoning of the gyrase-DNA complex, and the antidote action of CcdA. *J. Biol. Chem.* **1999**, *274*, 10936–10944. [CrossRef] [PubMed]
182. Kampranis, C.S.; Howells, A.J.; Maxwell, A. The interaction of DNA gyrase with the bacterial toxin CcdB: Evidence for the existence of two gyrase-CcdB complexes. *J. Mol. Biol.* **1999**, *293*, 733–744. [CrossRef] [PubMed]
183. Simic, M.; De Jonge, N.; Loris, R.; Vesnaver, G.; Lah, J. Driving forces of gyrase recognition by the addiction toxin CcdB. *J. Biol. Chem.* **2009**, *284*, 20002–20010. [CrossRef] [PubMed]
184. Smith, A.B.; López-Villarejo, J.; Diago-Navarro, E.; Mitchenall, L.A.; Barendregt, A.; Heck, A.J.; Lemonnier, M.; Maxwell, A.; Díaz-Orejas, R. A common origin for the bacterial toxin-antitoxin systems parD and ccd, suggested by analyses of toxin/target and toxin/antitoxin interactions. *PLoS ONE* **2012**, *7*, e46499. [CrossRef]
185. Ruiz-Echevarría, M.J.; Giménez-Gallego, G.; Sabariegos-Jareño, R.; Díaz-Orejas, R. Kid, a small protein of the parD stability system of plasmid R1, is an inhibitor of DNA replication acting at the initiation of DNA synthesis. *J. Mol. Biol.* **1995**, *247*, 568–577. [CrossRef]
186. Agarwal, S.; Mishra, N.K.; Bhatnagar, S.; Bhatnagar, R. PemK toxin of Bacillus anthracis is a ribonuclease: An insight into its active site, structure, and function. *J. Biol Chem.* **2010**, *285*, 7254–7270. [CrossRef]
187. Masuda, Y.; Miyakawa, K.; Nishimura, Y.; Ohtsubo, E. chpA and chpB, *Escherichia coli* chromosomal homologs of the pem locus responsible for stable maintenance of plasmid R100. *J. Bacteriol.* **1993**, *175*, 6850–6856. [CrossRef]
188. Kamphuis, M.B.; Monti, M.C.; van den Heuvel, R.H.; Santos-Sierra, S.; Folkers, G.E.; Lemonnier, M.; Díaz-Orejas, R.; Heck, A.J.; Boelens, R. Interactions between the toxin Kid of the bacterial parD system and the antitoxins Kis and MazE. *Proteins* **2007**, *67*, 219–231. [CrossRef]
189. Madl, T.; Van Melderen, L.; Mine, N.; Respondek, M.; Oberer, M.; Keller, W.; Khatai, L.; Zangger, K. Structural basis for nucleic acid and toxin recognition of the bacterial antitoxin CcdA. *J. Mol. Biol.* **2006**, *364*, 170–185. [CrossRef]
190. Walling, R.L.; Butler, J.S. Structural Determinants for Antitoxin Identity and Insulation of Cross Talk between Homologous Toxin-Antitoxin Systems. *J. Bacteriol.* **2016**, *198*, 3287–3295. [CrossRef]
191. Yang, M.; Gao, C.; Wang, Y.; Zhang, H.; He, Z.G. Characterization of the interaction and cross-regulation of three Mycobacterium tuberculosis RelBE modules. *PLoS ONE* **2010**, *5*, e10672. [CrossRef]

192. Połom, D.; Boss, L.; Węgrzyn, G.; Hayes, F.; Kędzierska, B. Amino acid residues crucial for specificity of toxin-antitoxin interactions in the homologous Axe-Txe and YefM-YoeB complexes. *FEBS J.* **2013**, *280*, 5906–5918. [CrossRef] [PubMed]
193. Nolle, N.; Schuster, C.F.; Bertram, R. Two paralogous yefM-yoeB loci from Staphylococcus equorum encode functional toxin-antitoxin systems. *Microbiology* **2013**, *159*, 1575–1585. [CrossRef]
194. Ahidjo, B.A.; Kuhnert, D.; McKenzie, J.L.; Machowski, E.E.; Gordhan, B.G.; Arcus, V.; Abrahams, G.L.; Mizrahi, V. VapC toxins from Mycobacterium tuberculosis are ribonucleases that differentially inhibit growth and are neutralized by cognate VapB antitoxins. *PLoS ONE* **2011**, *6*, e21738. [CrossRef] [PubMed]
195. Iqbal, N.; Guérout, A.M.; Krin, E.; Le Roux, F.; Mazel, D. Comprehensive Functional Analysis of the 18 Vibrio cholerae N16961 Toxin-Antitoxin Systems Substantiates Their Role in Stabilizing the Superintegron. *J. Bacteriol.* **2015**, *197*, 2150–2159. [CrossRef] [PubMed]
196. Tandon, H.; Sharma, A.; Wadhwa, S.; Varadarajan, R.; Singh, R.; Srinivasan, N.; Sandhya, S. Bioinformatic and mutational studies of related toxin-antitoxin pairs in Mycobacterium tuberculosis predict and identify key functional residues. *J. Biol. Chem.* **2019**, *294*, 9048–9063. [CrossRef]
197. Ramage, R.H.; Connolly, L.E.; Cox, J.S. Comprehensive functional analysis of Mycobacterium tuberculosis toxin-antitoxin systems: Implications for pathogenesis, stress responses, and evolution. *PLoS Genet.* **2009**, *5*, e1000767. [CrossRef]
198. Jin, G.; Pavelka, M.S., Jr.; Butler, J.S. Structure-function analysis of VapB4 antitoxin identifies critical features of a minimal VapC4 toxin-binding module. *J. Bacteriol.* **2015**, *197*, 1197–1207. [CrossRef]
199. Gucinski, G.C.; Michalska, K.; Garza-Sánchez, F.; Eschenfeldt, W.H.; Stols, L.; Nguyen, J.Y.; Goulding, C.W.; Joachimiak, A.; Hayes, C.S. Convergent Evolution of the Barnase/EndoU/Colicin/RelE (BECR) Fold in Antibacterial tRNase Toxins. *Structure* **2019**, *27*, 1660–1674. [CrossRef]
200. Goulard, C.; Langrand, S.; Carniel, E.; Chauvaux, S. The Yersinia pestis chromosome encodes active addiction toxins. *J. Bacteriol.* **2010**, *192*, 3669–3677. [CrossRef]
201. Sierra, S.S.; Giraldo, R.; Orejas, R.D. Functional interactions between chpB and parD, two homologous conditional killer systems found in the *Escherichia coli* chromosome and in plasmid R1. *FEMS Microbiol. Lett.* **1998**, *168*, 51–58. [CrossRef] [PubMed]
202. Santos-Sierra, S.; Giraldo, R.; Diaz-Orejas, R. Functional interactions between homologous conditional killer systems of plasmid and chromosomal origin. *FEMS Microbiol. Lett.* **1997**, *152*, 51–56. [CrossRef]
203. Zhu, L.; Sharp, J.D.; Kobayashi, H.; Woychik, N.A.; Inouye, M. Noncognate Mycobacterium tuberculosis toxin-antitoxins can physically and functionally interact. *J. Biol. Chem.* **2010**, *285*, 39732–39738. [CrossRef] [PubMed]
204. Pinaud, L.; Ferrari, M.L.; Friedman, R.; Jehmlich, N.; von Bergen, M.; Phalipon, A.; Sansonetti, P.J.; Campbell-Valois, F.X. Identification of novel substrates of Shigella T3SA through analysis of its virulence plasmid-encoded secretome. *PLoS ONE* **2017**, *12*, e0186920. [CrossRef] [PubMed]
205. Berman, H.M.; Westbrook, J.; Feng, Z.; Gilliland, G.; Bhat, T.N.; Weissig, H.; Shindyalov, I.N.; Bourne, P.E. The Protein Data Bank. *Nucleic Acids Res.* **2000**, *28*, 235–242. [CrossRef] [PubMed]
206. Burley, S.K.; Berman, H.M.; Bhikadiya, C.; Bi, C.; Chen, L.; Di Costanzo, L.; Christie, C.; Dalenberg, K.; Duarte, J.M.; Dutta, S.; et al. RCSB Protein Data Bank: Biological macromolecular structures enabling research and education in fundamental biology, biomedicine, biotechnology and energy. *Nucleic Acids Res.* **2019**, *47*, D464–D474. [CrossRef]
207. Li, G.Y.; Zhang, Y.; Inouye, M.; Ikura, M. Inhibitory mechanism of *Escherichia coli* RelE-RelB toxin-antitoxin module involves a helix displacement near an mRNA interferase active site. *J. Biol. Chem.* **2009**, *284*, 14628–14636. [CrossRef]
208. Takagi, H.; Kakuta, Y.; Okada, T.; Yao, M.; Tanaka, I.; Kimura, M. Crystal structure of archaeal toxin-antitoxin RelE-RelB complex with implications for toxin activity and antitoxin effects. *Nat. Struct. Mol. Biol.* **2005**, *12*, 327–331. [CrossRef]
209. Francuski, D.; Saenger, W. Crystal structure of the antitoxin-toxin protein complex RelB-RelE from Methanococcus jannaschii. *J. Mol. Biol.* **2009**, *393*, 898–908. [CrossRef]
210. Garcia-Pino, A.; Christensen-Dalsgaard, M.; Wyns, L.; Yarmolinsky, M.; Magnuson, R.D.; Gerdes, K.; Loris, R. Doc of prophage P1 is inhibited by its antitoxin partner Phd through fold complementation. *J. Biol. Chem.* **2008**, *283*, 30821–30827. [CrossRef]

211. Kamada, K.; Hanaoka, F. Conformational change in the catalytic site of the ribonuclease YoeB toxin by YefM antitoxin. *Mol. Cell* **2005**, *19*, 497–509. [CrossRef] [PubMed]
212. Miallau, L.; Jain, P.; Arbing, M.A.; Cascio, D.; Phan, T.; Ahn, C.J.; Chan, S.; Chernishof, I.; Maxson, M.; Chiang, J.; et al. Comparative proteomics identifies the cell-associated lethality of M. *tuberculosis RelBE-like toxin-antitoxin complexes. Structure* **2013**, *21*, 627–637. [PubMed]
213. Schureck, M.A.; Maehigashi, T.; Miles, S.J.; Marquez, J.; Cho, S.E.; Erdman, R.; Dunham, C.M. Structure of the Proteus vulgaris HigB-(HigA)2-HigB toxin-antitoxin complex. *J. Biol. Chem.* **2014**, *289*, 1060–1070. [CrossRef]
214. Hadži, S.; Garcia-Pino, A.; Haesaerts, S.; Jurėnas, D.; Gerdes, K.; Lah, J.; Loris, R. Ribosome-dependent Vibrio cholerae mRNAse HigB2 is regulated by a beta-strand sliding mechanism. *Nucleic Acids Res.* **2017**, *45*, 4972–4983. [CrossRef]
215. Talavera, A.; Tamman, H.; Ainelo, A.; Konijnenberg, A.; Hadži, S.; Sobott, F.; Garcia-Pino, A.; Hõrak, R.; Loris, R. A dual role in regulation and toxicity for the disordered N-terminus of the toxin GraT. *Nat. Commun.* **2019**, *10*, 972. [CrossRef] [PubMed]
216. Yoon, W.S.; Seok, S.H.; Won, H.S.; Cho, T.; Lee, S.J.; Seo, M.D. Structural changes of antitoxin HigA from Shigella flexneri by binding of its cognate toxin HigB. *Int. J. Biol. Macromol.* **2019**, *130*, 99–108. [CrossRef]
217. Krissinel, E.; Henrick, K. Inference of macromolecular assemblies from crystalline state. *J. Mol. Biol.* **2007**, *372*, 774–797. [CrossRef]
218. Pettersen, E.F.; Goddard, T.D.; Huang, C.C.; Couch, G.S.; Greenblatt, D.M.; Meng, E.C.; Ferrin, T.E. UCSF Chimera—A visualization system for exploratory research and analysis. *J. Comput. Chem.* **2004**, *25*, 1605–1612. [CrossRef]
219. Russell, A.D.; Hatfull, G.F. PhagesDB: The actinobacteriophage database. *Bioinformatics* **2017**, *33*, 784–786. [CrossRef]
220. Dienemann, C.; Bøggild, A.; Winther, K.S.; Gerdes, K.; Brodersen, D.E. Crystal structure of the VapBC toxin-antitoxin complex from Shigella flexneri reveals a hetero-octameric DNA-binding assembly. *J. Mol. Biol.* **2011**, *414*, 713–722. [CrossRef]
221. Lee, I.G.; Lee, S.J.; Chae, S.; Lee, K.Y.; Kim, J.H.; Lee, B.J. Structural and functional studies of the Mycobacterium tuberculosis VapBC30 toxin-antitoxin system: Implications for the design of novel antimicrobial peptides. *Nucleic Acids Res.* **2015**, *43*, 7624–7637. [CrossRef] [PubMed]
222. Korch, B.S.; Contreras, H.; Clark-Curtiss, J.E. Three Mycobacterium tuberculosis Rel toxin-antitoxin modules inhibit mycobacterial growth and are expressed in infected human macrophages. *J. Bacteriol.* **2009**, *191*, 1618–1630. [CrossRef] [PubMed]

13

Clostridium perfringens Sialidases: Potential Contributors to Intestinal Pathogenesis and Therapeutic Targets

Jihong Li [1], Francisco A. Uzal [2] and Bruce A. McClane [1,*]

[1] Department of Microbiology and Molecular Genetics, University of Pittsburgh School of Medicine, Room 420, Bridgeside Point II Building, 450 Technology Drive, Pittsburgh, PA 15219, USA; jihongli@pitt.edu
[2] California Animal Health and Food Safety Laboratory, San Bernardino Branch, School of Veterinary Medicine, University of California-Davis, San Bernardino, CA 92408, USA; fauzal@ucdavis.edu
* Correspondence: bamcc@pitt.edu.

Abstract: *Clostridium perfringens* is a major cause of histotoxic and intestinal infections of humans and other animals. This Gram-positive anaerobic bacterium can produce up to three sialidases named NanH, NanI, and NanJ. The role of sialidases in histotoxic infections, such as gas gangrene (clostridial myonecrosis), remains equivocal. However, recent in vitro studies suggest that NanI may contribute to intestinal virulence by upregulating production of some toxins associated with intestinal infection, increasing the binding and activity of some of those toxins, and enhancing adherence of *C. perfringens* to intestinal cells. Possible contributions of NanI to intestinal colonization are further supported by observations that the *C. perfringens* strains causing acute food poisoning in humans often lack the *nanI* gene, while other *C. perfringens* strains causing chronic intestinal infections in humans usually carry a *nanI* gene. Certain sialidase inhibitors have been shown to block NanI activity and reduce *C. perfringens* adherence to cultured enterocyte-like cells, opening the possibility that sialidase inhibitors could be useful therapeutics against *C. perfringens* intestinal infections. These initial in vitro observations should be tested for their in vivo significance using animal models of intestinal infections.

Keywords: *Clostridium perfringens*; intestinal infections; gas gangrene; toxins; sialidases; sialidase inhibitors

---

1. An Introduction to *Clostridium perfringens*

Clostridium perfringens is present throughout the environment, including soil, sewage, feces, foods, and the normal gastrointestinal flora of animals [1,2]. This Gram-positive, anaerobic, spore-forming bacterium is also a feared pathogen of both humans and other animals [2,3]. The most notable *C. perfringens* histotoxic infection is the rapidly-fatal human disease named clostridial myonecrosis (traumatic gas gangrene) [4,5]. This bacterium is also a preeminent cause of common, and sometimes lethal, infections originating in the intestines of humans or livestock [2,6]. Those intestinal infections often involve damage to the small intestine, or to both the small intestine and colon, which results in enteritis or enterocolitis, respectively [1,2]. *C. perfringens* intestinal infections can also progress to enterotoxemia, where a toxin(s) is produced in the intestines and then absorbed to affect extraintestinal organs such as the brain [2,7].

The virulence of this bacterium involves its ability to produce a vast toxin armory [2,3,8]. Currently ~20 different *C. perfringens* toxins have been identified, with more likely awaiting discovery [8–15]. Toxin production repertoires vary greatly among different *C. perfringens* strains, permitting classification of these isolates into five types (A–E), based upon an isolate's production of four typing toxins (alpha, beta, iota, and epsilon toxins) (Table 1) [9,10].

Table 1. *C. perfringens* typing table.

Type	Toxin Production [a]			
	α	β	ε	ι
A	+	−	−	−
B	+	+	+	−
C	+	+	−	−
D	+	−	+	−
E	+	−	−	+

[a] + indicates production of that toxin, while – indicates no production of that toxin.

C. perfringens type designations correlate with disease causation, as shown in Table 2. Two typing toxins, i.e., beta toxin (CPB) and epsilon toxin (ETX), have proven importance in *C. perfringens* intestinal infections of mammalian livestock [3,16,17]. *C. perfringens* produces other toxins that, while not used for typing classification, are nonetheless important for infections originating in the intestines of agriculturally-important animals. The foremost example is necrotic enteritis B (NetB) toxin, which is critical when *C. perfringens* causes avian necrotic enteritis in poultry [12].

Table 2. Diseases associated with the major types/subtypes of *C. perfringens*.

Toxinotypes [a]	Subtype	Most Significant Diseases [b]
A	No CPE [c] or NetB production	Human and animal myonecrosis (gas gangrene)
	NetB-producing	Necrotic enteritis of poultry
	CPE-producing	Human food poisoning and non-foodborne gastrointestinal disease
B		Necro-hemorrhagic enteritis of sheep (lamb dysentery)
C		Human enteritis necroticans (Darmbrand, pigbel); necrotic enteritis of neonatal individuals of several animal species (e.g., cattle, sheep, pigs)
D		Enterotoxemia of sheep and goats
E		Suspected association with gastrointestinal disease of cattle, sheep and rabbits

[a] All types of *C. perfringens* may also produce several other toxins, including, but not limited to, beta2 toxin (CPB2), perfringolysin O (PFO), and toxin *C. perfringens* large cytotoxin (TpeL); [b] Only diseases that have been confirmed to be associated with each type of *C. perfringens* and significant in terms of prevalence are included in this table; [c] CPE is *C. perfringens* enterotoxin.

With respect to human *C. perfringens* infections, type A strains are responsible for causing most histotoxic infections. During gas gangrene, alpha toxin (CPA) plays the major role in virulence. A non-typing toxin named PFO also contributes to this disease [4,5].

To date, only type A and C strains of *C. perfringens* have been conclusively linked to human diseases originating in the intestines [1–3,18]. Type C strains use their CPB to cause enteritis necroticans (EN), which was first described in post-World War II Germany, where it was referred to as darmbrand [18–20]. In the 1960s–1970s, EN, known locally as pigbel, was a major cause of death of children in the Papua New Guinea (PNG) Highlands [20,21]. Pigbel develops in children with reduced trypsin levels due to predisposing conditions, including malnutrition, a diet rich in sweet potato (which contains a trypsin inhibitor), and/or intestinal infections with pathogens producing a trypsin inhibitor [20,21]. Their low intestinal trypsin levels render these children susceptible to infection by type C strains because normal trypsin levels would otherwise easily inactivate CPB when it is produced in the intestines. Consequently, children suffering from pigbel develop CPB-induced necrotic enteritis or enterotoxemia and often die rapidly. The only treatment for pigbel is resection of the bowel; however, this surgical intervention is only effective if performed early after the onset of infection [20,21]. A vaccine introduced in the 1980s dropped the incidence of pigbel dramatically in PNG. Unfortunately, pigbel vaccination has since decreased and this illness may now be reappearing.

Although not used for typing classification, CPE is the toxin responsible for causing the gastrointestinal symptoms of *C. perfringens* type A food poisoning (FP) [1,22]. This FP is currently the 2nd most common bacterial foodborne disease in the USA, where one million cases occur annually and economic losses approach $500 million/year [1,23]. In people with fecal impaction or severe constipation due to side-effects from medications used to treat other pre-existing conditions, *C. perfringens* type A FP can be much more severe and is often fatal [1,24]. Studies with animal models suggest this may be due to absorption of CPE from the intestines, resulting in an enterotoxemia that involves the liver and kidneys [24]. CPE-producing type A strains also cause about 5%–15% of all cases of nonfoodborne human GI diseases, most notably antibiotic-associated diarrhea (AAD) [1,25].

As discussed later, CPE-associated AAD cases are more severe and longer lasting than typical cases of *C. perfringens* type A FP, which usually self-resolve within 24 h [1,25]. While the *cpe* gene can be either chromosomal or plasmid-borne in type A strains, most (~70%) type A FP strains carry a chromosomal *cpe* gene [1]. In contrast, nearly 100% of type A AAD strains carry a plasmid *cpe* gene [1]. There are also many other genetic differences between these two groups of type A *cpe*-positive groups. For example, type A strains carrying a chromosomal *cpe* gene produce a unique small acid soluble protein (SASP4) variant that provides their spores with greater resistance to heat, cold, and chemical treatment and probably facilitates survival of these bacteria in the food environment [1,26,27]. Other differences between type A chromosomal *cpe* strains and type A plasmid *cpe* strains, particularly with regard to sialidase gene carriage, will be discussed later.

In addition to its essential role during intestinal infections by type A *cpe*-positive strains, CPE may also contribute to some cases of human EN caused by type C strains [18]. However, CPB is clearly of critical importance for the pathogenesis of this disease [16].

2. *C. perfringens* Sialidases

Sialic acids are a carbohydrate family containing about 40 different nine-carbon relatives of neuraminic acid [28,29]. Under physiological conditions, sialic acids are negatively charged [29]. N-acetylneuraminic acid (Neu5Ac), whose amino group is acetylated, is the most widespread sialic acid [29,30]. Sialic acids are important components of the serum and mucus and represent the terminal sugar residue of many glycan chains on host cell surfaces, where they are involved in cell- cell recognition. Sialic acids can also stabilize enzymes or cell membrane proteins. Finally, due to their negative charge, sialic acids can mediate binding and transport of positively-charged molecules [28–30].

Sialidases, also referred to as neuraminidases (E.C.3.2.1.18), are key enzymes that hydrolyze the α-linkage of terminal sialic acids on various sialoglycoconjugates to generate free sialic acid [28–31]. Sialidases are made by certain viruses, microorganisms, and vertebrate animals, but not by plants. Included amongst the sialidase-producing viruses and bacteria are several prominent pathogens, e.g., influenza virus, *Vibrio cholerae*, *Streptococcus pneumoniae*, and *C. perfringens*. Sialidases can have a nutritional function for normal flora or pathogenic bacteria [28,29,31]. In addition, they often function directly as virulence factors during bacterial pathogenesis [30,32], as will be discussed later for *C. perfringens*.

C. perfringens produces three different sialidases, which are named NanH, NanI, and NanJ [33]. NanH (43 kDa) lacks a secretion signal peptide and thus has a cytoplasmic location in log-phase cultures [33,34]. In contrast, NanI (77 kDa) and NanJ (129 kDa) are secreted exosialidases ([33], Figure 1). The catalytic modules of all three sialidases show conserved amino acid sequence identity and belong to the family 33 carbohydrate binding module (CBMs, Figure 1) [35]. Compared to NanH, which only consists of a catalytic domain, NanI and NanJ also possess additional accessory carbohydrate-binding modules [35]. It is thought that these carbohydrate-binding domains increase the binding affinity between NanI and NanJ and their polyvalent substrates. NanJ has a complex multimodular structure comprised of a central catalytic module and five accessory modules [35]. The two N-terminal modules show amino acid sequence identity with family 32 and family 40 CBMs. NanI has a simpler structure consisting of a catalytic module and an N-terminal family 40 CBM [35].

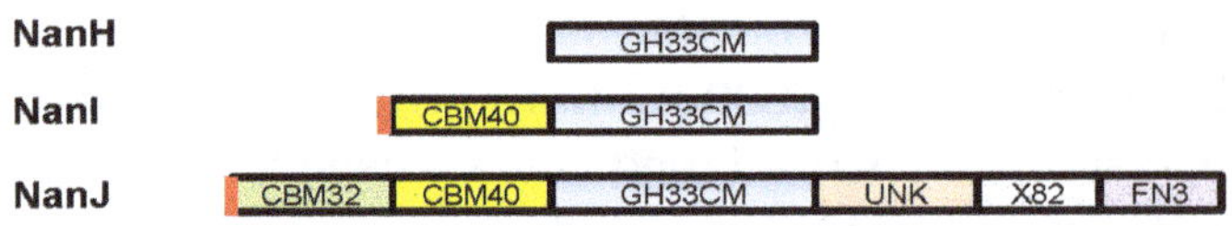

Figure 1. Modular organization of the *C. perfringens* sialidases. CBM32 is a carbohydrate binding module (CBM) belonging to the family 32 CBM; CBM40 is a module having an amino acid sequence identity with a family 40 CBM; GH33CM is a family 33 CBM; UNK is a module having unknown function; X82 is a family 82 "X module" of unknown function; FN3 is a module sharing distant identity with fibronectin type III domains. The red small boxes are secretion signals. Modified with permission from [35].

Most *C. perfringens* strains produce all three sialidases (further discussion below). However, as discussed in more detail later, some *C. perfringens* strains produce only one or two of the three sialidases. For strains producing all three sialidases, NanI is usually responsible for ~70% of total exosialidase activity [33,34,36,37].

To characterize the properties of the three *C. perfringens* sialidases, a recent study constructed a series of isogenic mutants, where two of the three sialidase genes present in *C. perfringens* type D strain CN3718 were inactivated [33]. This strategy created mutants that were each expressing, at their native levels, only NanJ, NanI, or NanH in a background free of contamination from the other two sialidases to allow a precise characterization of the enzymatic properties of each sialidase. NanI was found to be more heat-tolerant compared to NanJ or NanH, both of which exhibited greatly reduced sialidase activity at temperatures above 43 °C [33]. In this experimental system, all three sialidases worked best at low pH conditions (pH ~5). The enzyme activity of each sialidase was shown to vary in sensitivity to various metal ions [33]. Furthermore, unlike the sialidases from *Streptomyces* spp., *C. perfringens* sialidases were found to be sensitive to p-chloromercuribenzoate, which reacts with thiol groups in proteins [33]. Finally, the three *C. perfringens* sialidases showed different substrate preferences. NanI exhibited preferential activity in the order of α-2,3 > α-2,6 > α-2,8 sialic acid linkages and was responsible for most of the activity in CN3718 supernatants that was directed against those sialic acid linkages. NanJ showed a preference for α-2,6 > α-2,8 > α-2,3 sialic acid linkages. Finally, NanH activity was strongest for α-2,8 > α-2,3 > α-2,6 sialic acid linkages [33]. This diversity in linkage preferences suggests that, when present together, the three *C. perfringens* sialidases work in combination to generate free sialic acid, even from complex substrates [33].

3. *C. perfringens* Sialidases: Genetics and Regulation of Expression

All three *C. perfringens* sialidases are encoded by chromosomal genes, although those genes are located in different regions of the chromosome [38,39]. The sialidase-encoding ORFs in many *C. perfringens* strains have now been sequenced. Those sequencing analyses indicated that, amongst different *C. perfringens* strains, the NanJ sequence shares 96% to 100% identity, the NanI sequence has 98% to 100% identity, and the NanH sequence shares 93% to 100% identity [34,38,39].

As true for their toxin production, *C. perfringens* strains vary in their patterns of sialidase production. Most strains produce all three sialidases, with NanI usually being responsible for most of the sialidase activity in culture supernatants of those *C. perfringens* strains [34,36]. However, NanI production is not essential for *C. perfringens* growth since some strains of this bacterium naturally lack the *nanI* gene [36]. Interestingly, the *nanI* gene is consistently absent from the type A FP strains carrying a chromosomal *cpe* gene, as well as the genetically related type C darmbrand strains [36]. In contrast, the *nanI* gene is carried by most plasmid *cpe*-carrying type A AAD strains, *cpe*-negative type A normal human intestinal flora strains, and type C pig-bel strains [36]. Since NanI is usually the major sialidase of *C. perfringens*, it is not surprising that exosialidase activity is typically significantly lower for naturally *nanI*-negative strains compared to strains carrying a *nanI* gene [36]. The potential pathogenic importance of these differences will be discussed later.

Most type A FP strains with a chromosomal *cpe* gene lack the *nanJ* gene, as well as the *nanI* gene. However, those typical FP strains do carry the *nanH* gene. It should be noted that occasional strains of *C. perfringens* besides the type A chromosomal *cpe* FP strains or type C darmbrand strains also lack a sialidase gene. For example, Strain 13 is a *cpe*-negative type A strain that can cause gas gangrene yet it lacks the *nanH* gene [37,40].

The regulation of sialidase production by *C. perfringens* is complicated (Figure 2). A number of regulators influencing expression of one or more sialidase genes have been identified. For example, the VirS/VirR two-component signal transduction system was shown to upregulate *nanI* and *nanJ* expression [41]. Rather than a direct regulatory effect involving VirS binding to sialidase gene promoters, this VirS/R two-component system positively controls expression of the *vrr* gene, which encodes *virR*-regulated RNA (VR-RNA). This regulatory RNA then modulates sialidase gene expression [42]. ReeS is another sensor kinase whose presence also increases *nanI* and *nanJ* sialidase gene expression by a putative response regulator, referred to as ReeR [43]. In contrast, a transcriptional regulator named RevR has different regulatory effects on *nanJ* and *nanI* expression, possibly via a proposed sensor kinase named RevS. RevR increases *nanJ* expression, but negatively regulates *nanI* expression [44]. Finally, using a *codY* null mutant of *C. perfringens* type D strain CN3718, our group showed that CodY represses NanJ and NanH production, although it does not affect NanI production [45].

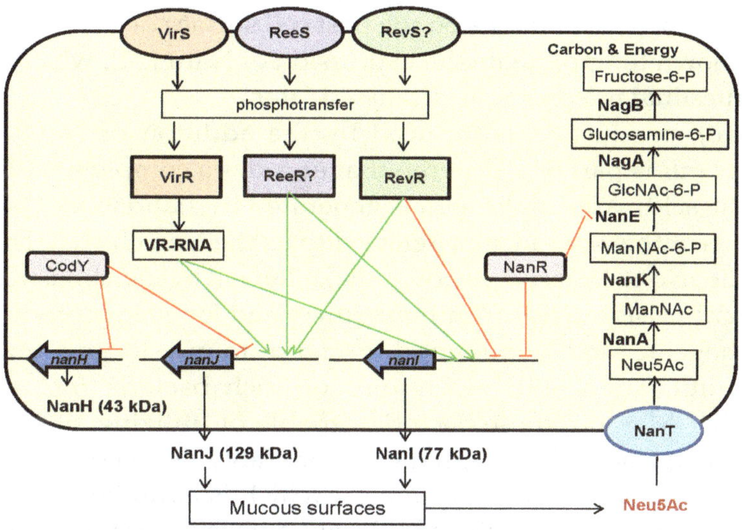

Figure 2. A proposed model for regulation of expression of sialidase genes and the pathway for sialic acid metabolism in *C. perfringens*. Exosialidases generate free sialic acid from mucus or host cell surfaces [33,40]. The free sialic acid is then transported into *C. perfringens* where it is metabolized to fructose-6-P [40]. There is evidence that VirS/VirR, RevR, ReeS, NanR, and CodY systems directly or indirectly affect sialidase production, although inter-relationships between these regulators are unclear [40,41,43–46]. The VirS/VirR two component system acts as a positive regulator of the *vrr* gene, which encodes VR-RNA. VR-RNA is then a positive regulator of *nanI* and *nanJ* expression [41,42]. The ReeS sensor kinase positively regulates *nanI* and *nanJ* gene expression, presumably by a putative transcriptional regulator named ReeR [43]. RevS positively regulates *nanJ* expression but negatively regulates *nanI* expression [44]. CodY represses *nanH* and *nanJ* expression [45]. Based on sequence homology comparisons with similar regulators in other bacteria, NanR may repress *nanI* expression and the sialic acid metabolism pathway, but there is no direct experimental evidence yet to support this hypothesis. Green lines indicate positive regulation while red lines indicate negative regulation.

No proven regulator has yet been identified that directly modulates sialidase production by binding to the promoters of *C. perfringens* sialidase genes. However, gel mobility shift assays have demonstrated high affinity binding of a purified protein named NanR to DNA from the promoter region of the *nanI* gene [40]. Since NanR has homology with the ribose-5-phosphate isomerase B regulator (RpiR) family of transcriptional repressors known to control sialidase production in bacteria such as *E. coli*, *Vibrio vulnifcus*, and *Staphylococcus aureus*, this result suggests that NanR may be involved in *nanI* expression regulation [31,40,47]. Consistent with its potential involvement in sialic acid generation and usage, NanR lies within a six-gene operon encoding the complete pathway for transport and metabolism of sialic acid by *C. perfringens* [40]. As discussed in more detail below, this operon also encodes NanE (epimerase) and NanA (sialic acid lyase) enzymes.

For the *nanI* and *nanJ* genes, primer extension analyses identified three or two putative transcription start sites, respectively [40]. These promoters are located within ~500 bp of the start codons of *nanI* and *nanJ*. This multiplicity of promoters may provide one explanation for why so many different regulators control sialidase expression, although (as mentioned) detailed understanding of regulator binding to these promoter sequences is currently lacking [40].

Only a limited number of bacterial species, mainly those having a close association with the sialic acid-rich environment of the host, can utilize Neu5Ac [48]. *C. perfringens* was actually the first bacterium demonstrated to be capable of utilizing a sialic acid (Neu5Ac) as a carbon source [31]. This result was later confirmed by another study showing that Neu5Ac can be used by *C. perfringens* when growing in a semi-defined medium [31]. It was also recently found that NanI and NanJ, but not NanH, can cause the release of sialic acid from Caco-2 cells [33]. This effect may contribute to pathogenesis, since these sialidases could help *C. perfringens* obtain nutrients in vivo by releasing sialic acid from glycolipids or glycoproteins on the host cell surface or in mucus. Interestingly, contact with Caco-2 cells was shown to upregulate the expression of NanI [34], which could potentiate NanI contributions during intestinal infections, as discussed later.

NanI, but not NanJ, production is induced by the addition of Neu5Ac to a medium [40]. As introduced earlier, NanR is part of an operon that encodes a complete pathway for the transport and metabolism of sialic acid. After sialic acid is generated by sialidases, it is then transported and metabolized by products of this operon, also referred to as the Nan cluster. This process involves an initial conversion of sialic acid to N-acetyl glucosamine, followed by metabolism of that carbohydrate to fructose-6-P, which *C. perfringens* can use as carbon sources and for energy production (Figure 2) [40,46].

Many other pathogenic and commensal bacteria found in the intestines carry a Nan cluster for sialic acid utilization [48]. Examples of such bacteria include *Vibrio cholerae* and *Salmonella enterica* [48,49]. Many of these bacteria capable of utilizing sialic acid also colonize the human or animal intestinal tract. This correlation is probably not a coincidence since sialic acid is a component of mucin, the major protein in mucus, which is abundant in the human and animal intestines [48]. However, to generate free sialic acid from mucus for uptake and metabolism inside the bacterial cell, extracellular sialidases must be present in the intestinal environment. Interestingly, some pathogenic bacteria, e.g., *C. difficile*, have a Nan cluster but do not produce their own sialidase. Instead, it is believed that *C. difficile* uses free sialic acid generated in the intestines by sialidases produced by other bacteria, such as *Bacteroides thetaiotaomicron* [50,51].

4. Possible Contributions of Sialidases to *C. perfringens* Diseases

Some pathogens use sialic acids to coat their cell surface, the flagellum, the capsule polysaccharide, or the lipopolysaccharide. This masks these bacteria so they can avoid the host immune system defense [28]. Whether *C. perfringens* coats its surface with host-derived sialic acid has not been studied, to our knowledge.

Sialidases can also promote in vivo growth and colonization of bacterial pathogens [30]. For extraintestinal infections, an example is the major human respiratory tract pathogen *Streptococcus pneumoniae*, which encodes up to three sialidases, named NanA, NanB, and NanC. NanA is the predominant sialidase that removes the sialic acid Neu5Ac from a variety of glycoconjugates. NanB is an intramolecular trans-sialidase producing 2,7-anhydro-Neu5Ac selectively from α2,3-sialosides, while NanC produces 2-deoxy-2,3-didehydro-*N*-acetylneuraminic acid (Neu5Ac2en), which can be hydrated to Neu5Ac. The three pneumococcal sialidases share a common catalytic mechanism up to the final product formation step, and all three sialidases are implicated in pathogenesis, including colonization, and are potential drug targets [52]. *S. pneumoniae* then takes up and metabolizes sialic acid using a similar pathway as present in *C. perfringens* [40].

The contribution of sialidases to growth and colonization during extraintestinal infections caused by *C. perfringens* is unclear. One study demonstrated that a *nanJ* and *nanI* double null mutant of *C. perfringens* strain 13 remains fully virulent in the mouse myonecrosis model [37]. This result could suggest that Neu5Ac metabolism is not essential for growth or colonization by *C. perfringens* in muscle. However, as noted in that study [37], this result does not necessarily preclude subtle contributions of sialidases to colonization or growth since the mouse myonecrosis model requires a massive inoculum that may mask such contributions.

Increasing evidence indicates that sialidases and sialic acid often play significant roles in growth and colonization of the intestines by bacterial pathogens [34,49,52,53]. For example, sialic acid appears to be an important source of carbon and energy for survival, growth and adherence of *E. coli* in the gastrointestinal system [54]. Similarly, catabolism of sialic acid by *V. cholerae* plays a significant role in both in vitro and in vivo colonization and growth [49]. In addition, *V. cholerae* sialidase enhances the binding and uptake of cholera toxin [55]. In *V. vulnificus*, a Nan utilization system is also important for bacterial colonization and growth in the intestines [56]. The sialidase NanS play a role in *Clostridium sordellii* adhesion and enhances Non-TcsL mediated cytotoxicity [53]. As mentioned above, *C. difficile* uses sialic acid generated by sialidases of other bacteria to grow and persist in the intestines [50,51].

As mentioned earlier, NanI may contribute to in vivo growth. Contact of *C. perfringens* with Caco-2 cells increases the production of this enzyme [34] and NanI can induce the release of free sialic acids from enterocyte-like Caco-2 cells [33]. Since *C. perfringens* possesses a sialic acid utilization system, similar NanI-generated sialic acid in the intestines should be useful for in vivo growth of this bacterium.

Earlier studies demonstrated that NanI production also impacts the production of several *C. perfringens* toxins [34,37,57]. Specifically, inactivating the *nanI* gene in *C. perfringens* type A strain 13 caused a slight increase in supernatant activities of alpha toxin and perfringolysin O (PFO), which are contributors to gas gangrene [37], but their role (if any) in most intestinal infections, particularly human infections, is unsettled [3]. However, it was later demonstrated that inactivating the *nanI* gene in type D strain CN3718 reduced ETX toxin production in vitro and that either genetic or physical complementation could recover this ETX production [57]. NanI effects on ETX production involved reductions in both *codY* and *ccpA* transcript levels, suggesting a model whereby NanI generates sialic acid release in the intestines and that free sialic acid then alerts a type D strain of its presence in the intestines, making it worthwhile for this bacterium to upregulate ETX production to induce disease.

There have been only limited studies to date on the effects of *C. perfringens* sialidase on toxin action, particularly on those toxins important for intestinal infections. An exception is ETX, where conclusions from early studies had been contradictory. One early study [58] reported that sialidases enhance ETX cytotoxicity towards Madin-Derby Canine Kidney (MDCK) epithelial cells, while a second study [59] reported that pretreating synaptosomal membranes with sialidases lowers their subsequent ETX binding levels. Recently a study used purified NanI, as well as *nanI*, *nanJ*, or *nanH* single null mutants, a *nanI/nanJ* double null mutant, and a triple sialidase null mutant, to study possible sialidase enhancement of ETX action [34]. Results obtained [34] conclusively showed that NanI sialidase increases the ETX sensitivity of MDCK cells and that this effect involves an increase in ETX binding levels. The mechanism of this enhancement is not yet clear but it could involve either NanI increasing the exposure of ETX receptors on the host cell surface or NanI modifying the host cell surface charge to increase binding of this toxin. Whether a similar effect extends to other *C. perfringens* toxins active during intestinal infections is not yet clear.

Only limited studies have addressed the mechanism of *C. perfringens* adherence to host cells and tissues [60–62]. For example, the adhesins used by *C. perfringens* in vivo remain unclear. However, some strains produce a collagen adhesion protein (CNA), and/or fibronectin binding proteins (FbpA, FbpB) that have been implicated in adhesion and colonization [60–63].

Sialidases are known to contribute to the ability of some bacterial pathogens to adhere to host cells, tissues, or mucosal surfaces in the airways or intestines [30]. A recent in vitro study suggests sialidase contributions to *C. perfringens* intestinal adherence [34]. That study [34] demonstrated that type D strain CN3718 attaches to cultured Caco-2 enterocyte-like cells. This adhesion was specific since this intestinal disease strain exhibited significantly less adherence to fibroblasts and kidney cells [34]. Interestingly, the attachment of CN3718 to Caco-2 cells is greatly facilitated by NanI production (Figure 3). Wild-type CN3718 exhibited much more adherence to Caco-2 cells compared to an isogenic triple mutant strain that does not produce any sialidase. Complementation studies showed that, of the three sialidases made by CN3718, restoring NanI production to the triple sialidase mutant yielded the greatest enhancement of adherence. That result indicated that NanI is of prime importance for sialidase enhancement of CN3718 attachment to Caco-2 cells [34].

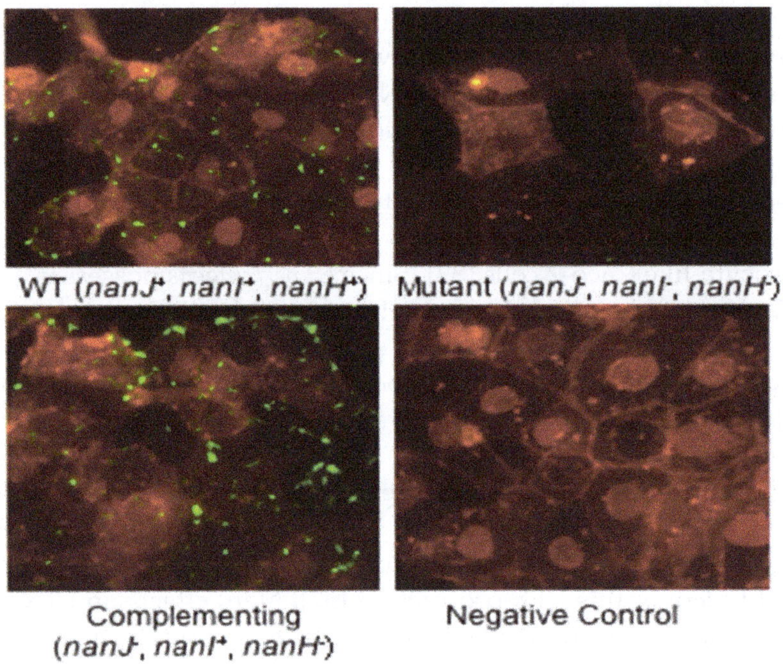

Figure 3. Adhesion of *C. perfringens* CN3718 to Caco-2 cells [34]. Caco-2 cells were incubated for 2 h at 37 °C under anaerobic conditions. *C. perfringens* CN3718 produces all three sialidases. This wild-type (WT) strain, and a complementing strain that produces only NanI, attach very well to Caco-2 cells as detected by immunofluorescence microscopy (600×). In comparison, only a few cells of an isogenic mutant with all three sialidase genes disrupted were able to attach to Caco-2 cells. Furthermore, when the triple mutant was complemented to produce only NanJ or NanH, those bacteria remained poorly adherent (not shown). Green: *C. perfringens*; Red: Caco-2 cells. Reproduced with permission from [34]. Creative Commons License 2011.

However, NanI itself does not appear to be a significant adhesin since the CN3718 *codY* null mutant shows 50% less adherence to Caco-2 cells compared to wild-type CN3718, despite this mutant producing the same levels of NanI as the wild-type strain [45]. This result suggests a two-step adherence mechanism. First, secreted NanI modifies the Caco-2 enterocyte-cell surface, which then allows the unknown adhesin present on the surface of *C. perfringens* to more easily bind to the still unidentified receptor(s) on the enterocyte-cell surface.

Sialidases may enhance *C. perfringens* adherence to host cells via both nonspecific and specific mechanisms. Sialic acids, typically present at the distal ends of carbohydrate chains, are negatively charged carbohydrates [28,30]. In addition, terminal sialic acids promote endothelial barrier integrity, so treatment of epithelial monolayers with *C. perfringens* sialidases may lead to barrier disruption and may increase access for *C. perfringens* so they can adhere [64]. Therefore, nonspecific effects of secreted NanI on both charge and epithelial barrier integrity could help to increase toxin binding and *C. perfringens* colonization. A *C. perfringens* sialidase (not specified) has, in fact, been shown to affect barrier resistance in some, but not all, host cells [64]. However, for some cell lines, nonspecific effects do not appear to completely explain NanI enhancement of *C. perfringens* adherence. As mentioned, *C. perfringens* CN3718 exhibits much greater adhesion for certain mammalian cells, such as Caco-2 and HT-29 intestinal cell lines, than for cell lines of nonintestinal origin [34]. Similarly, the toxins involved in *C. perfringens* intestinal infections only bind to and affect certain cells. Those observations suggest that NanI can also have specific effects in promoting *C. perfringens* adherence and toxin binding. This could involve specific effects of NanI sialidase on modifying host cell surface adhesins/toxin receptors and/or trimming back nearby molecules to better unmask the adhesin or toxin receptor on host cell surfaces.

When *C. perfringens* causes diseases originating in the intestines, its secreted proteins come into contact with host proteases, such as trypsin, that are present in the intestinal lumen. While some *C. perfringens* proteins, e.g., CPB, are highly sensitive to intestinal proteases [16], other proteins secreted by this bacterium are proteolytically-activated in the intestines. A prime example of a proteolytically-activated *C. perfringens* protein is ETX, which is initially produced as an inactive prototoxin and then activated by intestinal trypsin, chymotrypsin, and carboxypeptidases that remove N- and C-terminus residues from the prototoxin [65]. Similarly, trypsin or chymotrypsin treatment of CPE also increases cytotoxicity, in this case by removing N-terminal sequences from the native CPE protein to facilitate toxin oligomerization during pore formation [66]. Interestingly, NanI (but not NanJ or NanH) can also be proteolytically-activated by trypsin [33,34]. The increased enzymatic activity of trypsin-activated NanI was shown to be substrate-specific [33]. Collectively, these observations suggest that trypsin activation of NanI may contribute to *C. perfringens* intestinal diseases.

While NanI can enhance *C. perfringens* adherence and toxin binding, typical type A FP strains, i.e., those with a chromosomal *cpe* gene, do not carry the *nanI* gene. Furthermore, those typical FP strains possess very low exosialidase activities [36]. The limited production of sialidases by type A FP strains obviously does not hinder their ability to cause food-borne intestinal diseases. Unlike the typical FP strains, nearly all type A AAD strains do carry a *nanI* gene. When a *nanI* null mutant of a type A AAD strain was characterized, this *nanI* null mutant was shown to enter the sporulation cycle earlier and to produce more CPE than its wild-type parent. This finding supports the dispensability of NanI for the typical FP strains, which cause an acute disease. During the acute FP, NanI would not be needed for nutritional purposes or to enhance bacteria adherence or colonization since the typical FP strains are ingested in very large numbers in contaminated food, then quickly sporulate in the intestines to produce CPE, induce diarrhea via that CPE, and exit from the intestines via the flushing action of diarrhea (Figure 4) [36]. In contrast, type A AAD or sporadic diarrhea (SD) strains with a plasmid *cpe* gene cause chronic gastrointestinal disease lasting up to several weeks, so in vivo growth and colonization is important for their persistence. Those Type A AAD or SD strains may use NanI for their growth and colonization, as necessary to achieve the persistence required to cause a chronic diarrhea (Figure 4) [36].

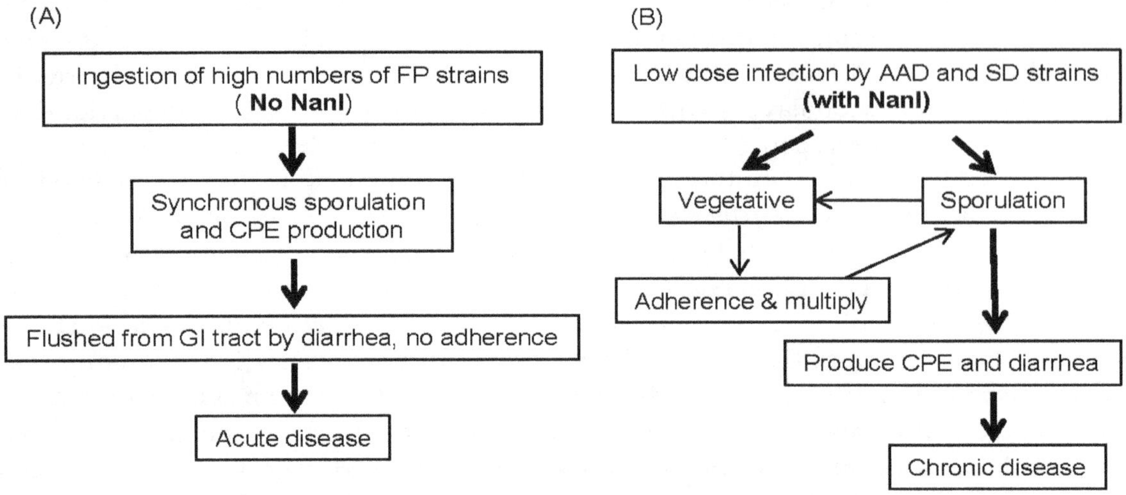

Figure 4. Possible models for acute *C. perfringens* type A FP vs. chronic gastrointestinal diseases like AAD. (**A**) *C. perfringens* type A FP does not require NanI production because strains are ingested in large amounts, sporulate in vivo to produce CPE and are then quickly removed from the intestines by diarrhea. (**B**) *C. perfringens* type A AAD does involve NanI production, which may promote adherence and colonization, as required for chronic diarrhea. Modified with permission from [36].

5. Sialidase Inhibitors: Potential Therapeutic Agents?

As described above, sialidases may offer several contributions to *C. perfringens* pathogenesis, particularly during intestinal infections. *C. perfringens* diseases are challenging to treat with antibiotics because already-synthesized toxins will continue to work even after the administration of antibiotics. Similarly, it is often difficult to treat or prevent *C. perfringens* infections with vaccines or neutralizing antibodies because some individual strains produce multiple (up to five) different toxins [2,3]. Those factors suggest that sialidase inhibitors represent interesting potential candidates for drug development.

Sialidase inhibitors have proven to be efficacious for the treatment of some infectious disease [67]. The most notable clinical application of sialidase inhibitors are several sialidase inhibitors (Zanamivir, Oseltamivir, Peramivir (Rapivab)) that are currently used for treating infections involving influenza virus [67]. However, there is also precedent for sialidase inhibitors interfering with bacterial growth and adhesion. Oseltamivir reduced the growth and adherence of *Tannerella forsythia*, which causes periodontitis, by two- to three-fold [68].

Two classic sialidase inhibitors, i.e., Siastatin B (SB) and N-acetyl-2,3-dehydro-2-deoxyneuraminic acid (NADNA), have been tested against *C. perfringens* [33,36,57,69,70]. SB, a broad-spectrum sialidase inhibitor isolated from a *Streptomyces* spp. culture, is an unusual 6-acetamido-3-piperidinecarboxy late [70]. NADNA is an analogue of 2-deoxy-2,3-didehydro-N-acetylneuraminic acid that is modified at the C-4 position. NADNA can inhibit the sialidase activity from influenza viruses A and B, parainfluenza 2 virus, *Vibrio cholerae*, *Arthrobacter ureafaciens*, *C. perfringens*, and sheep liver [69].

A recent study measured the effects of SB and NADNA on sialidase activities of overnight culture supernatants of *C. perfringens* CN3718 mutants ENanJ, ENanI, or ENanH, which contain natural levels of only NanJ, NanI, or NanH, respectively [33]. NanI was the most sensitive sialidase to both NADNA and SB, while NanH was the least sensitive sialidase to these inhibitors; IC_{50}s for these inhibitors are shown in Table 3 [33].

Table 3. Effect of sialidase inhibition on sialidase activity (IC_{50}).

Sample	NADNA (IC_{50})	Siastatin B (IC_{50})
CN3718	18.9 μM	42.2 μM
ENanJ	12.4 μM	15.1 μM
ENanI	13.4 μM	27.5 μM
ENanH	44.6 μM	50.9 μM

It was also shown that the adherence of *C. perfringens* AAD strain F4969 to Caco-2 cells can be reduced by either SB or NADNA [36]. In addition, both NADNA and SB efficiently inhibited sialidase activity in bacterial cell-free supernatants collected from *C. perfringens* type D strain CN3718. Consistent with those results, CN3718 cultures grown in the presence of the SB inhibitor also exhibited substantially reduced culture sialidase activity and ETX production. However, the NADNA sialidase inhibitor did not inhibit sialidase activity in CN3718 cultures, even at very high doses of the inhibitor. In the presence of NADNA, CN3718 cultures still possessed strong sialidase activity and also made the same amounts of ETX as wild-type CN3718 grown in the absence of any sialidase inhibitor [57]. Collectively, these results suggest that at least some sialidase inhibitors may be potentially useful therapeutics against *C. perfringens* infections.

6. Concluding Remarks and Future Directions

Most studies of *C. perfringens* virulence have, correctly, focused on toxin contributions. Only a few in vivo studies have thus far examined the role of exoenzymes in *C. perfringens* pathogenesis and their results have been equivocal. However, those studies examined the possible role of sialidases in gas gangrene. Recent in vitro studies suggest that sialidases, particularly NanI, may be important contributors to *C. perfringens* intestinal infections. Those contributions could include upregulated production of some toxins relevant to intestinal infections, enhanced binding and activity of some of those toxins, increased adherence of *C. perfringens* to host cells, and possibly generation of substrates for growth and metabolism [33,34,36,57]. There is a clear need to move these promising findings into appropriate animal models to firmly evaluate sialidase contributions to *C. perfringens* virulence.

In vitro studies also suggest that some sialidase inhibitors might be useful therapeutics for treating *C. perfringens* infections originating in the intestines [57]. This possibility also needs to be evaluated in animals. Identification of more potent inhibitors of *C. perfringens* sialidases may be helpful for those studies.

Acknowledgments: Preparation of this review was supported, in part, by a grant from the National Institute of Allergy and Infectious Disease (R21 AI125796-1).

References

1. McClane, B.A.; Robertson, S.L.; Li, J. *Clostridium perfringens*. In *Food Microbiology: Fundamentals and Frontiers*, 4th ed.; Doyle, M.P., Buchanan, R.L., Eds.; ASM Press: Washington, DC, USA, 2013; pp. 465–489.
2. McClane, B.A.; Uzal, F.A.; Miyakawa, M.F.; Lyerly, D.; Wilkins, T.D. The enterotoxic clostridia. In *The Prokaryotes*, 3rd ed.; Dworkin, M., Falkow, S., Rosenburg, E., Schleifer, H., Stackebrandt, E., Eds.; Springer: New York, NY, USA, 2006; pp. 688–752.
3. Uzal, F.A.; Freedman, J.C.; Shrestha, A.; Theoret, J.R.; Garcia, J.; Awad, M.M.; Adams, V.; Moore, R.J.; Rood, J.I.; McClane, B.A. Towards an understanding of the role of *Clostridium perfringens* toxins in human and animal disease. *Future Microbiol.* **2014**, *9*, 361–377. [CrossRef] [PubMed]
4. Rood, J.I. *Clostridium perfringens* and histotoxic disease. In *The prokaryotes: A Handbook on the Biology of Bacteria*, 3rd ed.; Dworkin, M., Falkow, S., Rosenberg, E., Schleifer, K.-H., Stackebrandt, E., Eds.; Springer: New York, NY, USA, 2007; Volume 4, pp. 753–770.

5. Titball, R.W.; Rood, J.I. *Clostridium pefringens*: Wound indections. In *Molecular Medical Microbiology*; Sussman, M., Ed.; Academic Press: London, UK, 2002; pp. 1875–1903.
6. Uzal, F.A.; Vidal, J.E.; McClane, B.A.; Gurjar, A.A. *Clostridium perfringens* toxins involved in mammalian veterinary diseases. *Open toxinol. J.* **2010**, *2*, 24–42. [CrossRef]
7. Songer, J.G. Clostridial enteric diseases of domestic animals. *Clin. Microbiol. Rev.* **1996**, *9*, 216–234. [PubMed]
8. Hatheway, C. Toxigenic clostridia. *Clin. Microb. Rev.* **1990**, *3*, 66–76. [CrossRef]
9. McDonel, J.L. Toxins of *Clostridium perfringens* types A, B, C, D, and E. In *Pharmacology of Bacterial Toxins*; Dorner, F., Drews, H., Eds.; Pergamon Press: Oxford, UK, 1986; pp. 477–517.
10. Petit, L.; Gilbert, M.; Popoff, M. *Clostridium perfringens*: Toxinotype and genotype. *Trends Microbiol.* **1999**, *7*, 104–110. [CrossRef]
11. Animoto, K.; Noro, T.; Oishi, E.; Shimizu, M. A novel toxin homologous to large clostridial cytotoxins found in culture supernatant of *Clostridium perfringens* type C. *Microbiology* **2007**, *153*, 1198–1206. [CrossRef] [PubMed]
12. Keyburn, A.L.; Boyce, J.D.; Vaz, P.; Bannam, T.L.; Ford, M.E.; Parker, D.; Di Rubbo, A.; Rood, J.I.; Moore, R.J. NetB, a new toxin that is associated with avian necrotic enteritis caused by *Clostridium perfringens*. *PLoS Pathog.* **2008**, *4*, e26. [CrossRef] [PubMed]
13. Mehdizadeh Gohari, I.; Parreira, V.R.; Timoney, J.F.; Fallon, L.; Slovis, N.; Prescott, J.F. NetF-positive *Clostridium perfringens* in neonatal foal necrotising enteritis in kentucky. *Vet. Rec.* **2016**, *178*, 216. [CrossRef] [PubMed]
14. Yonogi, S.; Matsuda, S.; Kawai, T.; Yoda, T.; Harada, T.; Kumeda, Y.; Gotoh, K.; Hiyoshi, H.; Nakamura, S.; Kodama, T.; et al. BEC, a novel enterotoxin of *Clostridium perfringens* found in human clinical isolates from acute gastroenteritis outbreaks. *Infect Immun.* **2014**, *82*, 2390–2399. [CrossRef] [PubMed]
15. Irikura, D.; Monma, C.; Suzuki, Y.; Nakama, A.; Kai, A.; Fukui-Miyazaki, A.; Horiguchi, Y.; Yoshinari, T.; Sugita-Konishi, Y.; Kamata, Y. Identification and characterization of a new enterotoxin produced by *Clostridium perfringens* isolated from food poisoning outbreaks. *PLoS ONE* **2015**, *10*, e0138183. [CrossRef] [PubMed]
16. Sayeed, S.; Uzal, F.A.; Fisher, D.J.; Saputo, J.; Vidal, J.E.; Chen, Y.; Gupta, P.; Rood, J.I.; McClane, B.A. Beta toxin is essential for the intestinal virulence of *Clostridium perfringens* type C disease isolate CN3685 in a rabbit ileal loop model. *Mol. Microbiol.* **2008**, *67*, 15–30. [CrossRef] [PubMed]
17. Garcia, J.P.; Adams, V.; Beingesser, J.; Hughes, M.L.; Poon, R.; Lyras, D.; Hill, A.; McClane, B.A.; Rood, J.I.; Uzal, F.A. Epsilon toxin is essential for the virulence of *Clostridium perfringens* type D infection in sheep, goats and mice. *Infect Immun.* **2013**, *81*, 2405–2414. [CrossRef] [PubMed]
18. Ma, M.; Gurjar, A.; Theoret, J.R.; Garcia, J.P.; Beingesser, J.; Freedman, J.C.; Fisher, D.J.; McClane, B.A.; Uzal, F.A. Synergistic effects of *Clostridium perfringens* enterotoxin and beta toxin in rabbit small intestinal loops. *Infect Immun.* **2014**, *82*, 2958–2970. [CrossRef] [PubMed]
19. Ma, M.; Li, J.; McClane, B.A. Genotypic and phenotypic characterization of *Clostridium perfringens* isolates from darmbrand cases in post-World War II Germany. *Infect Immun.* **2012**, *80*, 4354–4363. [CrossRef] [PubMed]
20. Lawrence, G.W. The pathogenesis of enteritis necroticans. In *The Clostridia: Molecular Genetics and Pathogenesis*; Rood, J.I., McClane, B.A., Songer, J.G., Titball, R.W., Eds.; Academic Press: London, UK, 1997; pp. 198–207.
21. Johnson, S.; Gerding, D.N. Enterotoxemic infections. In *The Clostridia: Molecular Biology and Pathogenesis*; Rood, J.I., McClane, B.A., Songer, J.G., Titball, R.W., Eds.; Academic Press: London, UK, 1997; pp. 117–140.
22. Freedman, J.C.; Shrestha, A.; McClane, B.A. *Clostridium perfringens* enterotoxin: Action, genetics, and translational applications. *Toxins* **2016**, *8*. [CrossRef] [PubMed]
23. Centers for Disease Control and Prevention (CDC). Posting date. CDC estimates of foodborne illness in the United States: *Clostridium perfringens*. Available online: http://www.cdc.gov/foodsafety/diseases/clostridium-perfringens.html (accessed on 28 July 2016).
24. Caserta, J.A.; Robertson, S.L.; Saputo, J.; Shrestha, A.; McClane, B.A.; Uzal, F.A. Development and application of a mouse intestinal loop model to study the in vivo action of *Clostridium perfringens* enterotoxin. *Infect. Immun.* **2011**, *79*, 3020–3027. [CrossRef] [PubMed]
25. Carman, R.J. *Clostridium perfringens* in spontaneous and antibiotic-associated diarrhoea of man and other animals. *Rev. Med. Microbiol.* **1997**, *8* (Suppl. S1), S43–S45. [CrossRef]

26. Li, J.; Paredes-Sabja, D.; Sarker, M.; Mcclane, B. *Clostridium perfringens* sporulation and sporulation-associated toxin production. *Microbiol. Spectr.* **2016**, *4*. [CrossRef]
27. Li, J.; McClane, B.A. A novel small acid soluble protein variant is important for spore resistance of most *Clostridium perfringens* food poisoning isolates. *PLoS Pathog.* **2008**, *4*, e1000056. [CrossRef] [PubMed]
28. Severi, E.; Hood, D.W.; Thomas, G.H. Sialic acid utilization by bacterial pathogens. *Microbiology* **2007**, *153*, 2817–2822. [CrossRef] [PubMed]
29. Traving, C.; Schauer, R. Structure, function and metabolism of sialic acids. *Cell. Mol. Life Sci.* **1998**, *54*, 1330–1349. [CrossRef] [PubMed]
30. Lewis, A.L.; Lewis, W.G. Host sialoglycans and bacterial sialidases: A mucosal perspective. *Cell. Microbiol.* **2012**, *14*, 1174–1182. [CrossRef] [PubMed]
31. Vimr, E.R.; Kalivada, K.A.; Deszo, E.L.; Steenbergen, S.M. Diversity of microbial sialic acid metabolism. *Microbiol. Molec Biol. Rev.* **2004**, *68*, 132–153. [CrossRef]
32. Rohmer, L.; Hocquet, D.; Miller, S.I. Are pathogenic bacteria just looking for food? Metabolism and microbial pathogenesis. *Trends Microb.* **2011**, *19*, 341–348. [CrossRef] [PubMed]
33. Li, J.; McClane, B.A. The sialidases of *Clostridium perfringens* type D strain CN3718 differ in their properties and sensitivities to inhibitors. *Appl. Environ. Microbiol.* **2014**, *80*, 1701–1709. [CrossRef] [PubMed]
34. Li, J.; Sayeed, S.; Robertson, S.; Chen, J.; McClane, B.A. Sialidases affect the host cell adherence and epsilon toxin-induced cytotoxicity of *Clostridium perfringens* type D strain CN3718. *PLoS Pathog.* **2011**, *7*, e1002429. [CrossRef] [PubMed]
35. Boraston, A.B.; Ficko-Blean, E.; Healey, M. Carbohydrate recognition by a large sialidase toxin from *Clostridium perfringens*. *Biochemistry* **2007**, *46*, 11352–11360. [CrossRef] [PubMed]
36. Li, J.; McClane, B.A. Contributions of NanI sialidase to Caco-2 cell adherence by *Clostridium perfringens* type A and C strains causing human intestinal disease. *Infect Immun.* **2014**, *82*, 4620–4630. [CrossRef] [PubMed]
37. Chiarezza, M.; Lyras, D.; Pidot, S.J.; Flore-Diaz, M.; Awad, M.M.; Kennedy, C.L.; Cordner, L.M.; Phumoonna, T.; Poon, R.; Hughes, M.L.; et al. The NanI and NanJ sialidases of *Clostridium perfringens* are not essential for virulence. *Infect Immun.* **2009**, *77*, 4421–4428. [CrossRef] [PubMed]
38. Myers, G.S.; Rasko, D.A.; Cheung, J.K.; Ravel, J.; Seshadri, R.; DeBoy, R.T.; Ren, Q.; Varga, J.; Awad, M.M.; Brinkac, L.M.; et al. Skewed genomic variability in strains of the toxigenic bacterial pathogen, *Clostridium perfringens*. *Genome Res.* **2006**, *16*, 1031–1040. [CrossRef] [PubMed]
39. Shimizu, T.; Ohtani, K.; Hirakawa, H.; Ohshima, K.; Yamashita, A.; Shiba, T.; Ogasawara, N.; Hattori, M.; Kuhara, S.; Hayashi, H. Complete genome sequence of *Clostridium perfringens*, an anaerobic flesh-eater. *Proc. Natl. Acad. Sci. USA* **2002**, *99*, 996–1001. [CrossRef] [PubMed]
40. Therit, B.; Cheung, J.K.; Rood, J.I.; Melville, S.B. NanR, a transcriptional regulator that binds to the promoters of genes involved in sialic acid metabolism in the anaerobic pathogen *Clostridium perfringens*. *PLoS ONE* **2015**, *10*, e0133217. [CrossRef] [PubMed]
41. Ohtani, K.; Hirakawa, H.; Tashiro, K.; Yoshizawa, S.; Kuhara, S.; Shimizu, T. Identification of a two-component VirR/VirS regulon in *Clostridium perfringens*. *Anaerobe* **2010**, *16*, 258–264. [CrossRef] [PubMed]
42. Ohtani, K.; Shimizu, T. Regulation of toxin gene expression in *Clostridium perfringens*. *Res. Microbiol.* **2014**, *166*, 280–289. [CrossRef] [PubMed]
43. Hiscox, T.J.; Harrison, P.F.; Chakravorty, A.; Choo, J.M.; Ohtani, K.; Shimizu, T.; Cheung, J.K.; Rood, J.I. Regulation of sialidase production in *Clostridium perfringens* by the orphan sensor histidine kinase ReeS. *PLoS ONE* **2013**, *8*, e73525. [CrossRef] [PubMed]
44. Hiscox, T.J.; Chakravorty, A.; Choo, J.M.; Ohtani, K.; Shimizu, T.; Cheung, J.K.; Rood, J.I. Regulation of virulence by the RevR response regulator in *Clostridium perfringens*. *Infect. Immun.* **2011**, *79*, 2145–2153. [CrossRef] [PubMed]
45. Li, J.; Ma, M.; Sarker, M.R.; McClane, B.A. CodY is a global regulator of virulence-associated properties for *Clostridium perfringens* type D strain CN3718. *mBio* **2013**, *4*, e00770–e00713. [CrossRef] [PubMed]
46. Walters, D.M.; Stirewalt, V.L.; Melville, S.B. Cloning, sequence, and transcriptional regulation of the operon encoding a putative N-acetyl-mannosamine-6-phosphate epimerase (nanE) and sialic acid lyase (nanA) in *Clostridium perfringens*. *J. Bacteriol.* **1999**, *181*, 4526–4532. [PubMed]
47. Olson, M.E.; King, J.M.; Yahr, T.L.; Horswill, A.R. Sialic acid catabolism in *Staphylococcus aureus*. *J. Bacteriol.* **2013**, *195*, 1779–1788. [CrossRef] [PubMed]

48. Almagro-Moreno, S.; Boyd, E.F. Bacterial catabolism of nonulosonic (sialic) acid and fitness in the gut. *Gut Microbes* **2010**, *1*, 45–50. [CrossRef] [PubMed]
49. Almagro-Moreno, S.; Boyd, E.F. Sialic acid catabolism confers a competitive advantage to pathogenic *Vibrio cholerae* in the mouse intestine. *Infect. Immun.* **2009**, *77*, 3807–3816. [CrossRef] [PubMed]
50. Ng, K.M.; Ferreyra, J.A.; Higginbottom, S.K.; Lynch, J.B.; Kashyap, P.C.; Gopinath, S.; Naidu, N.; Choudhury, B.; Weimer, B.C.; Monack, D.M.; et al. Microbiota-liberated host sugars facilitate post-antibiotic expansion of enteric pathogens. *Nature* **2013**, *502*, 96–99. [CrossRef] [PubMed]
51. Ley, R.E. Harnessing microbiota to kill a pathogen: The sweet tooth of *Clostridium difficile*. *Nat. Med.* **2014**, *20*, 248–249. [CrossRef] [PubMed]
52. Brittan, J.; Bucheridge, T.; Finn, A.; Kadioglu, A.; Jenkinson, H. Pneumococcal neuraminidase A: An essential upper airway colonization factor for *Streptococcus pneumoniae*. *Mol. Oral Microb.* **2012**, *27*, 270–283. [CrossRef] [PubMed]
53. Awad, M.M.; Singleton, J.; Lyras, D. The sialidase NanS enhances non-TcsL mediated cytotoxicity of *Clostridium sordellii*. *Toxins* **2016**, *8*. [CrossRef] [PubMed]
54. Chang, D.E.; Smalley, D.J.; Tucker, D.L.; Leatham, M.P.; Norris, W.E.; Stevenson, S.J.; Anderson, A.B.; Grissom, J.E.; Laux, D.C.; Cohen, P.S.; et al. Carbon nutrition of *Escherichia coli* in the mouse intestine. *Proc. Natl. Acad. Sci. USA* **2004**, *101*, 7427–7432. [CrossRef] [PubMed]
55. Galen, J.E.; Ketley, J.M.; Fasano, A.; Richardson, S.H.; Wasserman, S.S.; Kaper, J.B. Role of *Vibrio cholerae* neuraminidase in the fuction of cholera toxin. *Infect. Immun.* **1992**, *60*, 406–415. [PubMed]
56. Jeong, H.G.; Oh, M.H.; Kim, B.S.; Lee, M.Y.; Han, H.J.; Choi, S.H. The capability of catabolic utilization of N-acetylneuraminic acid, a sialic acid, is essential for *Vibrio vulnificus* pathogenesis. *Infect. Immun.* **2009**, *77*, 3209–3217. [CrossRef] [PubMed]
57. Li, J.; Freedman, J.C.; McClane, B.A. NanI sialidase, CcpA, and CodY work together to regulate epsilon toxin production by *Clostridium perfringens* type D strain CN3718. *J. Bacteriol.* **2015**, *197*, 3339–3353. [CrossRef] [PubMed]
58. Shimanoto, S.; Tamai, E.; Matsushita, O.; Minami, J.; Okabe, A.; Miyata, S. Changes in ganglioside content affect the binding of *Clostridium perfringens* epsilon-toxin to detergent-resistant membranes of Madin-Darby canine kidney cells. *Microbiol. Immun.* **2005**, *49*, 245–253. [CrossRef]
59. Nagahama, M.; Sakurai, J. High-affinity binding of *Clostridium perfingens* epsilon-toxin to rat brain. *Infect. Immun.* **1992**, *60*, 1237–1240. [PubMed]
60. Jost, B.H.; Billington, S.J.; Trinh, H.T.; Songer, J.G. Association of genes encoding beta2 toxin and a collagen binding protein in *Clostridium perfringens* isolates of porcine origin. *Vet. Microbiol.* **2006**, *115*, 173–182. [CrossRef] [PubMed]
61. Hitsumoto, Y.; Morita, N.; Yamazoe, R.; Tagomori, M.; Yamasaki, T.; Katayama, S. Adhesive properties of *Clostridium perfringens* to extracellular matrix proteins collagens and fibronectin. *Anaerobe* **2014**, *25*, 67–71. [CrossRef] [PubMed]
62. Katayama, S.; Tagomori, M.; Morita, N.; Yamasaki, T.; Nariya, H.; Okada, M.; Watanabe, M.; Hitsumoto, Y. Determination of the *Clostridium perfringens*-binding site on fibronectin. *Anaerobe* **2015**, *34*, 174–181. [CrossRef] [PubMed]
63. Wade, B.; Keyburn, A.L.; Haring, V.; Ford, M.; Rood, J.I.; Moore, R.J. The adherent abilities of *Clostridium perfringens* strains are critical for the pathogenesis of avian necrotic enteritis. *Vet. Microbiol.* **2016**, *197*, 53–61. [CrossRef]
64. Cioffi, D.L.; Pandey, S.; Alvarez, D.F.; Cioffi, E.A. Terminal sialic acids are an important determinant of pulmonary endothelial barrier integrity. *Am. J. Physiol. Lung Cell. Mol. Physiol.* **2012**, *302*, L1067–1077. [CrossRef] [PubMed]
65. Freedman, J.C.; Li, J.; Uzal, F.A.; McClane, B.A. Proteolytic processing and activation of *Clostridium perfringens* epsilon toxin by caprine small intestinal contents. *mBio* **2014**, *5*, e01994–e01914. [CrossRef] [PubMed]
66. Hanna, P.C.; Wieckowski, E.U.; Mietzner, T.A.; McClane, B.A. Mapping functional regions of *Clostridium perfringens* type A enterotoxin. *Infect. Immun.* **1992**, *60*, 2110–2114. [PubMed]
67. Streicher, H. Inhibition of microbial sialidases-what has happened beyond the influenza virus? *Curr. Med. Chem. Anti-Infect. Agents* **2004**, *3*, 149–161. [CrossRef]
68. Roy, S.; Honma, K.; Douglas, C.W.; Sharma, A.; Stafford, G.P. Role of sialidase in glycoprotein utilization by *Tannerella forsythia*. *Microbiology* **2011**, *157*, 3195–3202. [CrossRef] [PubMed]

69. Holzer, C.T.; von Itzstein, M.; Jin, B.; Pegg, M.S.; Stewart, W.P.; Wu, W.Y. Inhibition of sialidases from viral, bacterial and mammalian sources by analogues of 2-deoxy-2,3-didehydro-*N*-acetylneuraminic acid modified at the C-4 position. *Glycoconj. J.* **1993**, *10*, 40–44. [CrossRef] [PubMed]
70. Knapp, S.; Zhao, D. Synthesis of the sialidase inhibitor siastatin B. *Org. Lett.* **2000**, *2*, 4037–4040. [CrossRef] [PubMed]

14

mRNA Interferase *Bacillus cereus* BC0266 Shows MazF-Like Characteristics Through Structural and Functional Study

Sung-Min Kang [1,†], Ji Sung Koo [1,†], Chang-Min Kim [1], Do-Hee Kim [2] and Bong-Jin Lee [1,*]

1 The Research Institute of Pharmaceutical Sciences, College of Pharmacy, Seoul National University, Gwanakgu, Seoul 08826, Korea; men0528@snu.ac.kr (S.-M.K.); koojisung@snu.ac.kr (J.S.K.); kt912@snu.ac.kr (C.-M.K.)
2 College of Pharmacy, Jeju National University, Jeju 63243, Korea; doheekim@jejunu.ac.kr
* Correspondence: lbj@nmr.snu.ac.kr.
† These authors contributed equally to this work.

Abstract: Toxin–antitoxin (TA) systems are prevalent in bacteria and are known to regulate cellular growth in response to stress. As various functions related to TA systems have been revealed, the importance of TA systems are rapidly emerging. Here, we present the crystal structure of putative mRNA interferase BC0266 and report it as a type II toxin MazF. The MazF toxin is a ribonuclease activated upon and during stressful conditions, in which it cleaves mRNA in a sequence-specific, ribosome-independent manner. Its prolonged activity causes toxic consequences to the bacteria which, in turn, may lead to bacterial death. In this study, we conducted structural and functional investigations of *Bacillus cereus* MazF and present the first toxin structure in the TA system of *B. cereus*. Specifically, *B. cereus* MazF adopts a PemK-like fold and also has an RNA substrate-recognizing loop, which is clearly observed in the high-resolution structure. Key residues of *B. cereus* MazF involved in the catalytic activity are also proposed, and in vitro assay together with mutational studies affirm the ribonucleic activity and the active sites essential for its cellular toxicity.

Keywords: toxin–antitoxin system; mazF; type II; toxin; mRNA interferase; X-ray crystallography

Key Contribution: Structural and functional study on putative mRNA interferase BC0266 revealed that BC0266 is a *B. cereus* MazF toxin in type II TA system. This is the first reported toxin structure in type II TA system of *B. cereus*.

1. Introduction

Toxin–antitoxin (TA) systems are prevalent in prokaryotes and function to regulate cellular growth in response to stress such as antibiotic exposure, nutrient starvation, heat shock, and DNA damage. Initially, TA systems were discovered as a part of the plasmid maintenance system, in which only the daughter cells harboring the vertically transferred TA operon can survive [1–5]. In addition to the primary maintenance function, TA systems are important for bacterial survival. The TA systems are involved in many cellular processes, including cell growth, cell persistence, cell dormancy, biofilm formation, antibiotic resistance, DNA replication, translation, cell division, cell wall synthesis, and cell apoptosis [6–11].

TA systems are a two-component system, composed of a toxin and an antitoxin usually sharing the same operon. Currently, TA systems can be categorized into six phenotypes (I-VI) by the nature of each component and the regulatory mechanisms of the antitoxins to its cognate toxins. In the case of RNA antitoxins, direct inhibition of mRNA encoding toxins and protein toxins constitute the type I and type III systems, respectively. Similarly, protein toxins are directly inhibited by their counterpart protein antitoxins in type II TA system. As for indirect regulatory mechanisms, protein antitoxins counteract the activity of protein toxins in type IV TA system and cleave mRNA encoding protein toxins in type V TA system. Lastly, in type VI TA system, protein toxins are degraded by specific proteases and lose their toxicity as a result of complex formation with protein antitoxins [12–16].

Typically, in type II TA system, the rather stable toxins have much more invariant characteristics than their labile and flexible cognate antitoxins. Upon degradation of antitoxins, free toxins can function in postsegregational cell killing, abortive infection, and even bacterial persistence. Thus, it has been regarded that bacterial persistence would indeed be closely related to type II TA systems due to the cellular process-specific nature of type II toxins [17–21]. One of the best characterized type II toxins is mRNA interferase MazF toxin, belonging to the MazEF family. Comprehensive bioinformatic analyses have shown that MazF is widely distributed in both Gram-positive and Gram-negative bacteria, allowing an in-depth understanding of the consequences of stress and MazF activation on the physiological effects of the bacteria [6,22,23]. In addition, understanding of the cutting specificity of MazF on its target RNA is well established. Specifically, the MazF toxin acts as a ribonuclease and cleaves mRNA in a sequence-specific and ribosome-independent manner, but in some cases also cleaves rRNA at specific sequences, exerting its toxicity [17,24,25].

Bacillus cereus is a Gram-positive, facultatively anaerobic, and rod-shaped bacterium that is well known for its association with foodborne illness [26]. Recently, the pathogenicity potential of the bacteria gained attention due to its ability in causing serious and fatal nongastrointestinal infections [27]. As TA systems are involved in several cellular responses such as stress response, bacterial persistence, and biofilm formation, there is a clear linkage between TA systems and the virulence of the bacteria [28]. For this reason, research was conducted on the putative MazF toxin, mRNA interferase BC0266.

Here, we report 2.0 Å X-ray crystal structure of mRNA interferase MazF as the first toxin structure in the TA system of *B. cereus*. Similar to that of other MazF toxins, *B. cereus* MazF adopts a PemK-like fold and also has an elongated loop between the β1 and β2 strands. Further sequence alignment with several other MazFs supported the key residues in ribonucleic catalysis. Finally, in vitro ribonuclease activity test together with site-directed mutational studies demonstrated that *B. cereus* MazF exhibits a clear ribonuclease activity and its key residues play a significant role in exerting its role as a ribonuclease. In conclusion, we show that mRNA interferase BC0266 belongs to the MazEF family and is the first reported MazF toxin structure in the TA system of *B. cereus*. This is an important study to investigate the diversity of the TA system among many different bacterial strains.

2. Results and Discussion

2.1. *B. cereus MazF Adopts a PemK-Like Fold*

Crystals of *B. cereus* MazF were diffracted to high resolution (2.0 Å) and the final structure was determined using the molecular replacement method with the MazF structure from *Mycobacterium tuberculosis* (PDB code 5XE2) [29]. The structure was refined to R_{work} factor 18.9% and R_{free} factor 22.5%. All 116 amino acids were present and lie within the Ramachandran favored region. Crystals had $P3_121$ space group with the unit cell dimensions of $a = b = 60.648$ Å, $c = 76.247$ Å, and $\alpha = \beta = 90.0°$, $\gamma = 120°$. Detailed crystallographic data collection and refinement statistics are summarized in Table 1.

There are three α-helices and seven β-strands with the order of β-barrel arrangement in a monomer of *B. cereus* MazF (Figure 1A). Among those β-strands, five β-strands (β1, β2, β3, β6, β7) and two β-strands (β4, β5) form a β-sheet antiparallel to each other. Two loops between the β1 and β2 strands, and the β3 and β4 strands are also clearly observed. The total solvent-accessible surface area of the monomeric structure is 6740 Å^2, and the contact area between two monomers is 1440 Å^2 (21.4% per monomeric subunit). The crystal structure shows that the dimeric interface displays a concave structure covered by the neighboring loop (Figure 1B). Calculations of surface and interface areas were conducted using the PISA server [30]. The oligomeric state of *B. cereus* MazF was predicted through size-exclusion chromatography with reference proteins in the Gel Filtration Calibration Kits (GE Healthcare, Chicago, IL, USA) (Figure 1C). Because theoretical molecular weight of the *B. cereus* MazF monomer containing N-terminal His-tag is 15.1 kDa, it is reasonable that oligomeric state of *B. cereus* MazF is homodimer (30.2 kDa) as it was eluted at almost the same time as that of carbonic anhydrase (29 kDa).

2.2. Comparison of B. cereus MazF with MazF Homologs

To relate the structural characteristics of *B. cereus* MazF with mRNA interferase MazF, comparative analysis was performed with previously reported MazF structures. Illustrations of sequence alignment were displayed with *Staphylococcus aureus* MazF (PDB code 4MZM) [31], *M. tuberculosis* MazF3 (PDB code 5UCT), [32]; MazF4 (PDB code 5XE2) [29]; and MazF7 (PDB code 5WYG) [33] (Figure 2A). The results showed that two key residues of Arg25 and Thr48 involved in catalytic activity are well conserved among multiple MazFs. Interestingly, Lys19 is conserved in *M. tuberculosis* MazF4 instead of arginine [29], and Ser51 is conserved in *M. tuberculosis* MazF7 instead of threonine [33]. Yet, mutations of these two residues significantly impaired the activities of corresponding toxins, implying their role in ribonucleic catalysis [29,33]. Although considerable structural similarities are shared between the compared MazFs, there are notable differences in flexibilities and structural variations in their β1–β2 and β3–β4 loops (Figure 2B). These loops are reported as a major binding site with its cognate antitoxin MazE and, therefore, it is supposed that the flexibilities of these loops have relations to interactions with its target RNA substrate depending on MazE binding [32,34,35].

Table 1. Structure data collection and refinement statistics.

(a) Data Collection Details	
X-ray source	BL44XU beamline of Spring-8, Japan
X-ray wavelength (Å)	0.899995
Space group	$P3_121$
Unit cell parameters: a, b, c (Å)	60.648, 60.648, 76.247
Unit cell parameters: α, β, γ (°)	90.0, 90.0, 120.0
Resolution range (Å)	50.0-2.00
Observed reflections (>1σ)	231077
Unique reflections	21852
$<I/\sigma(I)>$	10.74 (2.97) [e]
Completeness (%)	99.3 (95.5) [e]
Multiplicity [a]	10.57 (9.76) [e]
R_{merge} (%) [b]	12.1 (47.9) [e]
$CC_{1/2}$	0.997 (0.918) [e]
(b) Refinement statistics	
R_{work} [c] (%)	18.9
R_{free} [d] (%)	22.5
No. of atoms/average B factor (Å^2)	997/46.2
RMSD [f] from ideal geometry: Bond distance (Å)	0.008
RMSD [f] from ideal geometry: Bond angle (°)	1.108
Ramachandran statistics: Most favored regions (%)	96.49
Ramachandran statistics: Additional allowed regions (%)	3.51
PDB accession code	7BXY

[a] N_{obs}/N_{unique}, [b] $R_{merge} = \Sigma (I - \langle I \rangle)/\Sigma \langle I \rangle$, [c] $R_{work} = \Sigma_{hkl} ||F_{obs}| - k |F_{calc}||/\Sigma_{hkl} |F_{obs}|$, [d] R_{free} was calculated in the same manner as R_{work} with 5% of the reflections excluded from the refinement. [e] Values in parentheses indicate the highest-resolution shell. [f] Root mean square deviation (RMSD) was calculated using REFMAC.

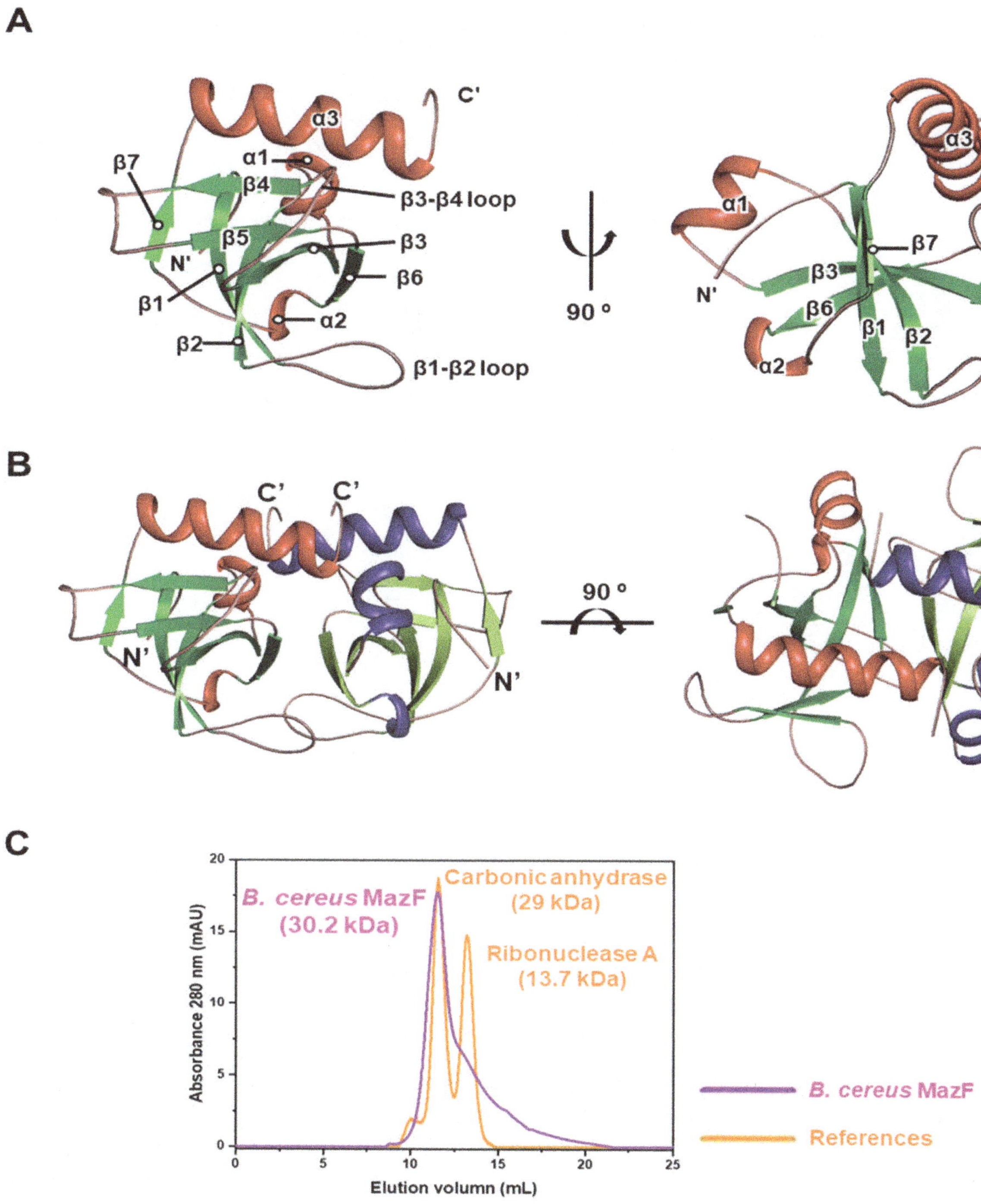

Figure 1. Overall structure of *B. cereus* MazF and gel filtration data with reference proteins. α helices are colored in red and blue. β strands are colored in green. (**A**) 90° rotational views on *B. cereus* MazF monomer. β1–β2 and β3–β4 loops are denoted. (**B**) 90° rotational views on *B. cereus* MazF homodimer. (**C**) Molecular weight estimates of *B. cereus* MazF obtained by size-exclusion chromatography with Superdex 75 10/300 gl column. Overlaid chromatograms were denoted with their names and molecular weights.

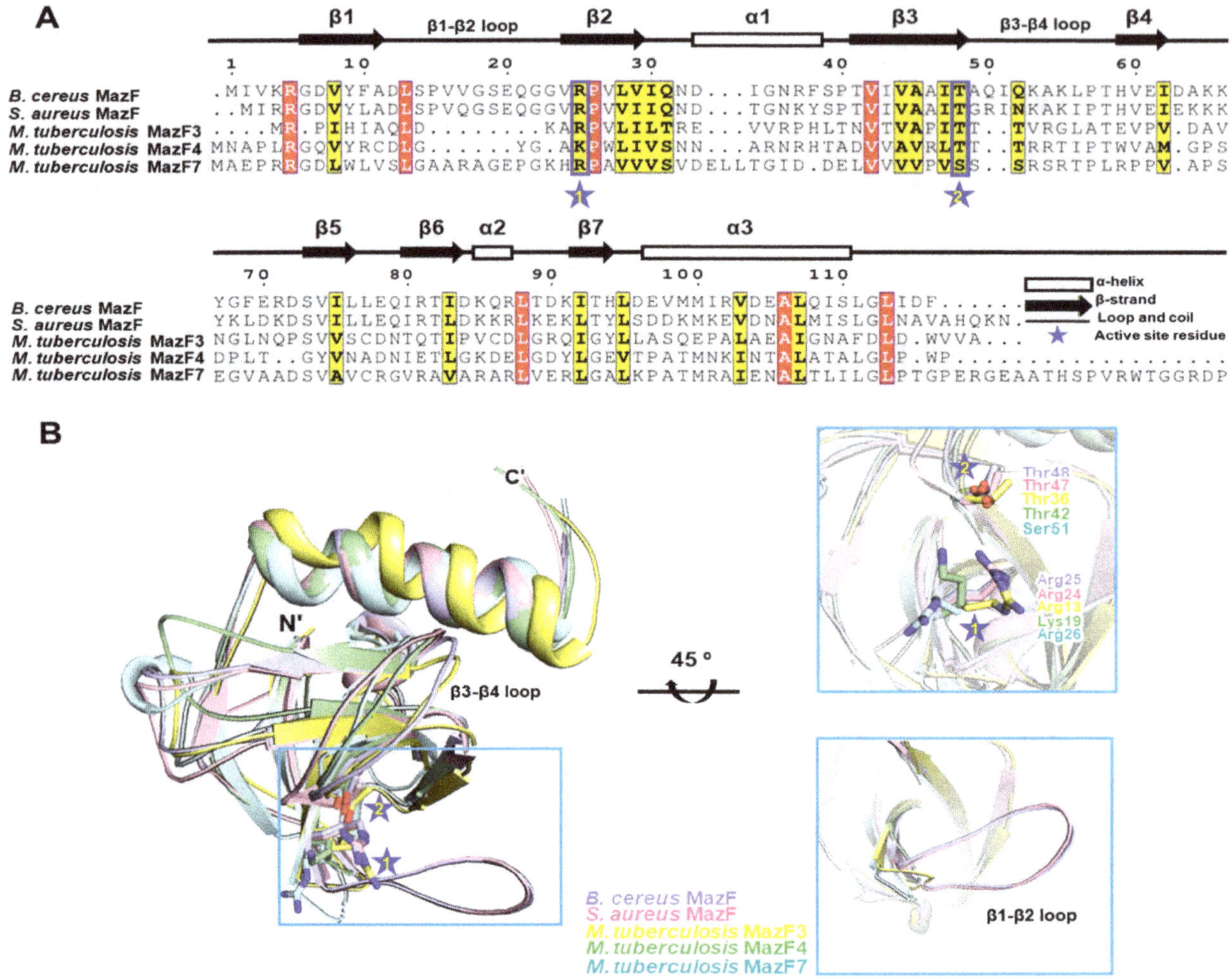

Figure 2. Comparative analysis of MazF toxins. (**A**) Sequence alignment of *B. cereus* MazF with other MazFs. Secondary structural elements are displayed above the alignment. Residues showing similarity are highlighted in red and yellow. Conserved active site residues are emphasized with star symbol. (**B**) Structural comparison of *B. cereus* MazF with other MazFs. Cartoon representations are employed to draw each structures. Conserved active site residues are shown in sticks. Conserved active site residues and variations in β1–β2 loop are illustrated by enlarged view in blue square.

Through early studies on *E. coli* MazEF, it was suggested that the β1–β2 loop in the MazF toxin undergoes conformational transition from disordered to ordered state upon binding to the MazE antitoxin [36]. During this transition, it is assumed that these loops are modulated by MazE, affecting the target RNA accessibility of MazF [36]. For instance, MazF4 and MazF7 toxins from *M. tuberculosis* with truncated β1–β2 loops had significant steric clashes within the β1–β2 loop region due to the high flexibility arising from the structurally unbound MazE antitoxin [29,33]. As for MazF3 toxin in *M. tuberculosis*, the length of the short β1–β2 loop may be insufficient to reach and recognize the downstream of the target RNA substrate [31,32]. Furthermore, it is predicted that absence of an elongated β1–β2 loop would cause even greater conformational transition of the β1–β2 loop, leaving the toxin structurally and functionally impractical. Therefore, it is suggested that target sequence specificity of truncated or short β1–β2 loops would be significantly reduced compared with MazFs having a long β1–β2 loop, such as that of *B. cereus* MazF. Therefore, it can be presumed that the long length of β1–β2 loop may be the most powerful factor that determines the target RNA specificity.

In contrast to the β1–β2 loop, β3–β4 loop shows an opposite conformational transition from ordered to disordered state upon binding to MazE antitoxin [29]. Furthermore, previous research has shown that aromatic residues located on the antitoxin MazEs are involved in essential MazEF interactions [29]. Specifically, His68 of *E. coli* MazE, Tyr61 of *B. subtilis* MazE, and Tyr76 of *M. tuberculosis* MazE4 interacted with the β3–β4 loops via van der Walls interactions [29,34,35].

Thus, considering the *E. coli* and *M. tuberculosis* model proteins, the *B. cereus* MazE antitoxin, BC0265, likely contains C-terminal aromatic residues that might interact with the β3–β4 loop in *B. cereus* MazF. Because of high structural unpredictability and lack of structural information of *B. cereus* MazE and other MazE antitoxins in general, an accurate prediction of which aromatic residue will interact with its cognate toxin is difficult. However, assuming the model is correct, one of the aromatic residues in *B. cereus* MazE, Tyr63, Phe78, Tyr82 or His86, likely engages in a hydrophobic interaction during complex formation.

2.3. Arg25 and Thr48 Act as Key Residues in Catalytic Activity

To assure that two key residues Arg25 and Thr48 are critical in catalytic activity, comparative analysis with *B. subtilis* MazF complexed with RNA (PDB code 4MDX) [34] (Figure 3A) and *E. coli* MazF complexed with DNA substrate analog (PDB code 5CR2) [35] (Figure 3B) was performed. Subsequently, in silico model of *B. cereus* MazF complexed with RNA and DNA was generated by superimposition of *B. cereus* MazF structure onto the 4MDX (left) and 5CR2 (right), respectively (Figure 3C).

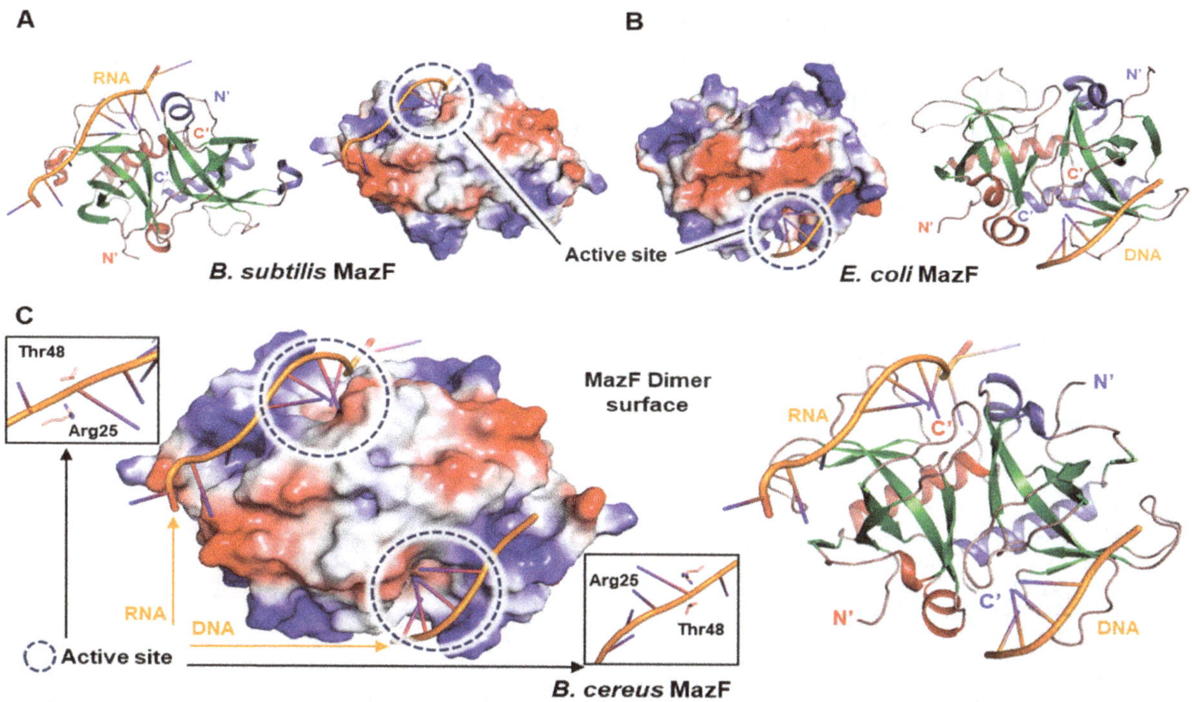

Figure 3. In silico model of previously reported MazFs and *B. cereus* MazF in complex with RNA and DNA. (**A**–**C**) Active sites for each proteins are illustrated in circles. (**A**) Ribbon representations and electrostatic surface potentials of *B. subtilis* MazF complexed with RNA. RNA is handled to be placed at left chain. (**B**) Ribbon representations and electrostatic surface potentials of *E. coli* MazF complexed with DNA. DNA is handled to be placed at right chain. (**C**) Superimposition of *B. cereus* MazF on *B. subtilis* MazF complexed with RNA (left chain) and *E. coli* MazF complexed with DNA (right chain). Close-up views of the interactions between active site residues and RNA/DNA are displayed.

In the in silico model of *B. cereus* MazF in complex with RNA and DNA substrate, Arg25 and Thr48 residues were well organized with RNA and DNA substrate (Figure 3C). Moreover, mutational study on *B. subtilis* MazF Arg25 and Thr48 resulted in complete inactivation of toxicity of MazF [34]. From these results, it can be assumed that Arg25 and Thr48 from *B. cereus* would also be involved in RNA catalysis. In addition, according to earlier work performed on MazF toxins, Arg25 is predicted to act as a general acid/base during RNA catalysis and Thr48 may serve as a charge stabilizer in transition state during RNA substrate binding [35,37,38].

Interestingly, charge distributions observed in the front surfaces of three proteins (Figure 3A–C) are similar but slightly different from each other, mainly due to the differences in the charged amino acid content. In *E. coli* MazF, negative-charged front surface is mostly formed by Asp16, Asp18, and Glu24 in the β1–β2 loop (residues 16 to 27). However, corresponding residues are substituted with different amino acids in both *B. subtilis* and *B. cereus* MazF. Because these front surfaces are known as the binding region of cognate antitoxin MazE, each MazE is thought to have a surface potential specific to its cognate MazF toxin to effectively bind and neutralize toxin surfaces. This seems to be the one of strong reasons why a specific toxin is neutralized only by its cognate antitoxin, although considerable structural similarities are observed within the same family of TA system [29,39,40].

2.4. *B. cereus MazF Shows Ribonuclease Activity by Two Key Residues*

MazF is a ribonuclease that is known to cleave RNA substrate [41]. Toxins such as VapC in the Type II TA system require metal ions as a cofactor to exhibit ribonuclease activity, but in the case of MazF, cofactor metal is not necessarily needed [42,43]. To demonstrate that *B. cereus* MazF shows ribonuclease activity like other MazF toxins, in vitro ribonuclease activity test was performed (Figure 4A). In this assay, a synthetic RNA strand of unknown sequence is used as substrate. A fluorophore is covalently attached to one end of the RNA strand while the quencher is located at the other end. If this RNA is cleaved by a ribonuclease, the quencher is detached and as a result fluorescence is detected depending on the amount of cleaved RNA. The result from in vitro ribonuclease assay showed an elevation of RFU (resulting fluorescence unit) as a function of time. *B. cereus* MazF exhibits a ribonuclease activity showing time and dose dependency, and hence shares the same functional activity with MazF toxins.

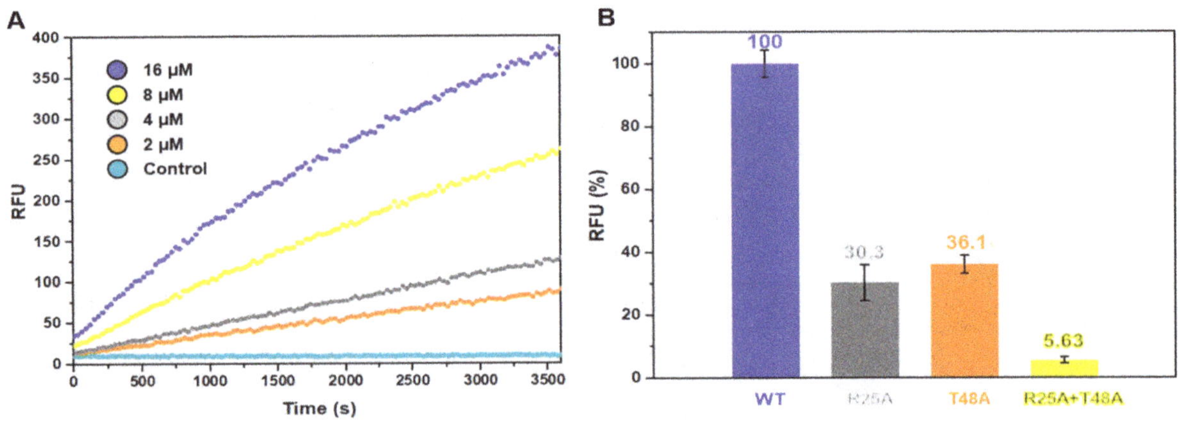

Figure 4. Ribonuclease activity test of *B. cereus* MazF toxin. (**A**) The in vitro ribonuclease activity test of wild-type *B. cereus* MazF. Fluorescence was measured as a function of time (s) during 1 h. Each curve is colored separately according to different *B. cereus* MazF concentrations. Concentrations of *B. cereus* MazF were increased by doubling from 2 to 16 μM. An equal aliquot of fluorescent RNA substrate was incubated with *B. cereus* MazF in different concentrations. Fluorescence was measured upon cleavage of RNA substrate. The control as well as different protein concentrations contained 20 mM Tris-HCl (pH 8.0), 500 mM NaCl, and 40 units of RiboLock™ (Thermo Scientific, Waltham, MA, USA) RNase inhibitor. (**B**) Ribonuclease activities of 10 μM wild-type *B. cereus* MazF and its mutants represented as bar graph with standard deviation derived from triplicate tests. Magnitudes of each reaction were assessed by subtracting the RFU (resulting fluorescence unit) of the starting point from RFU of the end point. The RFU obtained from wild-type *B. cereus* MazF was taken as 100%.

Furthermore, two single and one double site-directed mutagenesis on the two key catalytic residues of Arg25 and Thr48 were performed (Figure 4B). The results of the mutational studies showed that single mutations of either Arg25 or Thr48 resulted in severe reductions in the ribonuclease activity of toxin while double mutations of both Arg25 and Thr48 resulted in close to none ribonuclease activity. Therefore, in agreement with the results of *B. subtilis* MazF, Arg25 and Thr48 are key residues in *B. cereus* MazF, which is a characteristic observed consistently in MazF toxins.

3. Conclusions

We conducted an in-depth characterization and investigation on putative mRNA interferase MazF in *B. cereus*. High-resolution (2.0 Å) structure was obtained by X-ray crystallography, and overall structural folding of *B. cereus* MazF indicated β-barrel arrangement containing two independent β-sheets identical to that of MazF toxins. Through comparison analysis with previously reported structures of MazFs, amino acid residues Arg25 and Thr48 in *B. cereus* MazF were well conserved with other MazFs and were also confirmed as key residues in the catalytic activity of *B. cereus* MazF. Also, β1–β2 and β3–β4 loops were clearly observed in *B. cereus* MazF structure and their role in the recognition of target RNA substrate depending on the binding of its cognate antitoxin was discussed. From previously conducted studies, it can be presumed that the β1–β2 loop acts as a gateway to binding of its target RNA substrate and cognate antitoxin and the β3–β4 loop also acts as a major interaction site in antitoxin-binding via van der Walls interactions. Lastly, ribonuclease activity test via in vitro assay together with site-directed mutational studies on Arg25 and Thr48 showed a decrease in its ribonuclease activity, which was consistent in other MazF toxins. Altogether, our study provides a unified structural and functional basis that mRNA interferase BC0266 is indeed a MazF toxin. This is the first determined MazF and toxin structure in type II TA system of *B. cereus*. Our work may be a rational basis for understanding the TA systems regarding the stress-responsive toxin MazF and the general regulatory mechanisms of TA systems.

4. Materials and Methods

4.1. Cloning and Transformation

The *BC0266* gene was amplified by polymerase chain reaction (PCR). The following primers were used in PCR: forward, 5′- GGAATTCCATATGATGATTGTAAAACGCGGC-3′; reverse, 5′-CCGCTCGAGTTAAAAATCTATTAGTCCTAAAC-3′. *Nde1* and *Xho1* were used as restriction enzymes, and the cutting sites are underlined. The PCR products and pET28a vector were double-cleaved by the same restriction enzymes and ligated to each other. For purification, residual N-terminal His tag (MGSSHHHHHHSSGLVPRGSH) was attached to the gene of *B. cereus* MazF and was transformed into *E.coli* DH5α competent cells (Novagen, Madison, WI, USA). As the next steps, these cloned plasmids were re-transformed into *E. coli* Rosetta2 (DE3) pLysS competent cells (Novagen, Madison, WI, USA).

4.2. Protein Expression and Purification

Transformed cells were grown in Luria broth (LB) at 37 °C until the OD_{600} reached to 0.5. Overexpression of target protein was induced by 0.5 mM isopropyl 1-thio-B-D-galactopyranoside (IPTG). Expressed cells were further incubated for 4 h at 37 °C. These cells were harvested by centrifugation at 11,355× *g*. Harvested cells were suspended in a buffer A (20 mM Tris-HCl, pH 7.9, and 500 mM NaCl) containing 5% glycerol and lysed by ultrasonication. Lysed cells were centrifugated for 1 h at 28,306× *g*. After centrifugation, the supernatant containing soluble moiety of protein was loaded on a Ni^{2+} affinity chromatography column (Bio-Rad, Hercules, CA, USA), which was pre-equilibrated with buffer A. Loaded protein was washed with buffer A containing 50 mM imidazole to remove impurities. Then, the remaining protein, bound to the Ni^{2+} column, was eluted using an imidazole gradient (100–800 mM) and fractionized. The purity of eluted protein in each fraction was checked using sodium dodecyl sulfate polyacrylamide gel electrophoresis (SDS–PAGE). Finally, selected high-purity protein was loaded on a size-exclusion chromatography column HiLoad 16/600 Superdex 200 prep-grade column (GE Healthcare, Chicago, IL, USA), pre-equilibrated with final buffer (20 mM Tris, pH 8.0, and 500 mM NaCl). Final eluted protein was concentrated to 20 mg/mL using an

Amicon Ultra centrifugal filter unit (Millipore, Burlington, MA, USA). The purity of the final protein solution was verified by SDS-PAGE. Mutants of *B. cereus* MazF were expressed and purified by the same procedure as native ones.

4.3. Crystallization, Data Collection, and Processing

The final protein solution was screened for crystallization using crystallization kit Wizard (Rigaku Reagents, Bainbridge Island, WA, USA) and Index (Hampton Research, Aliso Viejo, CA, USA). Protein solution (1 μL) was mixed with 1 μL of each buffer solution in the crystallization kit. Crystals of *B. cereus* MazF were grown in the transparent 96-well plate using sitting-drop vapor diffusion method at 20 °C. Only one crystallized well contained the hit solution of 100 mM Sodium Citrate, pH 5.6, and 1.0 M ammonium phosphate monobasic. Cryo-protection was achieved by addition of 20% glycerol to the hit solution. After cryo-protection, crystals were immediately cooled in liquid nitrogen prior to data collection. The data were collected using an Rayonix MX-300 HE CCD detector at BL44XU of the Spring-8, Japan. All raw data of crystal were scaled and processed by XDS [44]. A set of data from a hit crystal was used to solve the structure at 2.00 Å resolution and to refine 7BXY (code name of *B. cereus* MazF). *PHENIX* [45] was first used to do molecular replacement and automatically build the model, and *COOT* [46] was utilized to yield the starting model for refinement. The R_{work}/R_{free} values [47] of the final model were obtained using *REFMAC* and *PHENIX* [45,48]. The validation of overall geometry was achieved using *MolProbity* [49]. PyMOL (PyMOL Molecular Graphics System, Version 1.2r3pre, Schrödinger, LLC, NY, USA) was used to generate figures in this study.

4.4. Mutiple Sequence Alignment

To conduct sequence alignment of *B. cereus* MazF and four MazFs whose structures have been identified, the sequence information of five proteins was browsed using Uniprot [50]. Alignments of amino acid residues were carried out using Clustal W [51] and visualized using ESPript 3.0 [52]. The consensus value was set to 0.85 and %Equivalent was used in similarity mode. In visualization, structural information of *B. cereus* MazF was used as top secondary structures and the description and sequence numbering on the topside correspond to *B. cereus* MazF.

4.5. In Vitro Ribonuclease Assay

To confirm the ribonuclease activity of *B. cereus* MazF, an RNase Alert Kit (IDT, Coralville, IA, USA) was purchased and used following the manufacturer's protocol. Equal aliquot RNA substrate (5 μL) was interacted with 2 μM, 4 μM, 8 μM, and 16 μM concentrations of *B. cereus* MazF in a final purification buffer. For ribonuclease activity test, 10 μM aliquots of wild-type *B. cereus* MazF and its mutants were prepared. Then, the resulting fluorescence units (RFU) were detected using a SPECTRAmax GEMINI XS spectro-fluorometer (Marshall Scientific, Hampton, NH, USA) at 37 °C by emission fluorescence at 520 nm upon excitation at 490 nm. Assay setting was performed on 384-well opaque plate, and all of experiments were performed in triplicate.

4.6. Site-Directed Mutagenesis

The mutations in *B. cereus* MazF (R25A, T48A, R25A+T48A) were performed using *B. cereus* MazF in pET28a (+) as template. Single-site mutations were performed in one step, and double-site mutations were performed in a step-wise manner. Reaction components were mixed and subjected to PCR machine guided by the manufacturer's protocol (EZchange Site-Directed Mutagenesis Kit, Enzynomics, Daejeon, Korea). Each plasmid was transformed into *E. coli* XL10-Gold competent cells (Agilent Technologies, Santa Clara, CA, USA), and the resulting inserted genes were verified through DNA sequencing.

Author Contributions: Conceptualization, S.-M.K. and J.S.K.; Methodology, S.-M.K. and J.S.K.; Data curation, S.-M.K. and J.S.K.; Formal analysis, S.-M.K., J.S.K and C.-M.K.; Writing, original draft preparation, S.-M.K., J.S.K.

and D.-H.K.; Writing: review and editing, S.-M.K., J.S.K., D.-H.K and B.-J.L.; All the authors approved the final version of the manuscript. All authors have read and agreed to the published version of the manuscript.

Acknowledgments: We thank the beamline (BL) staff members at Pohang Light Source (BL-5C, BL-7A and BL11C), Korea, and SPring-8 (BL44XU) under the Collaborative Research Program of Institute for Protein Research, Osaka University (proposal No. 2019A6972).

References

1. Page, R.; Peti, W. Toxin-antitoxin systems in bacterial growth arrest and persistence. *Nat. Chem. Biol.* **2016**, *12*, 208–214. [CrossRef]
2. Yamaguchi, Y.; Park, J.H.; Inouye, M. Toxin-antitoxin systems in bacteria and archaea. *Annu. Rev. Genet.* **2011**, *45*, 61–79. [CrossRef] [PubMed]
3. Lee, K.Y.; Lee, B.J. Structure, Biology, and Therapeutic Application of Toxin-Antitoxin Systems in Pathogenic Bacteria. *Toxins* **2016**, *8*, 305. [CrossRef] [PubMed]
4. Ogura, T.; Hiraga, S. Mini-F plasmid genes that couple host cell division to plasmid proliferation. *Proc. Natl. Acad. Sci. USA* **1983**, *80*, 4784–4788. [CrossRef] [PubMed]
5. Jurenaite, M.; Markuckas, A.; Suziedeliene, E. Identification and Characterization of Type II Toxin-Antitoxin Systems in the Opportunistic Pathogen Acinetobacter baumannii. *J. Bacteriol.* **2013**, *195*, 3165–3172. [CrossRef]
6. Pandey, D.P.; Gerdes, K. Toxin-antitoxin loci are highly abundant in free-living but lost from host-associated prokaryotes. *Nucleic Acids Res.* **2005**, *33*, 966–976. [CrossRef]
7. Sengupta, M.; Austin, S. Prevalence and significance of plasmid maintenance functions in the virulence plasmids of pathogenic bacteria. *Infect. Immun.* **2011**, *79*, 2502–2509. [CrossRef]
8. Kwan, B.W.; Valenta, J.A.; Benedik, M.J.; Wood, T.K. Arrested protein synthesis increases persister-like cell formation. *Antimicrob. Agents Chemother.* **2013**, *57*, 1468–1473. [CrossRef]
9. Goeders, N.; Van Melderen, L. Toxin-Antitoxin Systems as Multilevel Interaction Systems. *Toxins* **2014**, *6*, 304–324. [CrossRef]
10. Mutschler, H.; Gebhardt, M.; Shoeman, R.L.; Meinhart, A. A novel mechanism of programmed cell death in bacteria by toxin-antitoxin systems corrupts peptidoglycan synthesis. *PLoS Biol.* **2011**, *9*, e1001033. [CrossRef]
11. Unoson, C.; Wagner, E.G.H. A small SOS-induced toxin is targeted against the inner membrane in Escherichia coli. *Mol. Microbiol.* **2008**, *70*, 258–270. [CrossRef]
12. Goeders, N.; Chai, R.; Chen, B.H.; Day, A.; Salmond, G.P.C. Structure, Evolution, and Functions of Bacterial Type III Toxin-Antitoxin Systems. *Toxins* **2016**, *8*, 282. [CrossRef]
13. Lobato-Marquez, D.; Diaz-Orejas, R.; Garcia-del Portillo, F. Toxin-antitoxins and bacterial virulence. *FEMS Microbiol. Rev.* **2016**, *40*, 592–609. [CrossRef]
14. Harms, A.; Brodersen, D.E.; Mitarai, N.; Gerdes, K. Toxins, Targets, and Triggers: An Overview of Toxin-Antitoxin Biology. *Mol. Cell* **2018**, *70*, 768–784. [CrossRef]
15. Wang, X.X.; Lord, D.M.; Cheng, H.Y.; Osbourne, D.O.; Hong, S.H.; Sanchez-Torres, V.; Quiroga, C.; Zheng, K.; Herrmann, T.; Peti, W.; et al. A new type V toxin-antitoxin system where mRNA for toxin GhoT is cleaved by antitoxin GhoS. *Nat. Chem. Biol.* **2012**, *8*, 855–861. [CrossRef]
16. Aakre, C.D.; Phung, T.N.; Huang, D.; Laub, M.T. A Bacterial Toxin Inhibits DNA Replication Elongation through a Direct Interaction with the beta Sliding Clamp. *Mol. Cell* **2013**, *52*, 617–628. [CrossRef] [PubMed]
17. Yamaguchi, Y.; Inouye, M. mRNA interferases, sequence-specific endoribonucleases from the toxin-antitoxin systems. *Prog. Mol. Biol. Transl. Sci.* **2009**, *85*, 467–500. [CrossRef] [PubMed]
18. Aizenman, E.; Engelberg-Kulka, H.; Glaser, G. An Escherichia coli chromosomal "addiction module" regulated by guanosine [corrected] 3′,5′-bispyrophosphate: A model for programmed bacterial cell death. *Proc. Natl. Acad. Sci. USA* **1996**, *93*, 6059–6063. [CrossRef] [PubMed]
19. Marianovsky, I.; Aizenman, E.; Engelberg-Kulka, H.; Glaser, G. The regulation of the Escherichia coli mazEF promoter involves an unusual alternating palindrome. *J. Boil. Chem.* **2001**, *276*, 5975–5984. [CrossRef] [PubMed]

20. Kedzierska, B.; Lian, L.Y.; Hayes, F. Toxin-antitoxin regulation: Bimodal interaction of YefM-YoeB with paired DNA palindromes exerts transcriptional autorepression. *Nucleic Acids Res.* **2007**, *35*, 325–339. [CrossRef]
21. Brantl, S.; Muller, P. Toxin(-)Antitoxin Systems in Bacillus subtilis. *Toxins* **2019**, *11*, 262. [CrossRef] [PubMed]
22. Leplae, R.; Geeraerts, D.; Hallez, R.; Guglielmini, J.; Dreze, P.; Van Melderen, L. Diversity of bacterial type II toxin-antitoxin systems: A comprehensive search and functional analysis of novel families. *Nucleic Acids Res.* **2011**, *39*, 5513–5525. [CrossRef] [PubMed]
23. Anantharaman, V.; Aravind, L. New connections in the prokaryotic toxin-antitoxin network: Relationship with the eukaryotic nonsense-mediated RNA decay system. *Genome Biol.* **2003**, *4*, R81. [CrossRef] [PubMed]
24. Schifano, J.M.; Edifor, R.; Sharp, J.D.; Ouyang, M.; Konkimalla, A.; Husson, R.N.; Woychik, N.A. Mycobacterial toxin MazF-mt6 inhibits translation through cleavage of 23S rRNA at the ribosomal A site. *Proc. Natl. Acad. Sci. USA* **2013**, *110*, 8501–8506. [CrossRef] [PubMed]
25. Vesper, O.; Amitai, S.; Belitsky, M.; Byrgazov, K.; Kaberdina, A.C.; Engelberg-Kulka, H.; Moll, I. Selective translation of leaderless mRNAs by specialized ribosomes generated by MazF in Escherichia coli. *Cell* **2011**, *147*, 147–157. [CrossRef] [PubMed]
26. Ehling-Schulz, M.; Lereclus, D.; Koehler, T.M. The Bacillus cereus Group: Bacillus Species with Pathogenic Potential. *Microbiol. Spectr.* **2019**, *7*. [CrossRef] [PubMed]
27. Bottone, E.J. Bacillus cereus, a volatile human pathogen. *Clin. Microbiol. Rev.* **2010**, *23*, 382–398. [CrossRef] [PubMed]
28. Fernandez-Garcia, L.; Blasco, L.; Lopez, M.; Bou, G.; Garcia-Contreras, R.; Wood, T.; Tomas, M. Toxin-Antitoxin Systems in Clinical Pathogens. *Toxins* **2016**, *8*, 227. [CrossRef]
29. Ahn, D.H.; Lee, K.Y.; Lee, S.J.; Park, S.J.; Yoon, H.J.; Kim, S.J.; Lee, B.J. Structural analyses of the MazEF4 toxin-antitoxin pair in Mycobacterium tuberculosis provide evidence for a unique extracellular death factor. *J. Biol. Chem.* **2017**, *292*, 18832–18847. [CrossRef]
30. Krissinel, E.; Henrick, K. Inference of macromolecular assemblies from crystalline state. *J. Mol. Biol.* **2007**, *372*, 774–797. [CrossRef]
31. Zorzini, V.; Buts, L.; Sleutel, M.; Garcia-Pino, A.; Talavera, A.; Haesaerts, S.; De Greve, H.; Cheung, A.; van Nuland, N.A.; Loris, R. Structural and biophysical characterization of Staphylococcus aureus SaMazF shows conservation of functional dynamics. *Nucleic Acids Res.* **2014**, *42*, 6709–6725. [CrossRef]
32. Hoffer, E.D.; Miles, S.J.; Dunham, C.M. The structure and function of Mycobacterium tuberculosis MazF-mt6 toxin provide insights into conserved features of MazF endonucleases. *J. Biol. Chem.* **2017**, *292*, 7718–7726. [CrossRef]
33. Chen, R.; Tu, J.; Liu, Z.; Meng, F.; Ma, P.; Ding, Z.; Yang, C.; Chen, L.; Deng, X.; Xie, W. Structure of the MazF-mt9 toxin, a tRNA-specific endonuclease from Mycobacterium tuberculosis. *Biochem. Biophys. Res. Commun.* **2017**, *486*, 804–810. [CrossRef]
34. Simanshu, D.K.; Yamaguchi, Y.; Park, J.H.; Inouye, M.; Patel, D.J. Structural basis of mRNA recognition and cleavage by toxin MazF and its regulation by antitoxin MazE in Bacillus subtilis. *Mol. Cell* **2013**, *52*, 447–458. [CrossRef]
35. Zorzini, V.; Mernik, A.; Lah, J.; Sterckx, Y.G.; De Jonge, N.; Garcia-Pino, A.; De Greve, H.; Versees, W.; Loris, R. Substrate Recognition and Activity Regulation of the Escherichia coli mRNA Endonuclease MazF. *J. Biol. Chem.* **2016**, *291*, 10950–10960. [CrossRef] [PubMed]
36. Kamada, K.; Hanaoka, F.; Burley, S.K. Crystal structure of the MazE/MazF complex: Molecular bases of antidote-toxin recognition. *Mol. Cell* **2003**, *11*, 875–884. [CrossRef]
37. Guillen Schlippe, Y.V.; Hedstrom, L. A twisted base? The role of arginine in enzyme-catalyzed proton abstractions. *Arch. Biochem. Biophys.* **2005**, *433*, 266–278. [CrossRef] [PubMed]
38. Dunican, B.F.; Hiller, D.A.; Strobel, S.A. Transition State Charge Stabilization and Acid-Base Catalysis of mRNA Cleavage by the Endoribonuclease RelE. *Biochemistry* **2015**, *54*, 7048–7057. [CrossRef] [PubMed]
39. Gogos, A.; Mu, H.; Bahna, F.; Gomez, C.A.; Shapiro, L. Crystal structure of YdcE protein from Bacillus subtilis. *Proteins* **2003**, *53*, 320–322. [CrossRef] [PubMed]
40. Kang, S.M.; Kim, D.H.; Jin, C.; Lee, B.J. A Systematic Overview of Type II and III Toxin-Antitoxin Systems with a Focus on Druggability. *Toxins* **2018**, *10*, 515. [CrossRef]
41. Cook, G.M.; Robson, J.R.; Frampton, R.A.; McKenzie, J.; Przybilski, R.; Fineran, P.C.; Arcus, V.L. Ribonucleases in bacterial toxin-antitoxin systems. *Biochim. Biophys. Acta* **2013**, *1829*, 523–531. [CrossRef] [PubMed]

42. Kang, S.M.; Kim, D.H.; Lee, K.Y.; Park, S.J.; Yoon, H.J.; Lee, S.J.; Im, H.; Lee, B.J. Functional details of the Mycobacterium tuberculosis VapBC26 toxin-antitoxin system based on a structural study: Insights into unique binding and antibiotic peptides. *Nucleic Acids Res.* **2017**, *45*, 8564–8580. [CrossRef] [PubMed]
43. Kim, D.H.; Kang, S.M.; Park, S.J.; Jin, C.; Yoon, H.J.; Lee, B.J. Functional insights into the Streptococcus pneumoniae HicBA toxin-antitoxin system based on a structural study. *Nucleic Acids Res.* **2018**, *46*, 6371–6386. [CrossRef] [PubMed]
44. Kabsch, W. Xds. *Acta Crystallogr. D Biol. Crystallogr.* **2010**, *66*, 125–132. [CrossRef]
45. Adams, P.D.; Afonine, P.V.; Bunkoczi, G.; Chen, V.B.; Davis, I.W.; Echols, N.; Headd, J.J.; Hung, L.W.; Kapral, G.J.; Grosse-Kunstleve, R.W.; et al. PHENIX: A comprehensive Python-based system for macromolecular structure solution. *Acta Crystallogr. D Biol. Crystallogr.* **2010**, *66*, 213–221. [CrossRef]
46. Emsley, P.; Lohkamp, B.; Scott, W.G.; Cowtan, K. Features and development of Coot. *Acta Crystallogr. D Biol. Crystallogr.* **2010**, *66*, 486–501. [CrossRef]
47. Brunger, A.T. Free R value: A novel statistical quantity for assessing the accuracy of crystal structures. *Nature* **1992**, *355*, 472–475. [CrossRef]
48. Murshudov, G.N.; Skubak, P.; Lebedev, A.A.; Pannu, N.S.; Steiner, R.A.; Nicholls, R.A.; Winn, M.D.; Long, F.; Vagin, A.A. REFMAC5 for the refinement of macromolecular crystal structures. *Acta Crystallogr. D Biol. Crystallogr.* **2011**, *67*, 355–367. [CrossRef]
49. Chen, V.B.; Arendall, W.B.; Headd, J.J.; Keedy, D.A.; Immormino, R.M.; Kapral, G.J.; Murray, L.W.; Richardson, J.S.; Richardson, D.C. MolProbity: All-atom structure validation for macromolecular crystallography. *Acta Crystallogr. D Biol. Crystallogr.* **2010**, *66*, 12–21. [CrossRef]
50. UniProt, C. UniProt: A worldwide hub of protein knowledge. *Nucleic Acids Res.* **2019**, *47*, D506–D515. [CrossRef]
51. Thompson, J.D.; Higgins, D.G.; Gibson, T.J. CLUSTAL W: Improving the sensitivity of progressive multiple sequence alignment through sequence weighting, position-specific gap penalties and weight matrix choice. *Nucleic Acids Res.* **1994**, *22*, 4673–4680. [CrossRef] [PubMed]
52. Robert, X.; Gouet, P. Deciphering key features in protein structures with the new ENDscript server. *Nucleic Acids Res.* **2014**, *42*, W320–W324. [CrossRef] [PubMed]

Permissions

All chapters in this book were first published by MDPI; hereby published with permission under the Creative Commons Attribution License or equivalent. Every chapter published in this book has been scrutinized by our experts. Their significance has been extensively debated. The topics covered herein carry significant findings which will fuel the growth of the discipline. They may even be implemented as practical applications or may be referred to as a beginning point for another development.

The contributors of this book come from diverse backgrounds, making this book a truly international effort. This book will bring forth new frontiers with its revolutionizing research information and detailed analysis of the nascent developments around the world.

We would like to thank all the contributing authors for lending their expertise to make the book truly unique. They have played a crucial role in the development of this book. Without their invaluable contributions this book wouldn't have been possible. They have made vital efforts to compile up to date information on the varied aspects of this subject to make this book a valuable addition to the collection of many professionals and students.

This book was conceptualized with the vision of imparting up-to-date information and advanced data in this field. To ensure the same, a matchless editorial board was set up. Every individual on the board went through rigorous rounds of assessment to prove their worth. After which they invested a large part of their time researching and compiling the most relevant data for our readers.

The editorial board has been involved in producing this book since its inception. They have spent rigorous hours researching and exploring the diverse topics which have resulted in the successful publishing of this book. They have passed on their knowledge of decades through this book. To expedite this challenging task, the publisher supported the team at every step. A small team of assistant editors was also appointed to further simplify the editing procedure and attain best results for the readers.

Apart from the editorial board, the designing team has also invested a significant amount of their time in understanding the subject and creating the most relevant covers. They scrutinized every image to scout for the most suitable representation of the subject and create an appropriate cover for the book.

The publishing team has been an ardent support to the editorial, designing and production team. Their endless efforts to recruit the best for this project, has resulted in the accomplishment of this book. They are a veteran in the field of academics and their pool of knowledge is as vast as their experience in printing. Their expertise and guidance has proved useful at every step. Their uncompromising quality standards have made this book an exceptional effort. Their encouragement from time to time has been an inspiration for everyone.

The publisher and the editorial board hope that this book will prove to be a valuable piece of knowledge for researchers, students, practitioners and scholars across the globe.

List of Contributors

Joanna Trylska
Centre of New Technologies, University ofWarsaw, ul. S. Banacha 2c, 02-097 Warsaw, Poland

Marcin Równicki
Centre of New Technologies, University ofWarsaw, ul. S. Banacha 2c, 02-097 Warsaw, Poland
Department of Bacterial Genetics, Institute of Microbiology, Faculty of Biology, University ofWarsaw, Miecznikowa 1, 02-096 Warsaw, Poland

Robert Lasek and Dariusz Bartosik
Department of Bacterial Genetics, Institute of Microbiology, Faculty of Biology, University ofWarsaw, Miecznikowa 1, 02-096 Warsaw, Poland

Angelika Kronhardt and Christoph Beitzinger
Rudolf Virchow Center, Research Center for Experimental Biomedicine, University of Würzburg, Versbacher Straße 9, 97078 Würzburg, Germany

Holger Barth
Institute of Pharmacology and Toxicology, University of Ulm Medical Center, Albert-Einstein-Allee 11, 89081 Ulm, Germany

Roland Benz
Department of Life Sciences and Chemistry, Jacobs-University Bremen, Campus-Ring 1, 28759 Bremen, Germany

María Moreno-del Álamo, Chiara Marchisone and Juan C. Alonso
Department of Microbial Biotechnology, Centro Nacional de Biotecnología, CNB-CSIC, 3 Darwin Str., 28049 Madrid, Spain

Goli Yamini, Nnanya Kalu and Ekaterina M. Nestorovich
Department of Biology, The Catholic University of America, Washington DC, WA 20064, USA

Laurita Klimkaitė, Julija Armalytė, Jūratė Skerniškytė and Edita Sužiedėlienė
Institute of Biosciences, Life Sciences Center, Vilnius University, LT-1025 Vilnius, Lithuania

Ilona Schelle, Janina Bruening, Mareike Buetepage and Harald Genth
Institute for Toxicology, Hannover Medical School, Carl-Neuberg-Str. 1, D-30625 Hannover, Germany

Anton Ambroa
Microbiology Department-Research Institute Biomedical A Coruña (INIBIC), Hospital A Coruña (CHUAC), University of A Coruña (UDC), 15006 A Coruña, Spain

Ines Bleriot, Lucia Blasco and Laura Fernandez-Garcia
Microbiology Department-Research Institute Biomedical A Coruña (INIBIC), Hospital A Coruña (CHUAC), University of A Coruña (UDC), 15006 A Coruña, Spain
Study Group on Mechanisms of Action and Resistance to Antimicrobials (GEMARA) the Behalf of the Spanish Society of Infectious Diseases and Clinical Microbiology (SEIMC), 28003 Madrid, Spain

Maria Lopez
Microbiology Department-Research Institute Biomedical A Coruña (INIBIC), Hospital A Coruña (CHUAC), University of A Coruña (UDC), 15006 A Coruña, Spain
Spanish Network for Research in Infectious Diseases (REIPI), 41071 Seville, Spain

Jesus Oteo-Iglesias
Study Group on Mechanisms of Action and Resistance to Antimicrobials (GEMARA) the Behalf of the Spanish Society of Infectious Diseases and Clinical Microbiology (SEIMC), 28003 Madrid, Spain
Spanish Network for Research in Infectious Diseases (REIPI), 41071 Seville, Spain
Reference and Research Laboratory for Antibiotic Resistance and Health Care Infections, National Centre for Microbiology, Institute of Health Carlos III, 28222 Majadahonda, Spain

Thomas K. Wood
Department of Chemical Engineering, Pennsylvania State University, University Park, PA 16801, USA

Alvaro Pascual and Felipe Fernandez-Cuenca
Study Group on Mechanisms of Action and Resistance to Antimicrobials (GEMARA) the Behalf of the Spanish Society of Infectious Diseases and Clinical Microbiology (SEIMC), 28003 Madrid, Spain
Clinical Unit for Infectious Diseases, Department of Microbiology and Medicine, Microbiology and Preventive Medicine, Hospital Universitario Virgen Macarena, University of Seville, Biomedicine Insititute of Seville (IBIS), 41009 Seville, Spain
Spanish Network for Research in Infectious Diseases (REIPI), 41071 Seville, Spain

Mercedes Delgado-Valverde and Ana Gual-de-Torrella
Clinical Unit for Infectious Diseases, Department of Microbiology and Medicine, Microbiology and Preventive Medicine, Hospital Universitario Virgen Macarena, University of Seville, Biomedicine Insititute of Seville (IBIS), 41009 Seville, Spain

Maria Tomas and German Bou
Microbiology Department-Research Institute Biomedical A Coruña (INIBIC), Hospital A Coruña (CHUAC), University of A Coruña (UDC), 15006 A Coruña, Spain
Study Group on Mechanisms of Action and Resistance to Antimicrobials (GEMARA) the Behalf of the Spanish Society of Infectious Diseases and Clinical Microbiology (SEIMC), 28003 Madrid, Spain
Spanish Network for Research in Infectious Diseases (REIPI), 41071 Seville, Spain

Keith Weaver
Division of Basic Biomedical Sciences, Sanford School of Medicine, University of South Dakota, Vermillion, SD 57069, USA

Leonie Schnell, Ann-Katrin Mittler and Holger Barth
Institute of Pharmacology and Toxicology, University of Ulm Medical Center, Albert-Einstein-Allee 11, 89081 Ulm, Germany

Andrea Mattarei
Department of Chemical Sciences, University of Padova, 35121 Padova, Italy

Domenico Azarnia Tehran and Cesare Montecucco
Department of Biomedical Sciences, University of Padova, 35131 Padova, Italy

Claude Gutierrez and Olivier Neyrolles
Institut de Pharmacologie et de Biologie Structurale, IPBS, Université de Toulouse, CNRS, UPS, 205 route de Narbonne, 31400 Toulouse, France

Kanchiyaphat Ariyachaokun
Institut de Pharmacologie et de Biologie Structurale, IPBS, Université de Toulouse, CNRS, UPS, 205 route de Narbonne, 31400 Toulouse, France
Department of Biological Science, Ubon Ratchathani University, 85 Sathollmark Rd.,Warin Chamrab, Ubon Ratchathani 34190, Thailand

Anna D. Grabowska
Institut de Pharmacologie et de Biologie Structurale, IPBS, Université de Toulouse, CNRS, UPS, 205 route de Narbonne, 31400 Toulouse, France
Department of Biophysics and Human Physiology, Medical University of Warsaw, 02-091Warsaw, Poland

Simona Kavaliauskiene, Anne Berit Dyve Lingelem and Tore Skotland
Department of Molecular Cell Biology, Institute for Cancer Research, Oslo University Hospital, N-0379 Oslo, Norway
Center for Cancer Biomedicine, Faculty of Medicine, Oslo University Hospital, N-0379 Oslo, Norway

Kirsten Sandvig
Department of Molecular Cell Biology, Institute for Cancer Research, Oslo University Hospital, N-0379 Oslo, Norway
Center for Cancer Biomedicine, Faculty of Medicine, Oslo University Hospital, N-0379 Oslo, Norway
Department of Biosciences, University of Oslo, N-0316 Oslo, Norway

Chih-Han Tu, Michelle Holt, Shengfeng Ruan and Christina Bourne
Department of Chemistry and Biochemistry, University of Oklahoma, Norman, OK 73019, USA

Jihong Li and Bruce A. McClane
Department of Microbiology and Molecular Genetics, University of Pittsburgh School of Medicine, Room 420, Bridgeside Point II Building, 450 Technology Drive, Pittsburgh, PA 15219, USA

Francisco A. Uzal
California Animal Health and Food Safety Laboratory, San Bernardino Branch, School of Veterinary Medicine, University of California-Davis, San Bernardino, CA 92408, USA

Sung-Min Kang, Ji Sung Koo, Chang-Min Kim and Bong-Jin Lee
The Research Institute of Pharmaceutical Sciences, College of Pharmacy, Seoul National University, Gwanakgu, Seoul 08826, Korea

Do-Hee Kim
College of Pharmacy, Jeju National University, Jeju 63243, Korea

Index

www.ingramcontent.com/pod-product-compliance
Lightning Source LLC
Chambersburg PA
CBHW080623160326
41279CB00005B/490